WORKING WITH GIANTS

JOHN NORMAN
WORKING WITH GIANTS

Fire Engineering®
BOOKS & VIDEOS

Copyright © 2021 by
Fire Engineering Books & Videos
110 S. Hartford Ave., Suite 200
Tulsa, Oklahoma 74120 USA

800.752.9764
+1.918.831.9421
info@fireengineeringbooks.com
www.FireEngineeringBooks.com

Senior Vice President: Eric Schlett
Operations Manager: Holly Fournier
Sales Manager: Joshua Neal
Managing Editor: Mark Haugh
Production Manager: Tony Quinn
Developmental Editor: Chris Barton
Cover Designer: Jared Hood
Book Designer: Robert Kern, TIPS Technical Publishing, Inc., Carrboro, NC

Library of Congress Cataloging-in-Publication Data

Names: Norman, John, 1952- author.
Title: Working with giants : a memoir / John Norman.
Description: Tulsa, Oklahoma, USA : Fire Engineering Books and Videos,
   [2021] | Includes index. | Summary: "Memoirs of prominent New York City
   firefighter and author John Norman"-- Provided by publisher.
Identifiers: LCCN 2021017793 | ISBN 9781593704346 (paperback)
Subjects: LCSH: Fire extinction--New York (State)--New York. |
   Fires--Casualties--New York (State)--New York. | Fire fighters--New York
   (State)--New York. | Norman, John, 1952---Friends and associates.
Classification: LCC TH9505.N5 N67 2021 | DDC 363.37092 [B]--dc23
LC record available at https://lccn.loc.gov/2021017793

Printed in the United States of America

1 2 3 4 5   25 24 23 22 21

This book is dedicated to the memory of Lieutenant Dennis Mojica (left), Firefighter Joseph Angelini (middle), and the 92 other members of the Special Operations Command of the New York City Fire Department who made the supreme sacrifice while protecting life and property in the City of New York on September 11, 2001; and to Steve Casani (right), my friend, former boss, and later coworker, who is responsible for so much of my career; and to all the members of this nation's fire and rescue services who have perished while conducting fire or rescue operations.

To the truly unsung heroes, the engine company firefighters who put their heads down and their backs into the line when their brain is screaming at them "Please, no more!" and yet they take one more step closer to the red devil so that "the wet stuff" truly goes onto "the red stuff." As the FDNY Training Bulletin Engines 1 taught me: "More lives are saved by properly positioned and operated hoselines than by all other means combined."

Contents

Prologue ix

1 September 11th 1

2 Vacants 7

3 Being "Broken In" 17

4 The Death of a Building (and a Neighborhood) 35

5 Father's Day on the Pier 41

6 Harlem, USA 53

7 The Major Deegan Expressway 69

8 Railroad Pizza 75

9 I'll Make You a Detective! 81

10 The Southside 93

11 A Tree Grows in Brooklyn 115

12 Flight 5050 125

13 Some Really Important Steps 145

14 The Winter Offensive 159

15 My Turn to Be Schooled 171

16 The What If's? 179

17 Kingston Is Not Just in the Caribbean 189

18 The Troy Avenue Boys 201

19 The Soundest Sleeper You Ever Met! 207

20 The Thirsty Corrections Officer 219

21 Sometimes, I'd Rather Be Lucky Than Good 227

22 Only the Lucky Ones Get to Retire! 241

23 An Unusual Smile 247

24 "Precious" 257

25 Planes . . . Something That Big Isn't Supposed to Move,
Let Alone Fly! 267

26 When You've Got a Great Act, Take It to Broadway! 291

27 30 Rock 313

28 Call the Veterinarians 325

29 Popes, Presidents, and Other Assorted Dignitaries 335

30 V.I.B.s—Very Important Buildings 343

31 Harlem II 359

32 The World Trade Center 367

33 The Rebirth 435

Epilogue 443

Glossary 453

Index 465

About the Author 491

Prologue

The idea for this book came shortly after my textbook, *Fire Officer's Handbook of Tactics*, was published in 1991. That's an educational book now in its fifth edition. However, many readers expressed to me their appreciation of my realistic, life-and-death description of what firefighters face. Many readers had suggested I write "just a book of stories." I toyed with the idea and began writing down interesting things I had experienced. Many times, we would come back from fires, explosions, or other disasters, and everyone would say, "Somebody ought to make a movie about that; that was incredible!" But then, someone else would doubtfully reply: "Nobody outside the firehouse would ever believe that it happened."

In the aftermath of the September 11, 2001 attack on the World Trade Center, the public finally woke up to a little of what firefighters do, making this book possible. These stories necessarily are written from my perspective because I was there. I can only speak of what I know. There are only two incidents related in these pages that I was not personally involved in or did not actually witness. They both happened to my unit, and I made my notes after hearing the actual participants tell their stories, often multiple times, and accompanied by rollicking laughter (see Chapter 28: Call the Veterinarians).

This book is about all firefighters, the men and women who are absolutely the salt of the earth: everyday members of extraordinary institutions whose sole mission is to protect the lives and property of their neighbors.

Guys and gals who would give you the shirt off their backs, not because it will get them any recognition, or even any points with the "Big Man" upstairs, but because it's the right thing to do. Guys like Timmy Higgins, one of four firefighter sons of a fire captain, each of them better than the next. Guys like Mike Esposito and Billy Lake, pure bulldogs, who would never let go of a person who was trapped: guys that I would want to come to get me if I were ever trapped. Guys like Pete Martin, Dennis Mojica, Billy Henry, Ray Downey, and all the rest of the men who have proven over and over what the definition of courage means. It has been said that a firefighter performs his or her greatest act of courage when they take their oath of office, after that everything else is part of the job. That may be so, but the people who go out every day, laying their lives on the line, display raw courage daily, a courage that is rarely seen off the battlefield. This is their story.

Figure P-1. On a brutally hot August afternoon, while operating at a scaffold collapse 750 feet above Times Square, the author was visited by several of the Giants of the FDNY. They set an example for all humanity in their service to their fellow man. From left, Chief of Department Peter Ganci, BC John Keenan, Assistant Chief Frank Cruthers, Chief of Special Operations Ray Downey, Captain Norman. They set an example for all humanity in their service to their fellow man.

In the 18 years that have elapsed since the attack on the World Trade Center, I have struggled with writing two sections of this book: this prologue, and the last two chapters. Everything else was pretty much already written. The sections that follow have a specific mission in mind: first to set the tone of the book. It's not an attempt to cash in on the tragedy of others, which is why I waited 18 years before taking it to my publisher. Rather, it is meant to celebrate the lives of a very special group of people. The second purpose is to introduce the firefighters to the American public. Since September 11, all kinds of people have been speaking about firefighters as "America's Heroes." I truly believe they are heroes in the classical sense of the word, not in the casual, modern context that involves celebrity and making the rounds of the talk show circuit. A hero to me means someone who protects others at great risk to themselves, who does so because it's the right thing to do, and carries on with self-respect, providing a role model to others.

The time came for me to move forward with this manuscript in 2019, after I suffered a life-threatening injury that made me seriously consider what was left of my life. I felt it was time—this story needed to be told.

• • •

The loss of the heroes of September 11th and every other day is not only a tragedy for the nation, but a personal tragedy as well. Each of the 343 heroes that the FDNY lost that day was a real human being, loved by many, and perhaps disliked by a few. These people had passions and plans, hopes and desires for themselves, their families, and their loved ones. Peter Martin of Rescue 2 always talked of retiring to Vermont, to be far from the crush of humanity that made up the Crown Heights section of Brooklyn. Kevin Dowdell didn't wait until he retired. He set about claiming his own little bit of heaven right here on earth, moving his family into a house that he rebuilt with his own hands in the "Irish Riviera," Breezy Point. Ray Downey set about creating a better world, helping to establish the United States Urban Search and Rescue Teams to aid victims of severe disasters.

As you meet these individuals in the coming chapters, understand that this is real. The language may be coarse and the descriptions may be graphic.

This isn't Hollywood hype or public relations spin. Almost every one of the narratives is documented in official FDNY reports, citations, and other evidence: often photographic, video, or other means. If anything, most of the acts are understated, because these people always tend to downplay their involvement in heroic acts. "I was just doing my job," they say. Yes, they were. Thank God!

• • •

These stories are often about complex events that involve multiple people playing their roles in a pitch black, hot, dangerous environment, with many players in close proximity to others, but not visible to each other. As a result, some people who were at the same event may have different recollections of exactly how a situation played out from their perspective. I can only relay what I saw or heard. If I leave anyone out, it is totally my fault, and I assure you it is simply that I did not see or hear of your actions at the time they occurred. There are many people involved who could write multiple volumes about these same or similar events. I have tried to do justice to all of you. I was truly blessed to work with giants in their field. I know I owe so much to many of you whom I did not even get to mention some of what we have shared. My apologies in advance.

CHAPTER 1

September 11th

September 11, 2001 is a date which, like December 7, 1941, will live in infamy. Forty years from now people will be able to remember the event as it unfolded, thanks to live worldwide television coverage. For many of the men you will meet in the following chapters, the attack and the ensuing hour and a half until the collapse of the North Tower, was the rest of their lives. Three hundred and forty of New York City firefighters, two fire department paramedics, and a fire department chaplain who responded to the attack, would never spend another day serving the citizens of New York. *Attack*—I still can't bring myself to call it anything else; it was not some "tragic" event. It was an attack on our country unlike any other in history. Three hundred and forty-three men of the FDNY, 37 Port Authority police officers, and 23 New York City police officers responded with one focus in their minds: to save lives. And they never returned.

The number 343 is seared in our souls. It represents more firefighters lost in New York City than in the prior 50 years. It is nearly half the previous total of 778 in the 136-year history of the FDNY to that date. Of the 343, 94 were members of the Special Operations Command. Virtually every on-duty member that responded in the five rescue companies, five squad companies, and Haz Mat Company 1 were wiped out, except one member of the Haz Mat company: the resource man, who had been directed to move the unit's two apparatuses to a more effective location. To compound matters, the timing of the attack, occurring just prior to the 9:00 a.m. shift change, caused an even greater loss; some companies responded with both the on-duty platoon, as well as members of the incoming day tour. In retrospect, 20-20 hindsight, that was not a wise decision. Twenty men perished who did not have to. At least technically, they didn't have to. I'm here to tell you that there is no way they could have done anything different.

At Special Operations Headquarters on Roosevelt Island, John Paolillo was the on-duty battalion chief in the SOC Battalion, and John Moran was still in the office after finishing reports and working out after his shift. Ray Downey, the overall Commander of Special Operations, was in his office preparing for a budget meeting that afternoon, seeking extra funding to prepare for NBC defense: Nuclear, Biological, and Chemical attacks. When the first plane struck, the initial response included the SOC Battalion. Both John Paolillo and John Moran responded. As the initial radio reports came in and the magnitude of the incident became clear, Downey put aside the fiscal concerns and headed for his car. Joe Angelini was working in the office on light duty. Joey Angelini had spent over 41 years working in Manhattan high-rises and he knew what this attack would mean. He also knew that his son Joe Jr. was working that day in Ladder 4 in Manhattan. Joe Sr. hitched a ride with the Tactical Support Unit, which was responding on the 10-60 (major aircraft emergency) signal. He wasn't going to leave Joe Jr. alone on this one. Battalion Chief Charley Kasper responded from his home on Staten Island.

Captain Terry Hatton was at home in his apartment in lower Manhattan with his fire department radio scanner on keeping tabs on his unit, Rescue 1. His wife Beth was at work at City Hall as Mayor Giuliani's personal assistant. Terry responded directly to the scene. He somehow rendezvoused with Lt. Dennis Mojica, who had eight members of Rescue 1 with him, and started walking the 95 flights of stairs to the plane's impact site. They were joined en route by Joey. The handie-talkies were buzzing constantly with reports from the field communications unit relaying the calls to the dispatcher's office. "Eighty people trapped in a conference room on the 105th floor, northwest corner." "Fifty people trapped on the 100th floor with fire burning right underneath them." "Watch out for the jumpers! Watch out for the jumpers!"

As they climbed, they passed thousands and thousands of terrified office workers going in the opposite direction. Some were badly burned and suffering from trauma. They offered direction and encouragement, but kept on plodding upward. The burden they carried caused them to take a slow pace. The bunker gear and mask with 60-minute air bottles were bad enough, but the needed forcible entry tools and spare air bottles made it nearly unbearable. By the time they reached the 20th floor, they were out of radio contact with the lobby command post. The repeater system, which had been built into the complex to boost power so the upper floors could be contacted, wasn't working in the North Tower: neither was the power to the elevators. When they reached the 30th floor they started coming across downed firefighters: guys from the first arriving units who were suffering the effects of their climb. Several members were experiencing symptoms of heart attacks and other cardiovascular illnesses. From the bombing of the Trade Center in 1993, we knew that it would take about three hours to walk the 110 stories to the top floor. The trick is to pace yourself and discard any and all non-essential weight. It took the members almost 50 minutes to walk 35 flights. They were on pace, but it was an agonizingly slow one, with thousands of people trapped above them.

At the 31st floor, Rescue 1 stopped to treat several firefighters with chest pains. They gave their oxygen cylinders to other firefighters who are there. Police officers from the Emergency Services Unit, including my brother David, arrived with additional oxygen supplies and begin to assist. Just then the entire building shook. Tower 2 had just collapsed. The cops' radios started crackling with orders from their command post to evacuate the building. On the way down, many of the surviving rescuers recall that the previous flood of civilians had slowed to a mere trickle. Virtually all the ambulatory civilians below the point of impact had completed their escape. That still left others who were not able to walk the dozens of flights on their own, as well as the thousands trapped above the points of impact.

Tower 2, the South Tower, had been the second tower hit, but it was the first to collapse, probably because the plane hit lower in the building (adding tens of thousands of additional tons of weight to the melting steel) and sheared away more of the critical load-bearing members in the angled path that it took through that building. The firefighters working in that building, like Battalion Chiefs Orio Palmer and Eddie Geraghty, and Fire Marshal Ronny Bucca—the last two being former Rescue 1 firefighters—had absolutely no indication that they were about to be killed. All the firefighters in that building, an entire 5th alarm assignment, died nearly instantly. The collapse took 14 seconds from the time it began until the 110–story building stood only as a five-story pile of debris.

The collapse of Tower 2 spewed thousands of tons of steel and flaming debris in all directions. A very large section fell due west, landing on the 22-story Marriott Hotel. The impact had sheared the hotel into two remaining spires; the entire middle section was flattened. Many firefighters had been staged in the lobby of the hotel, awaiting orders to ascend into Tower 2 when the collapse began and they found themselves trapped. Chief of Department Peter Ganci ordered the evacuation of Tower 1 and prepared his staff to move the command post north to Vesey Street, but the urgent maydays coming from the Marriott drew him and several of his top commanders like magnets. Chief of Special Operations Ray Downey, and Battalion Chiefs Larry Stack and Brian O'Flaherty of the Safety Operating Battalion were among a group of personnel Chief Ganci dispatched to oversee evacuation of the Marriott. Several of the trapped members were dug out of their potential tombs, but more were still trapped, including several who were severely pinned. Time was running out for them as well as the men in Tower 1. The chiefs ordered all the survivors that could move to make their way away from the scene, taking many non-ambulatory casualties with them.

At 10:28 a.m., Tower 1 collapsed in only 12 seconds. The falling south wall of Tower 1 flattened the remaining sections of the Marriott hotel. The west wall landed on the command staff that had remained behind. Chief Ganci, First Deputy Fire Commissioner Bill Feehan, Ray Downey, Larry Stack, 11 members of Rescue Co. 1, and hundreds of others died in those short 12 seconds (fig. 1–1).

In hindsight, as I have said, the loss of the off-duty men was a tragedy that in some ways did not have to occur. Some people say they shouldn't have responded. That's nonsense. If the buildings hadn't collapsed, we'd have needed every one of those guys and more. The fact of the matter is they

Figure 1–1. The collapses of three modern steel framed high-rise buildings on September 11 left a hole in the heart of America, as the sacrifices of so many brave individuals came to light. Firefighters across the nation learned that nothing is "fireproof." (*Courtesy of Andrea Boohor*)

went into those buildings knowing that death was a real possibility. Death from being killed by jumpers, just like the first firefighter to die that day, Danny Suhr of Engine 216. Death from heart attacks after walking 110 flights, and then joining the battle with the red devil. Death from smoke and fire, and even partial collapse. But nobody anticipated total structural failure on the magnitude that occurred. We had no reason to. This was an absolutely unprecedented event, the total collapse of a modern high-rise. These were not the towers of Babel. They were the most modern structures ever created at the time. We had a track record with these kinds of things.

A World War II bomber had crashed into the 79th floor of the Empire State Building in 1945. That fire was only a third-alarm and there was not even a hint of a collapse. The same is true of another bomber that had crashed into another high-rise, just blocks away from the World Trade Center at 70 Pine Street. Several high-rise buildings had suffered severe multi-floor fires without collapse, including one in Sao Paolo, Brazil that saw over 20 floors fully involved at once. Fires in Philadelphia (One Meridian Plaza) and Los Angeles (First Interstate Bank) in the 1980s and 1990s had each burned five floors or more, and they both stood. Hell, the World Trade Center's architects and structural engineers had even told everybody who would listen that the buildings were built to take the impact of a Boeing 707. That was in 1970 when the buildings were going up and the 707 was the biggest plane in the air. We believed them. Now over 30 years later, the buildings proved the designers technically correct. They did take the *impact*. It was that little bit of fire created by the thousands of gallons of jet fuel from the crashing planes that the structural engineers failed to take into consideration and brought the buildings down.

The architects and structural engineers used lightweight steel bar joist floors and a steel load-bearing wall, each of which were protected only by a light coating of spray-on "fireproofing" which was incapable of standing up to a multi-floor fire, even without the impact that knocked the fibers off the steel. The buildings were lucky to make it to 2001, since originally most of the office floors were not protected by sprinklers, and they had the added danger of open "access stairs" between floors that would allow very rapid fire to spread from floor to floor. By the 1990s, however, the sprinkler systems were completed and the buildings were "safe"—until September 11, 2001. That day taught all firefighters a horrific lesson: that there are fires that are beyond our capacity to control, no matter how much we want to, and there are buildings that are not going to stand up long enough for us to do our job, no matter how much we want to. Never forget!

CHAPTER 2

Vacants

It is just another piece of shit vacant building. On a late summer evening at about 7:00 p.m. in August 1980, we are on our way back to the firehouse from the third false alarm of the young night shift since the start of the tour at 6:00 p.m., when the dispatcher calls us over the radio—"Brooklyn to 290" (pronounced two nine oh). Engine Company 290, the busiest unit in the New York City Fire Department. Captain Fred Kopetz (lovingly called "Fred the Red" behind his back by his troops) picks up the receiver and replies "290, K": signaling that he is standing by for the dispatcher's message. "290, take in an ERS alarm for box 2006, on Bradford Street between Blake and Sutter Avenues, 290?" "290, 10-4," radios the captain as his foot touches the air horn pedal on the floor of the Mack CF pumper. George Reidy is the chauffeur tonight and he pulls the rig out of what was going to be a left turn, and continues north on Pennsylvania Avenue, one block to Blake Avenue where he swings out left around stopped traffic, in order to make the right turn onto Blake Avenue. As he makes the turn, the dispatcher calls again and reports, "Brooklyn to 290. It sounds like you're going to work. We're getting the address of 350 Bradford Street in a vacant building." Captain Kopetz calls back to us in the passenger compartment, "Pull 'em up!": meaning, pull up the tops of our hip boots, something you only do if you're going to a real fire, as opposed to the hundreds of false alarms, auto accidents, and minor fires we do every month.

George turns the pumper into Bradford Street and stops at a hydrant about 50 feet before we reach the fire building. Normally, this is not our routine, since it adds a lot of work to our hose stretching effort. The hosebed is on the back of the truck, and stopping before the building means we will have to pull off enough hose to reach the building and operate inside it, and then drag that hose around the rig to the building. Usually, we perform a back stretch, stopping in front of the fire building and removing our hose, and then letting the rig drive down the street to the nearest hydrant. George hasn't done that here, because on his last tour the company had another fire on this block, and when he went to the next hydrant up from this building, it was out of service—vandalized. The bonnet had been removed and all the brass parts stripped out: the operating nut, the 2½-in. nozzle, and the 4½-in. steamer connection, all gone to the scrap yard. Engine 231 had connected to the hydrant we're at now and gotten the first line on the fire, while George had had to continue down the block, testing two more hydrants before he found one that worked. Company 290 got beat out of the bulk of that fire and they had to repack an extra 500 ft of hose. George won't let that happen to us this time.

We smell smoke in the air, wood smoke, but we don't see anything really pushing. Captain Kopetz has a better view, however, and tells us "It's a vacant on the left," then radios the dispatcher: "290 to Brooklyn, 10-30 on Box 2006." As I climb down out of the right-side jump seat behind the officer, I still haven't seen the building, but the nozzle operator Lenny Deszcz is telling me to "Make sure you bring your mask, kid. This could be snotty." Now that I have my three folds of hose and have pulled back away from the back of the pumper, I finally see the building. It is a three-story brick building, attached to a similar building on the right side. All of the front windows and doors are sealed up with cinder blocks to keep vagrants out. Obviously, they haven't worked!

Lenny and Mike Cushing, his backup man, have each taken one length of hose and walked up to the front of the building. I am the "door man" and my job is to make sure that they have enough hose to reach our area of operations and ensure the hose keeps moving forward, and then to be ready to relieve Lenny and Mike on the nozzle.

Lenny and Mike are both old school 290 guys from the era before masks became light enough to carry easily and clear enough to see out of the face-pieces. When they started, the masks were kept in boxes inside compartments on the sides of the rigs and were saved for "really nasty" fires, like the cellars of stores or factories. They didn't use them for every tenement or house fire. By 1979, there was no need for this anymore. The FDNY, working with NASA and breathing apparatus manufacturers, had developed a lighter, more comfortable, more useful mask: one that was stored in a pull-out bracket right behind or next to each seat. Company 290 was having none of that, however.

The company policy says that the nozzle operator and backup man would start the line in without masks. They didn't want to waste the extra 10 seconds it would take to pull the masks out of the bracket, slip them over their shoulders, and connect and tighten the straps. The doorman and the control man would don their masks and, if the nozzle operator or back-up needed relief, they would be there to move up.

Looking at this tightly sealed building, I feel this might be one of those times. The vacant, cinder-blocked building produces a particularly foul smoke, very high in carbon monoxide, since the building does not allow the necessary oxygen to enter to support clean burning until firefighters have smashed the cinder blocks out with sledge hammers.

Since 290 was already on the road when the run came in, and only five blocks away, we have a real head start on the other companies that are responding: Engine 231, Ladders 103 and 120, and the 44th Battalion. The blocked-up entranceway is a problem though. We can't begin to operate until we get in. I have pulled two lengths of hose to the front of the building and now I ask the captain if he wants me to get our sledge hammer, since the truck companies are probably two or three minutes away. He says, "No, let me look at the rear first," and disappears down the vacant lot (where there used to be another building until it also burned down) on the left side of the fire building. We start to follow him, but Lenny stops at the edge of the building line, waiting for orders. A second after disappearing around the rear of the building, Kopetz is back in the alley, waving us to the rear. There is a fire escape on the back of the building and the drop ladder is not only in place, but it is down to ground level. (That's probably how the "torch" got in there to start the fire in the first place.) Having a rear fire escape, as well as a front fire escape, tells us there are at least two apartments per floor. We don't know much about conditions in the front because, as I said, all the windows are sealed up, but from the backyard we can see we have fire on the second and third floors. Kopetz radios George to "Give the 10-75 signal. We're probably gonna need help."

Lenny tucks the nozzle over his left shoulder and pulls the hose across his chest and under his right arm, then starts to ascend the vertical drop ladder. Mike Cushing has flaked out Lenny's spare hose, as well as his own, and now he too prepares to follow Lenny. Captain Kopetz tells me to stay on the ground until the nozzle team moves in off the fire escape. "I don't want to overload that thing with too many guys," he says. Plus he wants me to push hose up as they advance. Then he calls George on the handie-talkie and tells him to "charge our line" (fill it with water). Immediately, water starts surging through the empty hoseline, making it dance like a snake and become rigid. Richie Steyert, our control man and the newest guy in the company, joins me as Lenny opens the nozzle on the fire escape balcony. With the flames shooting from both windows on the fire escape, there's no room for Mike or the captain, so both now have to make their way off the drop ladder onto the platform to follow Lenny who has already climbed into the first room. Lenny is one tough piece of work. Going on 50 years old, he runs circles around men half his age and doesn't like to be held up by anything. Besides, the faster he puts this thing out, the sooner he gets back outside to fresh air.

After Mike and the captain climb into the back room, I scoot up the ladder and position myself to feed hose to the advancing nozzle team. The last thing a nozzle operator wants is to be bogged down, unable to advance because of a lack of hose. That means he has to stand still and absorb punishment, and not kill the fire that is pounding him. The door man's job is to make sure that never happens. Richie feeds me hose from the rear yard and I push it into the building, trying to keep about a three-foot high loop in the hose that's leaning against the wall inside. As Mike pulls the hose in, the loop goes down and I push more in to restore it. The nozzle is advancing smoothly now and conditions are clear enough in the back room for me to see that there is sufficient space in it for me to store the spare hose that we will need for the rest of the advance. I pull up a full length and loop it around the room, then call Richie and tell him to "come on up." The nozzle is still moving forward. I don't want them to get too far away in case Lenny or Mike needs relief. We crawl forward, moving our loop of hose in front of us. The truck companies are in now, but clearing the cinder blocks from the front windows will take some time, even working off the front fire escape and from 120's bucket at the same time. I have advanced through three rooms and now have come to the public hall. The roof bulkhead and skylight must have been vented during a previous fire, because the hall is relatively clear of smoke. That's not the case at the door to the front apartment though. I can't for the life of me figure out how Lenny and Mike are still going without masks. This stuff is bad and I'm 20 feet behind them. (Captain Kopetz has his mask on his back, in case he needs it, but as long as his nozzle team isn't using them, he doesn't either.) I move up across the public hall to the door of the front apartment. I have my light on, shining it into the apartment in case the members inside are trying to make their way back toward us, but the line keeps advancing. Just then I hear Gene DiMarco shout an urgent message over the handie-talkies: "Watch out below! Get everyone out of the stairs! The bulkhead is really shitty and looks like it's gonna collapse!" I turn around to see where Richie is and find he's right behind me. I say, "Let's move up and get out of the public hall." Richie and I crawl forward about six feet. Richie's heels just clear the threshold into the front apartment when there is a massive crashing sound behind us that sounds like a bomb just went off. Instantly, we are engulfed in a cloud of dark swirling dust. The bulkhead has just collapsed inward, taking the entire staircase as well as the second and third floor landings into the cellar. Ten seconds earlier, it would have taken Richie and me with it. Through my mask facepiece I ask Rich if he's all right and he says he's OK. I breathe deeply and exhale with a sigh of relief: only we're not done yet.

The collapse also chopped our hoseline in half and now the nozzle's out of water. Just then Captain Kopetz appears through the smoke. "Are you guys all right?" We both say, "Yeah, we're good, Cap." "Was anyone else in the stairs?" he asks. Richie replies, "I was the last guy up, Cap. There was nobody behind me." The cinder-blocked front door probably saved some lives, since if it had been open, we would have come up the stairs going to the top floor, as well as Engine 231 and the inside teams of both ladder companies. Timing is everything sometimes. Kopetz tells Richie and me to follow him up to the front. Most of the fire is knocked down (thank God!) and the trucks have almost completed opening one of the cinder-blocked windows. We climb out through the opening onto the fire escape and down a portable ladder to the street. Chief Christ (pronounced Krissed) of the 44th Battalion has ordered a line stretched into Tower Ladder 120, and they are getting ready to open up the Stang nozzle on their bucket into the third floor. Engine 231 is ordered down off the rear fire escape as the tower gets ready to operate. The chief is in no mood to fool around with this vacant piece of garbage. We just had a really close call and he doesn't want anyone to get hurt. Besides, we've had some very bad things happen to these companies lately.

In March of 1980 we had yet another fire in a vacant three-story frame house at Sutter Avenue and Linwood Street. The fire had a tremendous head start, and the chief of the 39th Battalion ordered everyone to stay out of the building. Engine 332 had a line at the door to a one-story rear extension, trying to protect a rear exposure, when the entire three-story building suddenly collapsed. 332 turned to run for their lives. On the nozzle were two probies, Patrick Quinn, and Steve Fillipelli, whom I had gone to Probie School with. Their boss was Lieutenant Robert Dolney, a seasoned fire officer. When the main building started to come down, Dolney told Pat and Steve to drop the nozzle and run. But before they got two steps, the one-story extension was coming down on top of them. Dolney pushed the two younger men ahead. The wall slammed Dolney to the ground, burying him completely. It caught Pat from the waist down and slammed into Fillipelli. When they lifted the wall off of Pat and Steve and Lt. Dolney, the members of Ladders 103 and 107 thought they were all dead. They were only slightly wrong. Pat was critically injured, crippled from the waist down, although eventually he would learn to walk again and live with the pain and the memories: Steve slightly less so. Lieutenant Robert Dolney was dead, however. (Pat would later become my brother-in-law.)

I came in to work that morning just as 103 backed into quarters from the run. The mood in the firehouse was unlike anything I'd experienced before. Gone was the gung-ho spirit, the desire to fight any fire no matter how dangerous and at what cost. Several members of Ladder 103 had been on the roof of the building at the time of the collapse and "rode the building down," escaping with relatively minor injuries. Several were shaking. One of the senior men kept saying to all who would listen, "A bullshit, one-story wood wall on a fucking vacant building crushed them like bugs!" (fig. 2–1).

Five months later, in August of 1980, we had a fire in yet another real piece of shit vacant frame on Hinsdale Street, just north of Livonia Avenue. There was a hydrant about 50 feet past the fire building that was pouring out a stream of water from the open 2½-in. nozzle, which we pulled into. Benny Piegari had the nozzle and I was the backup man, while Lt. Clem Wingert was the boss. The building had experienced so many previous fires that it lost a good part of the rear of the roof and every window was thoroughly cleared out. As we got booster tank-supplied water in our line and prepared to advance, control man Tommy Pelligrino, who was helping the chauffeur hook up to the hydrant, called out that the hydrant was broken—it wouldn't

Figure 2–1. The collapse of this vacant frame dwelling killed Lt. Bob Dolney and critically injured probationary firefighters Patrick Quinn and Steven Fillipelli. (*Courtesy of Deputy Chief (Ret.) Vincent Dunn*)

open or close. The top operating spindle just spun loosely. Lt. Wingert didn't want to advance on this heavy fire with just the 500 gallons we carry in the pumper's booster tank, so he ordered us to stand fast on the front stoop and hit the fire from there. We ran out of tank water without making a dent.

Engine 231 had arrived and began the process of stretching a 3½-in. hoseline into our pumper's inlet, with which they fed us water from another hydrant around the corner. Just then Battalion Chief Frank Tuttlemundo of the 44th Battalion arrived. He had a better idea. "Put that three and a half into 120's tower," he commanded. "I don't want anyone inside this thing: there are no civilians in there and it's too dangerous." The members of 290 were disappointed. This is a young crew working that night and we'd take all the action we could get—even a vacant building. Ladder 120's master stream at 600 gpm was more than a match for this 20 by 40 foot building. The fire was knocked down in less than three minutes. There was one section of the cockloft however that just wouldn't go out. It was right behind the front wall, which was made of brick. (The cockloft is a blind space between the top floor ceiling and the underside of the roof.) The fire continued to burn away in this sheltered space, because the wall and roof prevented the hose stream from hitting it. We repositioned the tower ladder several times in an effort to hit the fire from better angles, but we still couldn't get water on the last remnants of the blaze. The only solution was to take a handline inside and finish it off from there; otherwise we would have been there for many more hours waiting for the roof to burn completely off.

By then, 290 had connected to a working hydrant and had a fully charged line to work with. Chief Tuttlemundo told Lt. Wingert to take our line up to the second floor and finish this thing off. He added, "I only want two guys up there, Lieu. This thing is in bad shape." Chief Tuttlemundo had spent seven years in Ladder 103 as a firefighter and many more years in the area as a chief, and was intimately familiar with the dangers of the vacant buildings in the neighborhood.

Benny started up the stoop for the building entrance and I followed him: after all, I was the backup man. The chief was at the front door checking the interior stairway for danger as we approached. Benny went first with Lt. Wingert right behind. I was just about to enter when the chief stopped me. "I only want two men in there, son. This isn't worth it. If anything happens, I only want to have to look for those two men." I started to say, "But my boss is…" but he cut me off. "I'll tell your lieutenant that I told you to stay outside. Now go down off the porch here." As I turned and made my way down the stair, Battalion Chief Frank Tuttlemundo climbed the stair to the second floor to feed hose around the banister as I would have done, and also to make sure nothing bad happened to 290. He succeeded. In another five minutes we were pulling the line out of the building and packing it onto the hosebed. The fire was out.

That was my second night tour in a row. We had about 15 more runs that night, none of which I remember. But I remember that one. At 9:00 the next morning our shift was finished. Half of the platoon would come back again at 6:00 that evening, for their second night tour. BC Tuttlemundo was one of them. The following morning my wife woke me up with terrible news. "It's just on the television. There's been a collapse at a fire in Brooklyn. They said a fire chief is dead and a bunch of firefighters are critically hurt." I knew instantly, right in my heart, who that chief was without any confirmation: Frank Tuttlemundo. It was another vacant frame, this time on Osborne Street in Brownsville. But this time there were two squatters trapped by the blaze. There was heavy fire on two floors of the sagging structure, and Tuttlemondo knew they didn't have time to save them. He went inside to order Engine 227 to withdraw when the entire three-story frame building fell on top of them. Tuttlemundo was killed and 5 members of Engine 227 were severely injured. The squatters died as well. Finally, the sense of mortality was taking hold. Until that day, we all thought we were invincible. We had lost another good boss.

CHAPTER 3

Being "Broken In"

Little boys and girls see the gleaming red trucks wailing by enroute to a blaze and get the urge at some point. If you ask the child, "What do you want to be when you grow up?" the reply quite likely would be, "I want to be a firefighter." Those of us who are in this vocation might correct them: "I'm sorry, but you can't do both. You can't grow up *and* be a firefighter. It's one or the other." Nearly all firefighters share a youthful enthusiasm and desire to help that they never outgrow. They do not get as jaded as others by the injustices of the world or the pessimists around them. They know their job is to make things better. They are among the givers of this world, where so many others are takers. The world could use a lot more like them.

I don't know what makes men and women choose to become firefighters. I do know they certainly don't do it for the money. Hell, most of the firefighters in the United States are volunteers and don't get paid a dime! I know some career firefighters are doing it for the paycheck or the benefits. It used to be a career with good job security before firefighters were laid off along with cops, sanitation workers, and teachers, when city budgets were overspent. But they'll never get rich doing this job; in fact, most of them have to work two and three jobs, or have their spouses work, just to make ends meet. I know a lot of firefighters in New York City who used to be cops, but got tired of being punched, spat on, and shot at: tired of being ridiculed for doing their job, and tired of being portrayed as the "bad guys" by every activist and pandering politician with an agenda. I guess getting out of the radio car, even if they go to work in the very same neighborhood as a firefighter, is a relief for these civil servants. I can't tell you what motivates all 11,500 of us, but I can tell you about the few hundred that I have had the privilege to work closely with in the course of my 50 years as a firefighter, and about what got me started. In my case, the love of the job began early in life.

LOVE OF THE JOB

I grew up in the fire service. My father, John W. (Butch) Norman Jr., was a past chief with the Inwood Fire Dept. on Long Island, New York, where we lived, and on Saturdays and Sundays, he worked as a relief dispatcher from 7:00 a.m. to 7:00 p.m. As a result, my brother Warren and I would get to take him lunch and dinner. While Dad ate, we would play on the firetrucks, waiting until we could return home with the empty plate. I still remember the first fire I went to; it was 1957, and I was five years old. The Inwood Fire Department is an approximately 100-member, all volunteer organization of neighbors dedicated to helping neighbors. In the late 50s and early 60s, New York State was in the process of cutting a swath right through the middle of the community, in order to build a section of Interstate Highway 878, intended to speed access from New York City to the oceanfront communities of Long Beach and Atlantic Beach. To do this, hundreds of homes, including the one we were living in at the time, were to be demolished. During the period from the time when the occupants were evicted until the wrecking crew showed up, which could often take several months, many of these vacant buildings were the target of the arsonist's match. This was to be another of those nights.

The fire siren atop the firehouse sounded the alert: four long whines, a pause, followed by four more—a general alarm! My father, who had served his term as chief from 1952 to 1954, was itching to go to another blaze. The pace of the fires kept those men highly motivated. There was only one problem. He was babysitting me. My mother and younger brother Warren were away for the evening, so Butch was unable to respond. A few minutes later, the siren went off again: another round of four, followed by four. It might have been another alarm for a separate call but, more likely, it meant that the chief arrived at the first call and found a serious fire, known as a "working fire" or "signal ten." Butch was pacing the floor now, trying to figure out who he can call at this late hour to come over to watch me. Then in the distance, the distinct mooing sound of the neighboring community's (Lawrence-Cedarhurst Fire Department) fog horn alert could be heard sounding its own alert: four then two blasts, a request for mutual aid. They were called to send assistance to Inwood. That did it. Butch scooped me out of bed, wrapped me in a light jacket and picked me up in one arm. Out the door we went, running the four blocks to the firehouse with him carrying me the whole way.

When we arrived, we found the place deserted because everyone was already at the scene. There was only one rig left in the building: a 1934 Mack pumper. The department's other two pumpers, a hook and ladder truck, a floodlight truck, and an ambulance had all responded. My father placed me in the right-hand bucket seat and showed me the two hand holds, telling me to hang on tight. There were no doors or roof on this truck. Then he ran around and jumped up into the driver's seat. He slammed the floor-mounted starter button with his left foot while the right feathered the throttle. With a throaty roar, the little straight-six engine coughed to life and Butch engaged the clutch. We rolled forward, out through the big swinging brass doors into a deserted street. Dad pressed the siren anyway to make sure everyone knows we're coming. There was a huge orange glow and a massive, moving gray cloud of smoke silhouetted against the pitch black night sky as we rolled up Wanser Avenue. Turning north onto James Street, I caught my first sight of a raging inferno.

Flames had totally consumed one huge old wooden home on Maiden Lane, and spread to others on each side, as well as to surrounding trees. It looked like the whole world was on fire to me. While the other involved homes were also vacant, the one to the right was still occupied. The sky was dotted with flaming embers, like orange snowflakes falling on the other homes, yards, and garages to the north. Dad was told by a chief in a white helmet to take a hydrant on James Street near Mott Avenue. The chief shouted orders to a number of men, who descended on "my truck" and stripped it nearly bare of equipment. Hoses were pulled off nearly as fast as a man can run. My father was connecting the pump to the hydrant with the big heavy suction hose. I called down to ask if he wanted help. He shouted back, "Stay up there and don't move!" My head swiveled in a hundred directions at once, trying to take in all the sensations around me: the impressive colors of the flames leaping skyward. I could feel the heat on my face as I turned toward the flames, warm in the night chill even though we were hundreds of feet away. The pungent odor of the wood smoke mingled with the acrid pitch of burning asphalt shingles. The noise of shouting and running men dashing into blazing buildings, smashing windows to let the smoke out, chopping at the roofs with axes, more sirens and air horns of approaching rigs arriving to help their neighbors confine this conflagration, all served to command the attention of the helpless spectator perched atop the throbbing pumper. These were tremendously powerful images for a five-year-old. I was hooked for life (fig. 3–1)!

Six months later, I saw my next burning building. I was walking to school after being hustled out early by my mother, who was obviously very upset. The house directly across the street from my school was still a smoldering shell. Sometime in the middle of the night this occupied home had caught fire. This time my father went out without waking me. As we made our way along the sidewalk, past the cordon of police barricades, my teacher came out to take my hand and hurry me away from the other kids and into the

Figure 3–1. The members of Hose Co.1 of the Inwood Fire Department pose on Memorial Day 1953, with their 1934 Mack pumper. Chief John W. (Butch) Norman Jr. is first on the left of the front row, in a white hat.

side door, shielding me away from the scene. Later that day, my Aunt Jean and my great-grandmother, "Nana" Hicks, picked me up from school. Only the next day did I find out that my father had been severely injured fighting that fire. Two women, one in a wheelchair, were trapped on the second floor of the blazing home. Butch had fallen off a ladder trying to reach them. He broke his back and was in a full body cast for nearly six months. It was only years later that I discovered what made him so dedicated to firefighting that he went back at it, even after recuperating from the terrible back injury: his own mother had burned to death in their family home while he was in high school (fig. 3–2).

The Norman family's involvement with the fire service began on a horribly tragic note when my paternal grandmother Elizabeth was cleaning clothes with a flammable solvent one day. Something ignited the vapors and fire spread quickly to her own clothes. My father was in high school at the time and came home to find the Inwood Fire Department at his home and his mother burned to death inside. It was the middle of World War II. Shortly after, upon graduating Lawrence High School in 1943, my father enlisted in the U.S. Navy, serving as a Gunners Mate on two escort carriers where, in addition to being sunk by a U-boat in the Atlantic and coming

Figure 3–2. A Norman family affair commemorates Lillian Norman and Butch's 50th year of service with the Inwood Fire Department in 1996. From left on the backstep are ex-Captain Joseph; Nancy; ex-Chief John III (author); ex-Captain Warren; and ex-Captain David. All four sons would go on to become officers in the organization. On the side of the truck are grandkids (left to right) Conor, Ashley, Patrick, and John IV. Conor and John are now New York City firefighters, while Patrick became a volunteer firefighter.

under kamikaze attack in the Pacific, he was exposed to firefighting training to try to save his ship. In 1946 he was discharged and returned to Inwood. He decided to do what he could with that training, to help ensure no other child lost their Mom.

Thirteen years after responding to my first fire in June of 1970, it was my turn to officially be sworn into the Inwood Fire Department. It didn't take long to get my baptism by fire. In early July, I responded to a dinnertime fire in a large frame, private house on Morris Avenue. I responded directly to the scene from a block away, since I had to pass it to get to the firehouse. As we pulled up to the block, a woman came running out of the first floor, her hair and clothes ablaze, screaming that her husband was trapped upstairs. Another firefighter knocked her to the ground and started beating the flames out. I ran down to the corner to meet the first due engine which was stopped there, preparing to drop two 3" supply lines as it pulled down to the front of the home. I got dressed on the way, donning a 15-minute Scott "sling-pak" for the first time in a real fire. Normally, with less than a month in the department, I would have been one of the last people into the building, and probably just pulling hose more than anything else. But this situation was different. I was the only free pair of hands with a mask on his back. FDNY Lt. John Baal was leading the handline, and he ordered me to "Get upstairs and get the husband!"

I found the interior stair easily enough and climbed rapidly to the second floor. I had a large hand light and a halligan tool. Quickly I thought of the training that I had received regarding search. I knew that a search has to be systematic. I knew that if I put one hand on a wall and just followed that wall around, I would eventually come right back to where I started. I knew that a search also has to be completed as quickly as possible if the victim is to stand any chance of surviving in a low-oxygen, highly toxic atmosphere. I began my search with a door on the right at the top of the stairs. This bedroom door had been closed, and even though visibility in the hall was poor, this room was relatively clear of smoke. It took only a matter of 30 or 40 sec to search it thoroughly. I made my way back out to the hall and, continuing my pattern, made a right turn. Just about 2 ft away, I found another door to another bedroom.

This door was open, however, and searching it took considerably longer due to the smoke. The results were still negative. Continuing to my right, I came to another doorway. I looked in below the smoke with my light and saw one wall about 2 ft ahead and walls to each side: a closet, I thought to myself. I moved on. The next room faced the front of the building and, since someone outside had vented its windows, it too was relatively clear. At this point, my air bottle's alarm began to sound, so I made my way back to the staircase, where I met two more members arriving to search. I told them to go to the left; that I had covered all but the one room in the front, just to the left of the stairs.

When I got to the street, the situation had worsened. The fire had begun in the cellar, and little headway was being made. Gasoline stored there kept reigniting around the members, injuring several of the original hose crew. I was directed to get another air bottle and act as relief on the 2½" line that was needed to advance into the cellar against the heavy fire. Afterward, the more senior firefighters did not act very differently than I had ever seen them act, so I just assumed that every fire was like this, with trapped people, firefighters needing assistance, and lots of hard work before anybody gets a break. It set a good example, even if it was not a truly common event. It taught me incredibly early that we have to be prepared for the very worst conditions at every alarm. As for the missing husband, he had jumped out of a rear window when the fire broke out and was probably coming around the side yard while I was running down to meet the engine.

I was privileged to have some great leaders and teachers to break me in: guys like Butch Borfitz, Bobby Dorn, John and Joe Baal, and many others. It was a great place to be a young firefighter. The department was doing a lot of fire duty. In fact, one night in 1971, we did as much work as I have done on even my busiest nights in New York City. It was a Saturday night in the beginning of August. I was at the firehouse washing the trucks for a coming parade, when the first alarm came in for a jackknifed gasoline tractor trailer that was leaking. We set up foam hoselines, in case the gasoline ignited, and stood by while the oil company pumped the gasoline out into another truck. While we were operating there, an alarm came in for an abandoned house on Walter Avenue, three blocks away. The street was loaded with a number of vacant houses waiting to be demolished to make way for a large public housing project known as 385 Bayview Avenue. The area produced a lot of fires, and I swear the ghosts of those buildings were assimilated into the new garden apartment complex that replaced them, because the department continues to go to a lot of fires in these new buildings to this day.

When the alarm came in for this second incident, Deputy Chief Bobby Dorn took command with my company, Hose Company 318, responding "first due." We arrived to find a one-story bungalow, maybe 20 feet wide by 30 feet deep, with fire showing out of about six windows. Dominic "Chow" Mari was on the side of the pumper that the fire was on, so when the rig stopped in front of the building, Chow had the nozzle. I became the backup man. We moved through the fire easily. The vacant house had a low fuel load, and with all the windows vented, the smoke was not too bad. In those days we didn't wear masks for every fire. We didn't really need them for this kind of fire; you took a deep breath and put your face down as close to the floor as you could get. We called it "sucking the nails out of the floor boards." You advanced the nozzle toward the fire for 8–10 ft until you reached the next window. There, you paused long enough to hang your head out the window to take a couple of breaths of fresh air and then started moving forward again. You opened the nozzle only when you were hitting fire, not smoke, in order to avoid upsetting the thermal balance that keeps the hot smoke up near the ceiling, over your head, and out of your lungs. That smoke was primarily what firefighters call Class A combustibles: wood and paper products. While not good for you, it was at least tolerable if you did inhale it. The products of combustion that are encountered today often have a heavy concentration of plastic materials, so highly toxic and so extremely irritating, that even small exposures can kill. This fire was extinguished in about 10 minutes. It was then that we became aware of a large crowd gathered out in front of the house.

It was a large, rather raucous gathering, drawn to the excitement of the blaze. While in the process of washing down, soaking any smoldering embers around the window and door frames, Chow swept the hose stream across the window. The spray went out through the opening and soaked 10 to 12 teenagers on the sidewalk. This upset them greatly, not because they got wet (it was a very warm, August evening); rather it was the teasing of the others around them that got them agitated. The next thing you know, a few rocks and sticks flew toward the window and the youths yelled threats, "We're gonna get you honkies!" The police were called and the crowd dispersed quickly. This was not a serious riot, just a group of teens who were trying to show how tough they were in front of their friends, even though they looked like a group of drowned rats. Under the watchful eye of the Nassau County Police, the hoselines were rolled up to be replaced with clean dry hose on our return to quarters.

We were no sooner back into the firehouse when the alarm rang, reporting another vacant house fire, three doors down from the first one. We hurriedly threw four lengths of 1½ in. hose onto the hose bed for each of the two preconnected hoselines we carry. We coupled them together en route. This time, I was standing in front of the nozzle when we arrived, so it was my turn to be the nozzleman. Chow was my backup. This house was nearly identical to the first one, but the fire was out every window. Somebody must have used at least five gallons of gasoline to get this much fire going less than five minutes after we left the area. The gasoline burned off before we arrived, so it didn't really make the fire any more difficult. It did make the place reek of the gas fumes, though. Ronnie Spinelli and Frank Balzano stretched the second preconnect and crouched behind me and Chow as we waited for the hose to be charged with water. Ronnie told me he'd go to the left when we got inside and I should go to the right. This complicated the attack somewhat. We had to make sure we didn't have "opposing fields of fire" where one line drives the smoke, heat, and flame toward the men on the other line, but the added fire extinguishing capability of the extra 125 gallons of water per minute they would provide might be useful. And they would be our backup, in case we lose water in our line for any reason, like falling glass slicing through our hose, or flaming asphalt shingle tar burning through it.

This fire was no match for the two lines, and it too went out in a matter of minutes. Overhauling was a little more complicated this time, however. The vacant house was stripped of everything of value and the strippers punched holes in walls and ceilings to get at the copper wires and brass pipes that were hidden inside them. The holes allowed the fire to penetrate the plaster, reaching the wooden lath, studs, and ceiling joists more readily. We had to spend a considerable amount of time pulling the remaining ceilings and walls open to trace the paths this hidden fire took. It took us nearly an hour to fully extinguish the blaze. Again, we dragged the hose, tools, and ladders back to the street. The crowd was back again, this time being held back behind police lines away from the building, but their shouts reached our ears, "You'll be back, motherfuckers!" This time we packed the wet hose directly back onto the pumper. We believed them!

That night we did six working fires, and I had the nozzle or the back-up position on the first line for all of them. With the gasoline truck incident, a number of rubbish and brush fires, and an overturned automobile with people trapped, we responded to a total of 15 runs. While I've done more runs on other nights, and have had far tougher fires, that night equals any other in my fire service career for the number of first due working fires. For a 19-year-old, it was pretty exciting stuff. I loved being a firefighter. I had my heart set on a higher calling, however—I wanted to be a career firefighter, getting paid to do it full time.

A family friend and fellow Inwood firefighter, Joe Baal, had opened my eyes to the New York City Fire Department one evening in 1969 while I was still in high school. Joe was a firefighter in Engine 230 in Bedford-Stuyvesant, Brooklyn. One afternoon around 4:00 p.m., I was checking the tools on my pumper at the Inwood Firehouse when Joe walked in. I had asked him several times before what it was like to be a firefighter in such a busy neighborhood. His usual reply was, "There's nothing else like it. It's hard to describe." That afternoon I mentioned again how I wondered what it was like: what went on in the firehouse and the neighborhood. This time his reply was, "C'mon, I'll show you." He was on his way in to work the night tour and he took me with him.

On the way in, Joe gave me the "cook's tour" of Bushwick and Bed-Stuy, pointing out the scene of recent large fires; a church on Bushwick Avenue, a storefront on DeKalb Avenue, and a vacant theater under the elevated train tracks along Broadway. As we passed a large, open area along Park Avenue, Joe pointed off to the right. "One block up there is box 359, the box that strikes fear in the hearts of Brooklyn firefighters!" "Why is that?" I ask. He then tells me about the numerous multiple alarm fires that have occurred in the proximity of the alarm box. The area was full of large old four-story wood frame multiple dwellings. These fire traps were heavily occupied by large numbers of people, often immigrants with large families. Fires there put the lives of the dozens of occupants and the firefighters alike at great risk. Now the buildings were being torn down, part of an urban renewal project that would see the Woodhull Hospital complex rise from its ashes. Many of the area firefighters were glad to see them go.

That first night in a Brooklyn firehouse was a career-forming event. The 1930s era building housed two units, Engine 230 and Squad Company 3. The building was only one bay wide, however, which meant that the two trucks would be lined up one behind the other. Since the two companies did not always respond to the same alarms, this meant that at times, the truck in front would be blocking the unit that did have to respond. This produced a rivalry among the units to see if the one in the rear could beat the other company to the apparatus and be ready to respond before the front unit could get out of its way. It kept everyone on their toes, ready to run for the rigs the moment an alarm sounded. The problem was that at the time, in 1969, the alarms were sent to firehouses throughout the city by a telegraph system which used the number of the alarm box to alert all the units in the borough. In the event of an alarm, the dispatcher pressed a telegraph key which caused a bell to ring in every firehouse in that section of Brooklyn. The bell would "tap out" the number of the alarm box being received. One firefighter, called "the house watchman," was charged with counting these bells and then consulting a book that listed every box in the neighborhood. The book would tell this member which units were to respond and in what order, whether first due, second, or third due. Each unit also had large charts posted around the firehouse that listed the boxes that they responded to, up to the third alarm. On a busy night like this one when I was there, the bells kept up a frenetic pace. It was nearly nonstop. When a box came in, everyone counted the bells. If one of the units was assigned, those guys were up and moving for the door even before the house watchman could call out the assignment, shouting loudly throughout the apparatus floor—"3757, Varet and White Street, Squad goes!"—even as Lt. Carino climbed into the cab.

By the time the shift ended at 9:00 the following morning, my life's path was well planned out. I wanted to be one of these guys: a real busy urban firefighter.

While that might have been my plan, it was not my father's. To Butch, I was the best chance that a Norman would get to go to college. Not that my three brothers, Warren, Joseph, and David, or my sister Nancy couldn't do the college routine, it was just that we couldn't afford it. We were poor. Not in spirit or in the things that count like family, friends, love, caring, and fun, but certainly monetarily. I was actually showing some academic ability. I had received a New York State Regents Scholarship for any state school, and an honorable mention in the National Merit Scholarship competition. In addition, I had expressed an interest in the United States Naval Academy at Annapolis in order to become a U.S. Marine officer, and had begun the candidate selection process. Unfortunately, as soon as the Navy dentist looked in my mouth and saw that I had two chipped front teeth (a flying rock), and two molars that had been pulled in my early teens, and several cavities, I was ruled ineligible. That news came in April of 1970. I had no other plans for college and the fall was approaching quickly. I had not even applied to another school besides the Naval Academy. So my father asked Joe Baal and his brother John to talk to me. Both of them had attended Oklahoma State University (OSU) majoring in fire protection technology. While Joe was a firefighter, his brother John was a lieutenant working in Squad Company 4 at the time, the busiest fire company in the City of New York. I idolized these guys. If Oklahoma State was good enough for them, it damned sure was good enough for me. That August I arrived at OSU in Stillwater, Oklahoma.

My two and a half years in Stillwater were certainly time well spent, even if I did not graduate. I completed all of my fire protection classes, and lived in the campus fire station working as a student and then part-time paid firefighter with the Stillwater Fire Department. My path was clear in my mind. I was going to be a New York City firefighter. I eagerly participated in all the fire related classes and activities at OSU. I went through the week-long essentials training and was assigned a shift on the pumper responding out of our station. Within a week of this assignment, my experiences in Inwood proved useful.

At about 7:30 in the evening, we were dispatched to a report of a fire in a garden apartment complex under construction. En route, we could see a huge column of smoke. This was going to be some job! When we pulled up, there were at least 20 apartments burning. Fire was spreading down the rows of plywood-covered buildings at an alarming rate. The ladder company from our station took a position ahead of the fire to try to cut it off. The attack began with two 1½ inch lines, flowing a grand total of about 200 gallons of water per minute. The fire laughed at the effort. As I was riding the back step of the pumper coming into this scene, I was wishing we had one of Inwood's pumpers with their preconnected master streams. I had already used the "deck gun" on our 1958 Mack at a fire that looked smaller than this one. I had watched other Inwood firefighters put it to good use just a few months earlier when a very large wood frame church burned and threatened to spread to several nearby wooden church buildings, including the pastor's home only five feet away from the church. I knew that this fire was screaming out for that kind of water. After helping the driver connect the pumper to the hydrant, I reported to the attack crew. An assistant chief directed me to stretch another line and get ahead of the fire. At this point the two preconnected lines were stretched, both 1½-inch lines. The streams of water they were throwing were too short to even reach the fire and the tremendous radiant heat prevented the nozzlemen from approaching any closer. I turned back to the pumper to get a line. There was one additional 1½-inch line left, but this fire obviously required a much larger stream. I opened a compartment on the side of the apparatus and began searching through the equipment for a large old brass play pipe that I knew was stored there. When I was being oriented to the ways of the Stillwater Fire Department during the previous "Rookie Week," several of the upper classmen who did the orientation showed me the old nozzle and told me I "had better make sure it stayed polished. It was an antique! It would never be used at a real fire since the modern fog nozzles were so much more efficient." But I knew that with its 2½-inch inlet and a 1⅜-inch discharge behind the main 1⅛-inch tip, I could put it on a length of three-inch hose and flow almost 500 gallons per minute. Since we wouldn't be dragging this hose inside the building, I had planned on doing this by myself (everyone else was busy).

Just as I located the "antique," I heard the chief shouting behind me, "What are you doing? I told you to stretch another line!" "Chief, you said stretch *another* line, and that's exactly what I'm doing," I said as I turned to the rear hosebed with the play pipe. I pulled off three lengths of 3-inch hose, connected the nozzle to one end, and connected the other end to the discharge of the pumper, then told the driver to watch for my signal to give me water. I dragged the hose over the red Oklahoma clay and moved into a position in front of the fire's path. There I made a loop in the hose, just like the instructors had taught us last week, and yelled and waved my arm to signal for water.

The moment was caught on film by a local news photographer and ended up on the front page of the next day's paper. In the photo, as I am preparing for the 3-inch line to be charged, are two 1½-inch lines, not even flowing a drop of water, as their operators retreated while trying to shield themselves from the radiant heat. The large line did exactly what it was supposed to do. It provided enough volume and reach to hit the buildings ahead of the fire and prevent their ignition, while alternately hitting the head of the approaching fire, knocking it down enough to further reduce the threat of fire spread. What it didn't do was placate the chief. The next morning, the fire was the talk of the fire station. I had an 8:00 a.m. English class, and when I returned, the Chairman of the Fire Protection Department of the University, Dave Ballenger, a huge bear of a man, was waiting for me. He took me back into the sprinkler system laboratory behind the firehouse, where we would learn the design and operation of fire sprinkler systems, to give me another, more imminent lesson (fig. 3-3).

"Son, what do you know about following orders?" he asked. I knew right away where this was going. Professor Ballenger was part of the University faculty and I was, technically, his responsibility. The fire department officers had control over us at operations and training and while on duty every third night, but discipline was handled through the University. "Sir, I know the chief was upset, but he did not specifically order me to take that inch and a half. He said, 'stretch another line,' and that's what I did. It was obvious that the inch and a halves couldn't do anything," I explained. Ballenger was in a little bit of a bind. The fire department must have made a complaint to him that he had to address, but he was also a professor of fire protection, and

Figure 3-3. Stillwater, Oklahoma: firefighters find themselves greatly outgunned at this first garden apartment conflagration. The two 1½-inch lines at the right center are being driven back by radiant heat. The 3-inch line is about to begin operations immediately in front of the chief in the white coat and helmet. (*Courtesy of Stillwater News-Press*)

he knew that I was right. Hell, there was even a picture on the front page to prove it and the photographer dropped off a whole series of shots that showed the entire operation from start to finish. "Try not to get in the chief's way for a while," he counseled. I promised to do my best to avoid problems, but I was sweating bullets: in trouble already!

That evening at dinner, my mood was lifted greatly by a brief conversation with one of the associate professors, Frank Gorup. Frank was very highly regarded by all of the personnel involved in the fire protection program, from the students to faculty, to the administration, as well as the members of the Stillwater Fire Department. He had earned his reputation as a very knowledgeable, street-smart kind of guy who was as tough a firefighter as he was an instructor. Frank liked things to be done right.

I was seated on the long wooden bench at the underclassmen's table, I was jabbering with my new best friend, David Rose of Bayou La Batre,

Alabama (who would go on to become a district chief with the Mobile, AL fire department) about the events of the previous evening, when Professor Gorup entered the dining room. "Which one of you guys is Norman?" he asked. I slowly raised my hand and waited for it. I knew I was about to get reamed again. He made his way directly over to me and placed his hand on my shoulder. "I hear you did a good job last night, kid. Keep up the good work." Then he was off, pausing to say hello to a few of the upperclassmen. All of the new freshmen at our table were as surprised as I was. To have Frank Gorup say you did well meant a lot to us. To me, it was a sign of vindication for my not following the exact intent of the chief's direction the previous evening, even if I did comply with what he actually said.

The fixation on small hoselines by some of the members of the Stillwater Fire Department was not unusual in that era. In fact, Stillwater was light years ahead of many fire departments, especially departments in rural areas, who often relied on ¾-inch booster lines that only flowed from 10 to 15 gallons of water per minute. Stillwater was flowing 95 gpm, to them a huge improvement, which indeed was sufficient 95% of the time when dealing with a typical one-bedroom house fire. The department was handicapped by their apparatus layout; there were no preconnected 2½-inch lines nor preconnected deck guns like we had in Inwood, and the officers didn't get to see very many really big fires, so they had a hard time recognizing when the 1½-inch should be left on the truck and replaced by a 2½-inch line or a master stream. I had no trouble making that distinction. I had been "broken in" by guys like Joe and John Baal, after all.

The following spring we were assigned to a "garage" fire on the west side of town just north of Sixth Street. As we were responding, we could clearly see the huge column of smoke in the sky, telling us this would be serious. By this time, I had been assigned to ride the squad: the first attack pumper. I and a number of the other serious students of firefighting had taken it upon ourselves to increase the punch that this unit carried. We were not allowed to do what many of us wanted, which was to mount a large master stream on top of the truck, but we were able to create an attack bed out of the 3-inch supply hose, to which we attached the 2½-inch playpipe nozzle, now outfitted with a 1¼-inch tip. It wasn't preconnected, but it removed several steps from the process of putting a big line in operation. We were ready this time. The only problem was this was not really our fire department.

On arrival, we found four garages, all grouped together like a square at the rear of the driveways, and all ablaze. The fire extended to the rear of the four houses that they abut, two on our block and two directly behind them facing the next street. A wind from the west was blowing the fire directly between the homes toward us. After charging the hydrant supply line, I returned to the apparatus and bumped right into the assistant chief, who promptly ordered me to stretch—you guessed it—"another line!" A quick look around the side of the squad showed that the first two 1½-inch lines were stalled in the front of the houses, unable to advance due to the tremendous quantity of heat and flames being blown at them. The nozzle firefighters were actually shielding themselves by hiding behind the front corners of the houses. This time we were able to put the line in operation much more quickly, so that by the time the chief realized what I was doing, I was already calling the driver to charge the line. When the firefighters on the first 1½-inch line saw the big line in operation, he put his 1½-inch line down and moved over onto the larger line. Where previously the two smaller lines could not advance, and their streams were being vaporized before hitting the seat of the fire, the 325 gpm line now made fast work of this blaze. In less than two minutes, the exterior of our two houses and the fire in the four garages was knocked down. Now the smaller lines could advance. They made their way into the rear of the houses and finished mopping up the inside fires.

This time the chief was much less upset. Again, the big line had proven its value in a most impressive manner. The nozzle operators on the smaller lines were all quick to point out the difficulty they were having and the way conditions changed so rapidly once the larger line went to work. After that, it was a little easier to get the big line pulled when the situation called for it. Other people were being "broken in."

CHAPTER 4

The Death of a Building (and a Neighborhood)

East New York is a neighborhood that has been ravaged by an arson epidemic that has lasted more than 50 years. The neighborhood in the 1930s and 40s was home to many Eastern European Jews and Italians. Many of the brick homes still had gold leaf lettering bearing the names of the various doctors and lawyers who lived there, etched on the glass transoms over the entrance doors. During the late 1960s, the neighborhood was transformed by new immigrants, and the wave of arson began, as the old landlords fled their neighborhood, abandoning their buildings to those left behind. The causes of this wave of arson were myriad. The growing social unrest of the turbulent time saw many rebel against all whom they saw as authority figures; arson was a tool to get at "the man." The fact that the fires burnt out the stores and houses where people worked, shopped, and lived was lost on these morons. Yeah, they chased the store owners out, but it made their lives and the lives of their parents and grandparents that much worse.

It doesn't take much of a fire to kill a building when the landlord doesn't care about the building anymore. One evening in November of 1980, the companies from the Sheffield Avenue firehouse, Engine 290 and Ladder 103, along with the neighboring units, Engine 231, Ladder 120, and Battalion 44, responded to an oil burner fire right around the corner from the firehouse on Pennsylvania Avenue. As these things go, this was not a big deal. The burner had been malfunctioning and a puddle of unburned oil had accumulated in the bottom of the firebox and leaked out of the box into the boiler room. The next time the burner ignited, the pooled oil also caught fire, making large quantities of black smoke and burning out the control wiring for the burner motor.

Engine 290 stretched a foam handline and extinguished the fire in relatively fast order. I say relatively because setting up for a foam operation is different from our usual stretch of hose. We had to change nozzles, stretch the hose to the correct location, then break the connection, put a foam eductor on the discharge, reconnect the 1¾ inch hose to the eductor discharge, get out the cans of protein foam concentrate, open them, and insert the foam educator pickup tube—all before we could start flowing water. That took added time. In the meantime, Ladder 103 had vented the cinder block-covered boiler room windows, which were needed to keep the heroin junkies from breaking into the boiler room and stealing all the copper wires and brass pipes. After the fire was out, the boiler was out of service. We had to kill the electricity to the oil burner motor to prevent another fire. The motor itself would require replacement, as would some of the valves, the sight glass on the boiler, and the light fixtures on the boiler room ceiling, as well as the cinder block itself. The total cost of the repairs would be somewhere around $3,000 to $4,000.

This was a rather large building—four stories high, 80 feet wide by 100 feet deep, "O" shaped, and housing 40 families. Those 40 families woke up the next morning with no heat or hot water, screaming at the super and cursing the absentee landlord (who almost certainly had lots of heat and hot water in his home) as they headed off to work. The landlord apparently decided that the return on his investment would not be there, so he refused to spend the money on the building. (Part of the problem is the city's rent control laws, which prevent landlords from raising rents. This keeps their profit margin so low that many landlords never recover financially from a minor catastrophe like this one, and simply abandon the building.) The spiral downward had begun.

Up until the time of the oil burner fire, this was a good, solid, relatively clean building, located very near a shopping strip and the Livonia Avenue elevated train line of the subway system. With the power off and the window open, the junkies descended like a pack of jackals, stripping everything that can be sold as scrap metal. Wiring was ripped out, as was the piping and light fixtures. Aluminum, copper, brass, and cast iron all ended up in stolen shopping carts and pushed down Pennsylvania Avenue to the scrap metal dealers along Flatlands Avenue. The cost of repairs grew by leaps and bounds. The building was dying.

The lack of heat and hot water drove the families from the building. By the end of January, there were only a half dozen families left, struggling to keep from freezing to death with what we call "creative heating" methods: turning on all the burners and the oven of the gas stove; plugging multiple portable electric space heaters into electric circuits that were never meant to supply more than one; at times running several extension cords from adjoining, abandoned apartments or even the building next door to get more power, without having the use show up on their own electric meter; using illegal kerosene heaters; even building wood fires in bathtubs or in barrels set in the sleeping rooms on the wooden floors! This creative heating is one reason why fire activity increases so much during cold weather. Usually after a week or so of subfreezing weather, the spike begins. It's one reason that most firefighters don't want to be on vacation during this time of year. They don't want to miss the height of the "winter offensive."

The city is also part of the problem. These families, trapped in a building with no heat, hot water, or other services, have few alternatives. The rent control laws have resulted in a shortage of apartments, since it is not very profitable to build a new building if you can't charge the rent you need to support the costs. The only place that many of these people can find is in public housing. The problem with that is the huge waiting list to get into these "housing projects." For people living in a building like the one on Pennsylvania Avenue, the years-long waiting list were unbearable. City policy made an exception to these lists, however. It gave immediate priority to a family who was left homeless because of a fire. Voila! Instant fire storm! Faster than you could say "flick your Bic," people began having fires. To make sure they did so, the city gave them even more incentives. Besides moving into a brand new apartment, the people were given new furniture and $1,500 for new clothing. All that was needed was a copy of a fire report listing your apartment as being damaged. To top it all off, at this time there were only about 50 fire marshals assigned to the Department's Bureau of Fire Investigation, and they were charged with investigating 40,000 fires a year! The situation was creating hundreds of thousands of fires, and it was known to everyone except the politicians. It was a common occurrence to drive by a building and see a moving truck out front in the late evening, and then come back to that same building later that night for a "job." These "midnight movers" might or might not let the other families in an apartment house know that "there's gonna be a fire!" Some families slept with a bag of clothes packed near the front door and another near the fire escape.

By the time I was assigned to Engine 290 in 1979, this situation had been going on for nearly 15 years. There had been so many fires, for so long, that there were thousands of blocks in the city with only one or two buildings left on a square city block. On some, there were none. On the firehouse block, which up until the 1970's had been solid four- and five-story tenements and apartment houses, sheltering thousands of occupants, the only building that still housed people was the firehouse. The shells of three five-story tenements on the next street towered over the two-story firehouse, sheltering rats and other vermin, and offering some of the locals a place from which to push old refrigerators off the roof and down onto the unsuspecting firefighters in the kitchen below (figs. 4–1 and 4–2).

The situation was even worse in certain sections of the neighborhood. In one 16-square block area between Blake Avenue and Belmont Avenue stretching from Miller Avenue to Cleveland Street, there were so few buildings remaining that we nicknamed the area "the airport" because as one old-timer put it, "You could land a 747 out there and never hit a building." Many of the buildings that did remain standing had been the scene of multiple fires and had been vacant for 10 to 15 years, exposed to the elements and in danger of collapse. That fact was brought home to the city's firefighters the hard way. The entire neighborhood was dying. Anybody with the resources to do so fled the area in order to protect their families from the slow-motion firestorm going on around them, as well as the rampant crime that large numbers of vacant buildings and the addicts they attract tend to create. The homicide rate in the neighborhood's local precinct (the

75) was so high that it was tops in the city for many years. The precinct even adopted its own motto, based on that of the local all-news radio station that boasted "You give us 22 minutes, we'll give you the world." The cops of the 75 rephrased it, not that inaccurately, to: "You give us 22 minutes, we'll give you a homicide!"

Figures 4–1 and 4–2. The destruction wrought on East New York during the 60s and 70s is evident in the contrast between these two photos. In the 1950s, East New York was a prosperous middle-class neighborhood, and apartment buildings like the one shown to the left of the firehouse completely filled the entire block. By 1979, the entire block had burned down, and was turned into vacant lots: "brick farms," except for three shells of vacant buildings directly behind the firehouse that were also eventually torn down.

CHAPTER 5

Father's Day on the Pier

Engine 290 and Ladder 103 have one of the largest response areas in the City of New York, running from Atlantic Avenue on the north, down to Jamaica Bay on the south. The companies respond from Brownsville on the west, to Cypress Hills on the east, and are assigned to nearly 400 first alarm boxes. As a result, these companies have been in the top 10 companies for runs every year for over 50 years. At the peak of their activity in the mid to late 1970s, Engine 290 was doing over 8,000 runs per year: an absolutely incredible pace. Activity was so high that a second section of Ladder 103 was created, Ladder 103-2, to provide additional ladder company coverage in the area, while Squad Company 4 was also assigned to the area, operating in 290's quarters every third night as an engine company to boost coverage. During this time frame, fire apparatus left the two-bay firehouse on Sheffield Avenue over 25,000 times a year!

While fires were the heart of the job on "Shitfield Avenue," as the local denizens were sometimes known to refer to their wonderful surroundings, they were by no means the only activity. Heroin overdoses, persons run over by trains on the Livonia Avenue and Canarsie elevated train platforms, as well as heart attacks, shootings, and stabbings, all made for interesting days. One evening, while I was sitting in the four- by eight-foot wooden house watch at the right front of the apparatus floor, there came a loud pounding on the apparatus door. Now one of the first things I learned on Sheffield Avenue was that "verbal alarms," (alarms received directly from the public who run to the firehouse) are usually serious. The pounding on the door had that ominous "verbal" tone and frequency to it. The second thing I learned on Sheffield Avenue though was to never just raise the apparatus doors without finding out what was going on first. The guy doing all the pounding might be the shootee, but the shooter might just be running up to finish what he started. So, to make sure we didn't get caught in any crossfire, the house watchman was supposed to go take a look by opening the side door which could be slammed shut much more quickly if bullets started flying than the overhead door. It was also in an alcove surrounded by a thick brick wall.

Most of the shootings are drug related—one dealer evicting another from what he sees as his turf, drug sales gone awry, or in retaliation for a previous assault. We call them "public service murders" since, for the most part, no innocent civilians are hit. Of course, as violence goes unchecked, the free use of guns spreads, so that the armed robbery victim is often shot just for the hell of it. On the middle section of the front wall of the firehouse is a memorial to members of the companies who have died in the line of duty. One of the plaques commemorates Fireman Thomas Hitter of Engine 290 who was killed by gunmen on Halloween 1938 while returning to the firehouse with the companies' payroll. East New York has been a tough neighborhood for a long time, long before the blacks and Hispanics moved in during the 1960s. The level and type of violence is astounding, as well as the organization of the groups involved. Every Sunday morning at 10:00 a.m., a group of militants marches up and down Livonia Avenue under the elevated train tracks in tight formation, dressed in camouflage fatigues and carrying wooden "rifles" at Port Arms throughout their exercises. Youths as young as five years old are recruited into this "army." Many go on to bigger and better things. You have to be ready for anything in East New York.

On this particular night, when I open the firehouse door and look out to see who is pounding, the coast looks clear so I call over to a teenager at the big door, "Can I help you?"

The caller, a very tall, extremely thin youth of about 15 is frantic. "My friend's been shot! My friend's been shot! Help me, man! Help me!"

"Where is he?" I ask, since I don't see anyone else on the deserted street.

"360, man, apartment 5E. C'mon, man, hurry!"

"360 what?" I ask.

The youth is already running north toward Dumont Avenue as he yells back over his shoulder, "360 Sheffield, man, 5E. Hurry, please!"

I close the door and run back into the housewatch where I punch the "turnout bell" five quick times as I simultaneously press the "All" switch on the intercom, activating the speakers throughout the firehouse. "Turnout Engine, two-nine-oh goes. A verbal for a shooting right up the block. Engine goes." I give a repeat of the five bells as I call the dispatcher on the voice alarm, telling him 290's responding to a verbal for a shooting, giving the address and asking to have the cops and an ambulance respond.

I climb into the jump seat as the rig sprints out the door. We roll north onto the southbound street, pulling up to number 360 less than a minute and a half after the young man first pounded on the door. I grab the trauma bag and Benny Piegari grabs the oxygen and we sprint into the building. Lt. Clem Wingert is calling to us to slow down and be careful because the shooter might still be around. As we enter the lobby, the kid who called us is in the lobby, holds the elevator door open, urging us to hurry even more. He is clearly distraught with tears flowing from his eyes as I pull up short of the elevator and ask, "Hold it. Where's the shooter?" "Don't worry man, he's gone!" followed by the constant, "Hurry!" was the reply.

The youth leads us into the tiny, creaking elevator which moves excruciatingly slow. It probably would have been faster to run up the four flights. I notice the young man's outfit; he is dressed all in bright red, from sneakers to his baseball cap. The elevator, like every elevator I've ever been in inside public housing developments, reeks of urine. Usually the smell doesn't seem so bad because when we're going to a fire, we have our fire gear on which overpowers the smells in the elevator. But tonight, responding to this EMS run, we leave our boots and coats on the rig. As the elevator door opens on five, the teenager practically explodes out of the elevator, his red shirt and pants making him look like the comic book character the Flash. He leads us to the apartment at the end of the hall, through a neat apartment, to what is obviously a teenager's bedroom. There, on the floor on the far side of the bed, lies another teen bleeding severely.

"Are you sure the shooter ain't coming back?" I ask again as I pull out a 5×9 trauma dressing and start cleaning away some of the blood so I can see where the stuff is coming from.

"No, man, don't worry 'bout the shooter. This was a accident, man. See, there's the gun over there," says Flash, pointing to the corner right behind me. I turn to look and there under the edge of the bed is the biggest handgun I've ever seen in my life: a large-frame, Dan Wesson 44 magnum, with what looks like about a 10-inch barrel. I swear this thing looks like a cannon. Lt. Wingert tells everybody not to disturb anything; the cops will want everything as it was for their investigation. After one or two swipes with the 5×9, the extent of the damage is clear.

The boy has been shot in the left side of his neck just below the ear. The huge slug has ripped through the throat, coming out the other side of the neck before blowing off the right shoulder. "Flash" is rocking back and forth on the heels of his tennis shoes, begging us to "Save him, please! Please save him!" and lamenting the fact that if his friend dies, "I'm gonna do some heavy time, man."

Lt. Wingert is not too happy about having the increasingly excited youth in such close proximity to the gun and, more importantly, to us! He tells the soon-to-be felon to wait in the living room so we can work, but the youth refuses. The sound of approaching police sirens from the street below quickly changes the shooter's mind. No one notices as the pleas to save the victim recede further down the hall. We are all preoccupied trying to do just as he was urging. Benny and Mike Pinsent are performing CPR on the shattered victim while I try to staunch the blood that flows out with each chest compression. We have yanked the blanket off the bed to give us something to soak up some of the blood on the floor just so we aren't sliding all over the tile. The trauma dressings that I have packed into the holes in the neck and shoulder are saturated with blood, as are our pants and our hands and arms. The flow of blood has just about stopped, more likely due to the fact that there is no more left in the body for the heart to pump than owing to my bandaging job.

The first cops arrive, two plainclothes guys from the 75th Precinct— "The Homicide Capital of New York."

"Where's the shooter?" the first one asks: a big strapping Irish looking guy with reddish gray hair and a redder face. Lt. Wingert says, "Oh damn! He was just here. You must have just missed him." "What did he look like?" asks the partner, a shorter, darker Mediterranean looking fellow, probably Greek or Italian. Wingert describes "Flash" to a "T"—red shirt and pants, red high-top sneakers, and baseball hat, six-foot, 150 pounds, 15–16 years old. The two cops grimace at each other. "That's the son of a bitch that was holding the elevator door for us," says the swarthy one, as his partner starts broadcasting the perp's description on his portable radio.

We're still doing CPR a half hour later when the cops walk "Flash" back into the room. The victim is obviously dead. The eyes have glazed over. There isn't even a pulse when we do chest compressions, since there's no blood left to pump, but we can't simply stop CPR because no one present can legally pronounce the guy dead. Only a doctor or paramedic can do that in 1980. Hell, if we had a paramedic there, he would have an ambulance with him, and the guy would be in the hospital. This is 1980, however, and the ambulance service in New York City is in shambles. It is nothing to wait two hours on a busy Friday or Saturday night for an ambulance, and that's for shootings and cardiac arrests!

The fire department, on the other hand, even in the face of drastic cuts made during 1974–75 fiscal crises, has maintained its ability to respond to each and every incident within an average of less than five minutes. The citizens of New York know this and expect nothing less and when they call the fire department, they know help is on the way. They count on help with fighting fires, delivering babies, stopping water leaks from clogged sinks, or pulling people out of the many rivers, bays, lakes, and ocean that surround and dot the city. That's how I came to be on the Canarsie Pier on Father's Day in 1982.

I am working that night with one of my favorite bosses, Lt. Steve Casani: a great officer for Engine 290 with its large number of new firefighters. Lt. Casani recognizes the importance of teaching the newer guys, and never misses an opportunity, whether at drill, on building inspection (BI or AFID-Apparatus Field Inspection Duty), or after a fire. He has only worked in Brooklyn for about two years now, having gotten promoted out of Rescue Company 1 in Manhattan, but he thrives on the activity on Sheffield Avenue. I would learn a tremendous amount from Steve, and my career would be guided by him for years to come. Also working this night is Gene DiMarco. The fact that Gene DiMarco was working in Ladder 103 usually makes for an eventful evening. Gene always liked to blow things up, which kept everyone on their toes. He was always experimenting with different fuel mixtures and fire, as well as pyrotechnics. Firefighters need to understand combustion and fuel behavior, and Gene likes to give hands-on demonstrations of the way fuels can behave.

Like the night about a week prior, when Gene got hold of some "block-buster-type" fireworks that had been confiscated at an apartment house. During drill that night, Lt. Wingert was discussing the way oil burners work, how the "gun" atomizes the relatively high flash point fuel oil, making it easier to ignite. This must have given Gene an idea, because as soon as the officers went upstairs, he was over emptying a plastic one-gallon milk container into every cup and glass he could find and drinking what was left. Then he came over with the empty jug and said, "Hey, John, gimme a hand. I want to try something," as he went out the door of the kitchen. He then went over to the truck's supply locker behind the housewatch where there was a five-gallon can of gasoline for the saws, generators, and other power equipment. "Here, hold this steady," he said, handing me the empty milk jug. He then proceeded to pour about a half gallon of gas into the plastic container. "This ain't such a good idea, Gene," I said, thinking how quickly gasoline dissolves plastic. "Oh, don't worry. Here, roll this sheet of newspaper up long ways." As I started to comply, Gene put the cap on the milk jug and pulled a roll of electrical tape out of the pocket of his turnout coat and taped one of the blockbusters to the bottom of the milk container. "C'mon," he said, leading the way out through the side door, around past the front of quarters, to the large vacant lot on the south side.

Actually, by the late 1970s, the entire firehouse was surrounded by vacant lots except for directly behind quarters, where the burnt-out shells of three tenements remained: too close to the firehouse to knock down without risking a collapse onto quarters. All the other buildings on the block, a block once solid with four- and five-story tenements, had been demolished:

victims of the fire storm that destroyed much of the rest of East New York and Brownsville in the late 1960s and 1970s. The lots are now Gene's testing grounds—the place where he developed the "Cremora Blowtorch," which consisted of a piece of pipe filled with the nondairy creamer and connected to a high pressure air hose. When the air hose is turned on, the fine powder is blown out of the pipe toward a candle or other open flame. The flame ignites the powder in a miniature dust explosion. This was usually done from just inside the apparatus doors as an interchange company pulled up. The sight of the fireball rolling out of the firehouse welcomed them to beautiful downtown "Shitfield" Avenue.

Out in the middle of the lot, Gene placed the fuse of the blockbuster on top of the rolled newspaper and told everybody to get back as he ignited the end of the paper. A minute later, the flame reached the blockbuster, which blew the plastic jug to pieces and ignited the gasoline, blasting it skyward. The effect was a rolling fireball worthy of a Hollywood special effects crew. The effect must have been particularly impressive when viewed from the engine office because, as we walked back into quarters, here came Lt. Wingert sliding the pole, yelling at the house watchman to turn out both companies. "There's been an explosion out on Livonia Avenue!" "Oh, shit," I thought, "we done it now!" Thinking quickly, Gene stepped out through the apparatus door as it went up and called in, "Where, Lieu? I don't see anything." Lt. Wingert stepped outside as he buckled his turnout coat, certain there must be a huge fireball coming out of the liquor store under the el tracks, and was startled when the entire scene was absolutely peaceful. He looked up and down the block but there was nothing amiss. He yelled at the house watchman to cancel the call to the dispatcher. Fortunately for us, the blockbuster had totally vaporized the gasoline—not even a smoldering ember could be seen in the rubble-strewn lot. The good lieutenant trudged-back up the stairs scratching his head and muttering aloud about "something fishy around here."

The night tour of this particular Father's Day has been typical of most others in the neighborhood: about a dozen runs by midnight, mostly rubbish and car fires, false alarms, and one mattress fire on the third floor of a building on Hinsdale Street, one of our busiest blocks. By midnight, things have settled down as, one by one, guys drift out of the kitchen heading up to the bunkroom or into the TV room to flop on a couch. At 1:45 a.m., the tone alert and the huge bell at the housewatch announce another run. The house watchman calls out the information. "Box 2199, Rockaway Parkway and the Belt Parkway, BARS." This is one of the longest runs in 290 and 103's response area. Both companies are second due to a pull box. Usually, this is a false alarm or maybe a car fire on the Belt Parkway: nothing to get excited about, as there is only one small maintenance building on the pier at that location. As we're going down Pennsylvania Avenue, the dispatcher calls the 58 Battalion and tells him they are now receiving reports of a car in the water off the Canarsie Pier. Rescue 2 and EMS are now being assigned. Lieutenant Casani leans back into the window separating the cab and the crew compartment and yells over the screaming siren and airhorn to me and Tommy Pellegrino to get out of our fire gear and get ready to go swimming. "And bring your masks!"

When we get to the pier, there are already a dozen emergency vehicles there: Engine 257, Ladder 170, a half-dozen or so police cars, a police harbor launch, and a couple of ambulances. The edge of the pier is lined with emergency responders and curious bystanders. Lt. Casani tells us that if anyone is in the car, he, Tommy, and I will have to try to get them. Our firefighting SCBA are not SCUBA-rated, but Casani knows that they can function underwater down to about 25 feet. As we reach the edge of the pier, we are told there's no car involved. It is just two men who fell off the pier. We can see them both floundering around in the water below. Lt. Casani hands me his mask and tells me to take the masks back to the rig. Ladder 170 has lowered a 20-foot ladder from the pier and secured it with a rope so that the two men can climb up. I return to the edge of the pier to watch. It appears that one man is trying to help the other onto the ladder but the second man is totally unresponsive. It turns out they are brothers, 30 and 32 years old. Both had been drinking heavily, and one apparently stumbled and fell off the pier. His brother jumped in to save him.

After several minutes, it is clear that the one man is incapable of climbing the swaying vertical ladder. The other brother is becoming increasingly desperate as he struggles to hold his now unconscious brother afloat. The water temperature, while not frigid, is cold enough to sap the man's strength in combination with his ongoing efforts and his state of inebriation. As I stand at the end of the crowd watching this unfolding drama, I can see the men's increasingly desperate plight. Ladder 170 and 103 have lowered one end of their life-saving ropes to the men, with instructions to "tie him in it and we'll pull him up." The one conscious man is unable to hold the other afloat and simultaneously loop the rope around him. Suddenly, there is nothing left in him. He releases his grip on his brother and the ladder, and both men slowly sink below the surface, being carried outward by the current.

I work my way closer to the spot where the men disappeared in the murky black water and jump off the edge of the pier. As I hurtle toward the surface 15 feet below, I think to myself I probably shouldn't be doing this. I don't know what is under the black water. I don't want to land on top of these guys and break their necks or push them deeper or, worse, land on a jagged piling that might be just out of sight below the water line. Out of the corner of my eye, I see a flash of red hair paralleling my flight. Suddenly, I hit the surface feet first. I had removed my shoes to get in to my boots back at the firehouse, and I had taken off my boots when Lt. Casani said to get ready to go in the water, so I am in my socks. Now, as my head goes under water, my foot strikes something below. The sharp pain tells me it is not one of the brothers; instead, it is a barnacle-encrusted piling sticking up from the bottom. "Shit, that hurt!" When I stop descending and start to rise toward the surface, I somehow force my eyes open underwater, looking for either of the men. Nothing. I twist from side to side, but still see nothing. I return to the surface for a breath.

From up on the pier, Lt. Tom Dunphy of Ladder 103 directs me out and to my left. I inhale deeply, put my head down and kick for the bottom. I am not a particularly good swimmer, and the long pants and shirt don't help. The water is really muddy and even with a bunch of floodlights directed from the pier, I can barely see a few inches in front of me. I grope vainly with my arms, much like when conducting a search in heavy smoke, as I swim deeper and deeper. Finally, when my lungs can take it no more, I turn and struggle for the surface. On my way up I think I see a flash of white off to my left. I reach for it but don't find anything. As I break the surface, I grab another breath and look around. The surface is unbroken except for a slight chop. Back down I go, convinced that flash of white was the T-shirt of one of the victims. Again, I swim in a type of breast stroke, trying to make my arms cover as much area as possible. Then suddenly there I am, face to face with the unconscious victim. His eyes are wide open and rolled up into his head. I almost swallow a quart of the polluted Jamaica Bay as the shock of bumping into this guy hits me. Then instinct takes over and I grab his arm and kick for the surface.

Breaking the surface with my victim, I see Gene DiMarco pushing the other brother toward the ladder. Neither man is conscious. We are going to need help here. I grasp for a rung of the ladder as I struggle to hold my victim's head above water. Gene does the same. The lifesaving rope is dangling there between us. I pull my victim closer and blow a breath into his open mouth, trying to begin resuscitation. Gene is much larger than I am and has looped one arm through a ladder rung. He tells me that he'll hold the two guys up while I tie the rope around them. I push my victim over to him and grab for the rope. My arms are getting tired now as I start to tie a bowline knot around the guy. Gene says to "Forget that! Just loop the rope around the guy's chest and snap the snaphook back onto the rope!" This will allow the rope to tighten around the guy's chest, which could break some ribs as he's hauled up, but at least the guy will survive. I do as I'm told and Gene gives the guys on the top the OK to haul. Fifteen people pull on the other end of the lifesaving rope, and the man practically flies up out of the water like a yellowfin tuna. In a couple of seconds, the rope is lowered back down and the process is repeated with the other brother. Then it's our turn. Gene tells me to go first. I start up the ladder, but when I put my weight on my left foot a pain shoots up my leg. I can't go up. My mind flashes back to the second after impacting the water's surface when my foot struck the piling. I must have hurt it more than I thought. Now that things have calmed down a little and the adrenaline has stopped pumping, the pain begins.

Gene calls for the rope again, this time for me, and even though I protested, I knew there was not much alternative. I yell up to the pier edge to go slow, I just need a little help. I hobble up the ladder, followed by Gene. At the top, what seem like a dozen pair of hands reach out to help pull us up over the edge. Lt. Casani directs two guys to help us over to the chief's car where Gene and I are wrapped in blankets. We are then driven to Brookdale Hospital, the same place our two victims were taken. We are given tetanus and gamma globulin shots for the ingestion of polluted water, and my foot is treated. It has several deep lacerations from the barnacles, and my ankle is sprained, but it could have been worse. I can't help but shudder to think what could have happened if I had jumped six inches or so to my left, where the piling would have come up between my legs. My son John might have been an only child. The fire department doctor shows up at Brookdale after about 2 hours, and I am placed on medical leave, while Gene returns to duty at the firehouse.

MEDAL DAY

When I return to work three weeks later, the excitement surrounding our rescue has passed. Lt. Casani and Lt. Dunphy have written Gene and me up for a meritorious act. Deputy Chief Emmanuel Skillings, of the 15th Division, recommended us for Class II awards, which indicates the members acted in the face of great personal risk, based on the eyewitness report of the 58 battalion chief who had seen the entire event. The report was forwarded to the Board of Merit for consideration. That September, department order #140 of 1982 announced that Fr. 1st Grade Gene DiMarco of Ladder 103 and Fr. 2nd Grade John Norman of Engine 290 had been awarded Class III awards for their actions at Box 2199. This helped restore some of the excitement to Sheffield Avenue again, since the receipt of a Class III award is still a pretty big deal, especially for a guy in an engine company. It practically guarantees that at the department's Medal Day Ceremony the following June, the member will be awarded one of the department's medals for valor: a really big deal. The key word here is "practically."

There are only 45 medals for valor awarded by the job, but no limit to the number of Class I, II and III awards, as well as Class A and B citations, that can be given in any one year. After the year ends, the Board of Merit meets again and reviews all of its previous awards. It then tries to evaluate the degree of danger each involved, compared to all the others, and awards the department medals in that order. Well in 1982, there were more meritorious acts performed than there were department medals to be presented. You have to be able to guess who didn't get one. Some say it was the fact that ours was a "water" rescue and not a "fire" rescue that doomed us. Others say it was the fact that two of us got awards for the same incident, since if the job gave one of us a medal, it would surely have to give one to the other guy as well, and with medals in short supply, well…who knows? Anyway, both Gene and I would have our turn at Medal Day another time. For now, it is just another way to pass a night tour on Sheffield Avenue.

The incident marked the end of a segment of my career in another way. It would be one of the last times I would work with Lt. Casani in 290, for shortly after the Medal Board decided that Gene and I would not be going to Medal Day, Lt. Casani received the news he had been awaiting for several months that he would be leaving us. He was being transferred to Rescue Company 3 up in Washington Heights. I was deeply saddened to see him go. He was a truly great fire officer.

CHAPTER 6

Harlem, USA

About three months after Lt. Steve Casani had transferred to Rescue Company 3, he called me back at Sheffield Avenue and asked if I'd be interested in transferring to Rescue Co. 3? I was just about astounded. Me, going to a rescue company? That's a dream come true. At the time, there were over 350 fire companies in the city of New York, but only four rescues. Rescue 1 covered Manhattan from its southern tip down at the Battery up to 110th Street. Rescue 2 covered all of Brooklyn and Staten Island. Rescue 4 covered Queens, and Rescue 3 covered Manhattan from 110th Street north, which includes all of Harlem, as well as the beautiful borough of the Bronx. The rescue companies respond to every working fire in these areas, plus every other life-threatening type incident such as subway crashes, building collapses, and major auto accidents. That means that the rescues go to a lot of fires, which makes them a very sought-after assignment.

Since there are nearly 100 fire companies (engines and ladders) for every one rescue, the selection process allows the captain of the rescue a lot of talent to choose their personnel from. Nearly all of the rescue captains use the same criteria when evaluating a candidate. They want a candidate with at least five years on the job, preferably with at least a few years in a busy ladder company: a person who has shown that they can perform on their own, someone with leadership and initiative. The member must also possess some skills that the company can use, such as being a rigger, welder, carpenter, or mechanic or someone that is good with tools and knows buildings.

Unfortunately for me, in June of 1983 when Steve called, I was a little light on some of these qualifications. I was pretty handy with a cutting torch, and was a NY State Hurst Tool instructor, and my seven years as a fire protection engineer had given me a very strong background in building construction of all types, but I only had about three and a half years on the job, and three of them had been in Engine 290. I'd only been in Ladder 103 for about six months. I was concerned that Captain Bill Ryan would be upset that some kid with just six months in a truck company thought that he could come up and play in the major leagues with the big boys. But Steve told me not to worry, that I had a lot of other things going for me. My status as an advanced emergency medical technician helped, since we deal with a lot of trauma and burns to firefighters and civilians, and my exploits on the Canarsie Pier the previous summer with Steve showed my willingness to jump in and do whatever needs doing. But the thing that apparently convinced Capt. Ryan that I was worthy of the opportunity was my certification as a New York State Hazardous Materials instructor.

Rescue 3 had just recently been designated as the backup hazmat unit. Rescue 4 was the primary hazmat unit for the city, but they needed additional people at a major incident, and a backup in case they're busy somewhere else. Ryan wanted a guy who could operate in these type incidents without having to spend a lot of time training him. I could basically hit the ground running, so that and Steve's recommendation helped convince "Capt. Billy" that I would be an asset. I got the nod about a month later and started at "Big Blue" just after Labor Day of 1983.

My first day tour in Rescue 3 happens to be the day of Dick "Pop" Hannon's 59th birthday. Now 59 might not be old in many professions, but 59 years old as a firefighter in a rescue company is positively *ancient*. This is a very physical job, one that can beat up even a young man. At roll call, Lt. Joe Spor assigns me as "Pop's" partner as part of our roof team. This team carries the Partner K-12 circular saw, the 150-ft long lifesaving rope, and the normal truck tools, Halligan hooks, and Halligan tools. Our assignment can vary, depending on the occupancy and type of building, as well as the location of the fire, but *the* fire problem in the Bronx in the 1970s and 80s is the "H-type" building. "H-types" are so named due to their shape when viewed from above. These can be absolutely huge build- ings, with lines of apartments in each of the vertical lines of the "H," and the stairs and elevators usually located in the horizontal segment, which we call the "throat" connecting the two "wings." A serious fire in these buildings can trap dozens of people above the fire. Every apartment must have two exits, one of which is usually an outside fire escape. But a fire can trap people in rooms where they can't get to either the door or the fire escape window. It is for these reasons that our team brings the lifesaving rope, informally known as the "roof rope."

Every year somewhere in the city, firefighters are tied into one end of the roof rope and lowered over the edge of the roof of a burning building to pick up civilians who are trapped at a window where there is no way to get a ladder to them. While I was in Engine 290, we went to a fire off Sutter Avenue where Joe Dirks of Ladder 103 made a *double* roof rope rescue. Joe was being lowered by Gene DiMarco, who had no substantial object to anchor to. A woman was trapped on the fourth-floor rear. Joe went over the side and Gene lowered him down to level with the windowsill. When he got there, Joe was shocked to see not one, but two women trapped! He tried to explain that he could only carry one person at a time and that someone else would be coming down to get the next victim, but they were having none of that. As Joe grabbed the first woman and began to be lowered, the second woman, driven to desperation by the approaching fire, decided to climb out onto the rope, on top of Joe and his charge and go for a ride. The victim couldn't grip the half-inch rope well, and slid down. The sudden extra weight nearly pulled Gene DiMarco off the roof, sending all four people attached to the rope for an express ride down four stories to the rear yard! Russ Bengston of 103 managed to jump on top of Gene just in time, and the two of them held on for dear life. As the three people on the travel- ing end of the rope passed the second floor, the "hitchhiker" finally lost her grip. Despite Joe's best efforts (he could only use one hand, he was holding the other woman with the other) the woman fell the remaining 12-14 feet into the debris strewn rear yard. She was bruised and beat up, but she was alive and happy to be so. When Engine 290 was dragging our line out of the building after the fire was out, she was still jumping up, down, and all around out in the street screaming about the "firefighter that saved me!"

Now I am given the rope and paired up with Pop Hannon. The guy who carries the rope gets to go over and make the rescue, which virtually assures the member of receiving a department medal the following June on Medal Day. Now I am not shy about these kinds of things, but out of respect for Pop Hannon's birthday, I ask him whether he would like to switch assignments. I'll take the Partner saw, if he wants the rope. (The saw and its roof kit weigh about 30 pounds, while the rope weighs nine. OK, so I *was* thinking that maybe a 59-year-old guy wouldn't want to carry an extra 30 lbs. if he could carry nine instead, but I'd never say that to Pop.) At first Pop asks, "Where'd you work, kid?" My reply of "290 and 103" doesn't seem to raise any eyebrows, so Pop says, "Yeah, I'd like that, thanks." Not 15 minutes later, we go out the door for a second alarm in the Fordham section of the Bronx.

The fire is on the top floor of a huge H-type off the Grand Concourse. Fire has extended into the cockloft and is threatening to burn the entire top floor off. The only way to stop it is to cut plenty of large holes over the fire to let it go up and out of the building, and to allow the engine companies to press home their attack with hoselines. I step off the rig with the saw and tool bag slung over my back and with a six-foot steel Halligan hook and Halligan tool in hand, searching for a way to reach the roof. This is a tall, seven-story building with a high decorative parapet wall. Ladder 59's 100-foot aerial looks to be fully extended, and going nearly straight up as I ascend to the turntable. I look down at Pop, who is shaking his head and muttering something like, "Why is it always the top fucking floor?" as I start up the ladder.

It takes me probably a minute and a half to scurry up the ladder to its pinnacle, and then clamber down from the parapet. I see Ladder 59 and 33's roof men are cutting away with their saws and have got one good hole about 8 feet by 8 feet over the fire. I ask what they want done, and 33's roof man says he thinks the one hole they've got and the one they're working on should be enough, but I should take my saw and start making some inspection holes (small triangular openings about eight inches long) over toward the throat, to see if fire is threatening to cross into the "A-wing." I unsling the saw and drop the saw bag with its spare blades and wrenches and a canteen full of gasoline off to one side out of the path of travel. I cast a quick glance around for Pop, but I don't see him through the smoke. I take the saw and make a quick set of three offset plunges of the blade into the thick tar and gravel roof, forming a triangle.

The roof here is built of so many layers, due to previous roofs being left in place and new layers added on top of them, that the four-inch deep blade doesn't cut all the way through! Usually, when you make one of these triangular holes, the piece either falls through into the hole on its own, or it is easily pushed through with the heel of one's boot. That is, of course, unless you have managed to cut right on top of a roof joist (Murphy's Law is always in play!). This one is different. The roof is so thick that the saw blade hasn't cut through the wooden roof boards. I'll need a Halligan to pry out the piece of roofing, and then I'll have to make a second, smaller triangle inside the first to get through the roof deck. I sure hope the fire hasn't traveled this far yet, because if we have to cut a vent hole or a trench over here, it is going to take a *very* long time. I look around for Pop, for some assistance, but I still don't see him, so I head over to the aerial to see if something has happened to him. Suddenly, a six-foot hook falls off of the parapet, and then just as suddenly, here comes the "roof rope." I look up and see Pop's helmet appear next, out of the banking and lifting clouds of smoke. "Whew, that was a long one!" he says, "Thank God you took that saw from me, kid. You're all right in my book." I start to explain about how thick the roof is and how I need a hand with cutting and pulling the inspection holes but Pop just looks around and pronounces, "Aahh, they got it, kid. They don't need no more holes." Smoke is still coming out of the two huge holes that 59 and 33 have made, but it is mixed with steam now, which is a good sign, and the hole closer to the throat is nearly all steam, meaning that Engines 43 and 75 are beating the hell out of "the devil" with their hose streams. I consider that this is good, that the fire is contained to a relatively small portion of the building and damage will be kept to a minimum, but Pop expresses why he considers this important. "Thank God, now we'll be able to go down the stairs. I really didn't want to have to go back down that damned aerial!" What a way to start a new assignment!

While the Bronx is famous for its H-types, the tenement is *the* fire problem in Harlem. Harlem tenements are large, non-fireproof, multiple dwellings, often seven stories high, but they are not subdivided into wings like the H-types. "New law" tenements, built after April 1901, make up the majority of Harlem's housing stock. Typically, they house four to six families per floor, and anywhere from 25 to 40 apartments. Like their larger cousins, the H-type, they have a common cockloft on the top floor that allows rapid horizontal fire spread, but since the buildings are usually only 50 feet wide by 75 to 100 feet deep, this is not as critical a problem as with the larger H-types. New law tenements are so named because they were meant to make life safer for the occupants compared to the "old law" tenements, by requiring noncombustible stairs and fire-resistant doors on apartments, as well as other fire safety improvements. That doesn't make them totally safe though. If the occupants of the fire apartment leave their door open, the stairway is still impassable, and fire can extend vertically through pipe chases and the space around the steel columns and "I-beams" used to support the wooden joists of the upper floors. The tenement is a very solid building, which can contain a lot of fire without being in danger of collapse. That enables firefighters to get inside and get close to "the devil." It also means we can have several fires in these buildings before they get knocked down. Once the cultural mecca and capital of black America, Harlem has a very high population density and some of the most deeply rooted abject poverty in a city known for its excesses of wealth.

Streets like Bradhurst Avenue and Edgecombe Avenue in the 1980s were like the streets of Beirut, thanks to the crack trade and the abdication of authority by the city's liberal government. Arson is one of the crimes this situation breeds. Rival drug gangs firebombing each other's "houses," absentee landlords looking to "sell" their buildings to the insurance company before they flee the neighborhood for good, and the aberrant behavior or just plain carelessness of people whose crack-scrambled brain has just about lost its grip on reality, make Harlem a very active area for fires. The fire companies of the 12th, 16th, and 25th battalions are among the city's most active and most experienced.

One night in the winter of 1986-87 leaves a vivid impression on me. For weeks there had been a serial arsonist running around Harlem instilling terror throughout the neighborhood. This guy has progressed from torching vacant buildings to stores closed for the night, until now he is so brazen that he is burning occupied tenements with dozens of people in grave danger. He is so audacious that he is setting one fire after another, sometimes within sight of each other! He even has a pattern of sorts: he picks a particular avenue, either Lenox, 7th, 8th, or Bradhurst, and works up and down that street for a night, branching out only slightly on the side streets on either side of the north/south avenues.

This particular night, it is 7th Avenue's turn. 7th Avenue above 110th St. is called Adam Clayton Powell Jr. Blvd. for the long-term Harlem congress-man, while 6th Avenue is Lenox, and 8th Avenue is Frederick Douglas Blvd. The firefighters still use 7th and 8th instead of the much longer formal names for speed as well as indicating direction; the lower numbers are to the east, higher to the west.

This evening's affair begins fairly early, just after the change of tours at 6:00 p.m., with a fire on 7th Avenue at 140th Street involving a vacant two-story commercial building. Engines 69 and 59 have each stretched 2½-in hand-lines and are beating the hell out of the fire, while Ladders 28, 30, and 23 are "opening up." Chief Meagher (pronounced Mar) of the 16th Battalion has special-called Ladder 14 as a tower ladder in case the fire breaks through the roof, since the building runs the length of the block from 139th to 140th Streets, and two large six-story tenements butt up against it, one on each street.

The first alarm companies are doing their usual bang-up job, so Chief Meagher tells my boss, Lt. Pete Lund, to "just stand fast, Rescue," meaning stand by to see what develops. Ladder 14 has been directed to do the same thing, since it looks like the fire won't get through the roof. As the "irons man" for this tour, I am carrying the traditional forcible entry tools of the FDNY: a Halligan tool and a heavy axe (eight-pounder, as opposed to the usual six pounds), as well as a new addition to our bag of tricks, the Rabbit Tool.

The Rabbit is a miniature set of hydraulic jaws designed to spread a door away from its jamb. The jaws are powered by a hand-operated hydraulic pump that is connected to them by a five-foot length of hydraulic hose.

The tool comes in a heavy duty, plastic-coated carrying bag with a shoulder strap. At 25 pounds, it's not light, but the ease and speed with which it can force most inward opening doors makes it invaluable at fires in tenements, with their high life hazard and large number of doors that must be forced. Rescue 3 got one of the Rabbit tools directly from its distributor, Clemens Fire Tools of Maryland, after one of our guys heard about how useful it was. Initially, most of us were skeptics. Hell, we have some of the best forcible entry guys in the country, often forcing 15–20 doors a night above the fire in zero visibility. "We don't need no stinking rabbit!" was one of the less tart descriptions. But after the guys used it a few times on some tough doors, they soon came to appreciate it as "the best thing to come along since canned beer." When I have "the irons," I like to take the Rabbit out of its bag and carry it with the pump slung under one arm on a nylon strap and the jaw draped by its hose up and over my shoulders and hanging on the other side. Carrying it this way makes it possible to have both hands free for climbing stairs or ladders, as well as making the jaws instantly accessible for use. Slung like this, it's possible to insert the jaws in place with one hand, while operating the hydraulic pump with the other. The jaws exert a force of 8,000 pounds, and open five inches, which is usually enough to force the most common residential locks in one try.

While we are standing fast in front of the vacant store fire, the captain of Ladder 14 has been eyeing the Rabbit slung around my neck and finally sidles over. "Is that the new tool I've been hearing about?" he casually asks. I reply, "Yeah, Cap, it's the Rabbit Tool. It's some piece of work." "The thing looks heavy, is it any good?" he asks. Lt. Lund replies, "It's the best tool we carry." The captain's skepticism is barely concealed as he asks, "Is it really that good?" By now, all of the guys in the Rescue are expressing their opinions of the tool's worth. They're unanimous in their support. "How do I get one of these things for my guys?" is 14's final question before we are told to "take up" by Chief Meagher. The answer to that one is not what the good captain wants to hear. We tell him that we got ours from the manufacturer on a trial basis and "the job" doesn't think we need them so "they won't buy any." You can tell that the captain is not exactly 100% sure that "the job" is wrong in its evaluation.

Shortly after midnight, we're on our way back down 7th Avenue, en route to another 10-75; this one is a fourth-floor fire in a six-story tenement on 131st Street. The fire is visible as we come down 7th Avenue, rolling out of the fourth and fifth floor windows in the rear. Lt. Lund calls Chief Meagher on the handie-talkie as we pull up on the wrong side of traffic on 7th Avenue. "Rescue to the 16" (spoken as one-six). We're in, chief. What can we do for you?" Before the chief can reply, Ladder 14, who had been special-called due to the volume of fire in this occupied building, interrupts. "14 to the 16 [one-four to the one-six], chief. Tell Rescue I need a hand on the top floor. I've got a lot of apartments to get into. And tell them to bring that new tool." It's show time!

We hustle down 131st Street, past the bedlam that surrounds a serious fire in an occupied building. Glass is crashing to the ground all around us as the chauffeurs of Ladders 30 and 40 perform VES on the front of the building, entering the upper floor front apartments to search for trapped occupants. Engines 37 and 36 are stretching another line to back up the one Engine 59 is operating on the fourth floor. Ladder 14 calls the 16 again as we enter the building to tell the chief he's going to need another line on the top floor. There's fire in at least one apartment up there already. Chief Meagher orders his aide to radio the dispatcher for a second alarm. We're going to need the help. At the base of the stairway, a member of one of the engines is dragging a guy into the lobby from the exterior courtyard. He's asking for help. This civilian is in about the absolute worst shape I've ever seen in a human being, and yet still be conscious. He must have been cut off from his escape routes by fire and found himself driven to a window in the shaft as his last resort. When the fire ignited his underwear, he dove right through the window into the shaft, the bottom of which is a half-story below grade, making his flight five-and-a-half stories. He must have landed in a crawling position, for the bones of his forearms, knees, and shoulders are all compound fractured, clearly visible, as is his skull through the badly charred skin. The guy is still talking though, or rather moaning "Help." Lt. Lund turns to order someone to stop and help the sole engine man who dragged the guy out of the way of the falling glass and flaming debris (although there is almost nothing that anyone short of a major trauma team and burn specialists can do for him, other than cool his burns).

As we pass the fourth floor, Engine 59 is doing a really good job on the original fire apartment and Engine 37's line is pushing the fire back into the adjoining apartment. That's a good sign. Ladder 30 is still working on the "primary" search of the two rear apartments. Their irons man is still working on forcing the door to the front right-side apartment. That's the one furthest from the fire, so the occupants there have a better chance of survival, plus they have a front fire escape that is not blocked by fire, so they should have been able to escape on their own. The irons man is working on this door by himself and as I pass by in the smoke- and steam-filled hall, I can make out his form swinging the Halligan tool like a baseball bat, trying to bury the point of the claw into the wooden door jamb. He grunts, coughs, and curses as he does. I think to myself, "This guy needs a Rabbit Tool." But the people that really need a Rabbit Tool are the occupants of the two top floor rear apartments. They're right over a lot of fire. The stairway was blocked by fire early, due to the apartment door being blocked open, and their secondary means of egress, the rear fire escape, is also blocked by flames venting out of the windows. They're in *trouble*.

As we pass the fifth floor, I see Ladders 40 and 28 working on doors. 40 has forced the two rear doors, and found fire in both apartments. They have expended their can and 28's can trying to get into the apartments for a search, while waiting for a hoseline. Now 40 has retreated to the public hall and closed the doors (making sure they don't re-lock), while 28 works on the doors to the front apartment. Lt. Lund waves me on while he stops to speak to Ladder 40's officer. He wants to let him know we're going above, so that 40 can advise us as well as 14 Truck if there's a problem with fire control on this floor. A face-to-face meeting underscores the importance of the contract that they create.

At the top floor landing, conditions are bad. Even with the roof bulkhead vented, the smoke and heat here are very heavy. We are, of course, above two floors of fire. The irons team of Ladder 14 has forced the door to the left rear apartment, and the captain has made a quick primary search of this apartment, while the two firefighters begin work on the door to the right rear apartment. He has just returned to the public hall as I arrive. I ask the "truckies" if they want a hand. "I've got a Rabbit Tool here." "Nah, we'll get it," they reply, but the captain has other thoughts. "Let him in there," he orders and the firefighters grudgingly crawl backwards a foot or two. I quickly insert the jaws with one hand, and begin pumping the handle. After only four or five seconds the door springs open. Ladder 14's captain orders his team to get inside and begin the search. Then, probably not fully convinced that his own guys hadn't done the vast majority of the work on that door, only to have the Rabbit take the credit, he slides over by me and says, "Let me see you do that again," pointing to the doors to the front two apartments.

I crawl quickly across the hall landing and insert the jaws into the door. This door is pretty tight, and I have to stand up momentarily to put my shoulder against the door, to create a large enough gap to fit the jaws in. The heat makes me want to be quick about it. It has the feel of steam, as opposed to solid smoke, condensing on my mask facepiece, and showing lighter in the beam of my flashlight. That's a good sign. We don't have to worry as much about fire venting up the stair from below us. Five quick pumps on the handle of the hydraulic pump cause the door to pop open. Lt. Lund and Ray Grawin crawl inside and begin their search.

"Do it again," commands the captain of Ladder 14. I slide over to the remaining door and again push the jaws deep into the gap. This one goes even easier than any of the others: only four pumps and it springs open. The captain crawls quickly through the doorway, and I follow him down the hall. He moves into the first room on the right, as I continue past to the next room. I'm in a bathroom. I quickly sweep my Halligan tool into the tub, listening for any splashing sound as well as waiting to bump into a body as I sweep. In most neighborhoods, I wouldn't bother with the Halligan, using my hand to probe instead. But in the most depressed neighborhoods, I've learned not to do this, after discovering the hard way that the tubs are, at times, used as toilets after the toilet has been hopelessly broken. A tub full of liquid (not to mention solids), is not always full of water when the apartment is occupied by squatters who are not paying rent. I use the Halligan for my initial probe.

This tub is empty, as is the rest of the bathroom. I meet the captain as I pass by the next room. The smoke condition is getting much lighter as we work our way toward the front windows. The chauffeur of one of the ladder companies has come up the front fire escape and has vented the front windows and begun the search of the apartment that we're in from that direction. I pass him as I enter the last room, and he calls out, "I've already done this, it's negative." I acknowledge with a simple "Gotcha," and continue over to the front windows. I stick my head out the fire escape window and glance over to my left, where I see Lt. Lund stick his head out and look over at me. I call, "This side's negative, Lieu," and he tells me to meet him back in the hall (fig. 6–1).

On my way back, the captain from Ladder 14 pulls me aside. I still have the Rabbit slung around my neck, and he now seems impressed with its

Figure 6–1. Members of Rescue 3, taking up from a 3rd alarm on 114th St. in Harlem. From left to right are FF. Ed Connelly, Lt. Pete "Lt. Vulcan" Lund, the author, and FF. Jerry Murtha. (*Courtesy of Matt Daly*)

performance. He wants to know how much it costs and if I know where to buy one. I have to tell him I don't, but it seems obvious that the Rabbit has made another convert. Lt. Lund promises to get him the information.

As we make our way back to quarters, it is nearly dawn. We pass by a building opposite the New York State office building on the corner of 125th Street and 7th Avenue; we are amused by graffiti on the wall. The first bit, reading "US get out of Nicaragua," where the American government was secretly (or not so secretly to the people of Harlem) backing the Contra rebels against the Communist Sandinista government. As the apparatus pulled forward, the next bit of graffiti came into view through the small side window on the crew box of the rescue rig: "US get out of Central America." So these protesters were widening the scope of their demands. But it was the last bit of sidewalk commentary that drove the firefighters to fits of howling laughter. It read "US get out of Harlem!" Welcome to a whole new world.

One of my most memorable nights in Harlem occurred shortly after this tour. The "torch" was still active. The fire marshals were busting their asses trying to catch him, but this guy was really brazen. For this particular tour, we had an officer who did not usually work in Rescue 3: a lieutenant from Ladder 28, working overtime. It was also the first night in the company for a new firefighter, John O'Connell. John had just been detailed to Rescue 3 for his "audition" from his old unit, Ladder 102 in Brooklyn. John had recently moved to upstate New York, and the commute to Brooklyn was too much for him. Rescue 3, in addition to being a great company for fire and rescue work, was also the first firehouse on the New York side of the George Washington Bridge, making it an especially desirable assignment for firefighters or officers who live in Orange County or Rockland County on the other side of the Hudson River.

On this night, I am again assigned the irons and teamed up with John, who has the hook and, if we're first due, the can. (Being first due in this neighborhood is an extremely rare event in R-3, since we are in the same firehouse on West 181 Street as Ladder 45, as well as Engine 93 and the 13th Battalion.) This night follows its usual routine: the evening hours are pretty quiet for R-3, and I use this time to give O'Connell the tour of the rig and its tools, as well as my take on how things operate in Harlem and the Bronx. Since I also came from Brooklyn, John's interested in my view of the quality of the companies we'll be working with, to hear if they live up to their reputations. Engine 93 goes in and out a few times for car fires and rubbish fires. Truck 45 goes to a few water leaks, and we all go to Box 1701, Columbia Presbyterian Hospital, down on 168th Street and Fort Washington Avenue. This box comes in at least three times a day: always a malfunction. Due to the extreme life hazard and huge size of the place (it's a complex of a dozen or so buildings running from 163rd to 168th Streets, on both sides of Ft. Washington), the mandatory assignment is three engines, two ladders, the rescue, and a chief. The constant false alarms breed complacency. Nobody expects a real fire when we go there. We're not disappointed this time either—same old story, 10-35, code 1, faulty alarm.

After midnight, things usually start to pick up for Rescue 3. We start out for an "All-Hands" out in Soundview, in the eastern Bronx. John Mulroy coaxes the "R" model Mack over the bridge into the Bronx, and as we start down the hill on the entrance ramp onto I-95, the Cross Bronx Expressway, the rig picks up some speed. The Mack needs the assist from the long downhill run to build up speed for a merge onto the highway. It's a 1974 model, and it's already paid its dues. Halfway out the Cross Bronx, just before the exit for the Sheridan Expressway, the 3rd Battalion announces that it's "using all hands, no special units required," meaning he does not want the Rescue, the "All Hands Chief," or the deputy chief to continue to respond. The fire will be handled with the three engines and two ladders of the first alarm. "Mullie" gets off the highway at the next exit, Bronx River Avenue, and swings back around to the westbound side. On the way back, the Bronx dispatcher advises us to "switch back to Manhattan, they've got a run for you." The lieutenant acknowledges and switches radio frequencies to the Manhattan channel. Immediately we know we've missed a job. The Manhattan dispatcher is giving the 16th Battalion a "run down" of the second alarm units for a box on Eighth Avenue. "Shit!" O'Connell exclaims, "We're missing a second alarm." Joe Cullen nonchalantly informs him, "Don't worry kid, there are plenty more where that came from." Even I get a chuckle out of this, even though I feel just as disappointed as O'Connell does. Cullen is a real old-timer who has seen more than enough fire to last a lifetime. He really doesn't care if we miss a fire or if he ever goes to another one again. Like many of the senior members of Rescue 3, he is still affected by the death of Larry Fitzpatrick five years earlier.

Larry Fitzpatrick was a real hero. A firefighter in Rescue 3, he was working on another summer night in Harlem when a firefighter from Ladder 28 found himself in deep trouble. "Jerry" Frisbee had been searching an apartment above the fire in a seven-story new law tenement, when he found himself out of air and cut off from the rest of his company. He retreated to a window in an air shaft and called for help. Billy "Arms" Murphy from Rescue 3 was the first to reach him. Billy had lowered himself from the roof on his ⅜-in. personal escape rope, and tried to help Jerry get himself oriented enough to make a crawl back into the smoke- and heat-filled apartment, and over one more room where there was a window that opened onto a fire escape. The smoke and heat had already taken its toll though, and Jerry was losing consciousness. Billy tried to help Jerry by removing the child guard gate that was blocking the bottom of the window, which would allow Jerry to get his head out the window more to get some fresher air, but removing the gate proved to be an impossible task while swinging from the rope in midair.

Larry was the next one over the parapet. He was even more powerfully built than "Arms," who was so nicknamed because his arms were thicker and more powerful than most men's thighs. Larry was suspended by Rescue 3's ⁷/₁₆-in. thick Life Saving Rope, which was rated to carry two people, as opposed to Murphy's single person rated "escape" rope. By the time Larry reached the window, Jerry was unconscious. Larry squatted down on the windowsill and managed to manhandle Frisbee up far enough to where he could get a grip around his chest. Struggling mightily, he pulled him up and over the window guard gate. As Larry pushed off from the windowsill, with Frisbee now wrapped safely in his bear-like hug, the rope suddenly snapped. Both men plunged 70 feet to the base of the rubble-strewn shaft. They died instantly. The impact blasted its way through the fire department like an atomic bomb. The rope that we use for roof rope rescues broke! Not only did it break, but in coming weeks it would be shown that the department and the manufacturer knew in advance that the rope was not really strong enough to hold two people. The resulting investigation ended the careers of two highly respected fire officers and ushered in an era within the job that made fearless fire officers afraid to make decisions about purchasing new tools and equipment. No one wanted their name on a piece of paper that might later turn up in court if a firefighter were injured as a result of using an item that wasn't time tested. CYA was the motto to live by for many years. Many compare the impact of this event on the fire department to the impact of Vietnam on the U.S. military. It took nearly a decade to get over the fact that we were not being equipped with the very best equipment to perform some extremely dangerous tasks. As a result of these deaths, however, the department formalized its safety division, including the research and development unit, mandating that every new piece of equipment be rigorously tested before being placed in service. That is Larry and Jerry's legacy. That and a new, thicker roof rope!

CHAPTER 7

The Major Deegan Expressway

By April of 1984, I had been in Rescue Co. 3 for about six months and had begun to feel pretty confident about my abilities and skill. Thanks to the drilling and training of some great bosses, guys like Capt. Bill Ryan, Lieutenants Steve Casani, Joe Spor, and Marty McTigue, and the senior men including John Mulroy, Billy Cody, and Pete Ferrante, I knew all of the tools in the gigantic tool box that we call a rescue truck, where they were stored, what the capabilities are of each, and how to use them. Unfortunately, nothing that I had learned so far had prepared me for what we were to face the night of April 20th.

At around 7:00 p.m., a young couple prepares to leave their north Bronx home for a night out. They have hired a young lady to watch their five-month-old son for them while they are out, probably giving the age-old admonitions that all parents give when entrusting their precious child to a babysitter. Then they are gone. The next time they see their child, he will have gone through a life-shattering experience, owing his survival to God and a resourceful group of firefighters.

At about 11:30 p.m. that evening, the babysitter's boyfriend stops by to visit his girl with a bunch of friends. They are all hungry and in a mood to party, but the babysitter remembers the instructions of the parents and decides against bringing the kids inside. But her boyfriend doesn't give up easily. If they can't come in and party, they can all go to the local White Castle (home of the infamous "belly bomber"), and she can bring the baby with her. Since the parents are not due home for several hours yet, the babysitter relents, and out the door she goes with the baby in her arms. Everything must have been happy and festive as she climbs into the back seat next to her boyfriend and another couple. The driver's date is seated in the front seat of the sports car as they pull away from the curb on what is supposed to be a brief jaunt up the expressway for a couple of burgers, shakes, and fries.

The exhilaration of a powerful car coupled with female companionship and youth itself must be at work, or maybe it is simply driver error or inexperience. Whatever the cause, the results are the same. As the vehicle speeds down the entrance ramp to a depressed section of the Major Deegan Expressway at 230th Street, the driver attempts to cut into traffic in front of an oncoming tractor trailer loaded with 80,000 pounds of bananas. The rear bumper of the car catches on the front bumper of the truck as it goes past, pulling the car into the truck's path. The car spins in front of the truck, causing the trucker to lose control. The two vehicles career along at 60 miles an hour, locked together like a rampaging Goliath crushing a defenseless David. Eventually the car slides past the front of the truck, probably to the relief of the terror stricken passengers, but they are not safe yet. On the contrary, they have just gone from the frying pan into the fire, quite literally. The car is now sliding along backwards on the driver's side of the truck, hooked to it at the trailer's set of dual rear axles. The drag on the driver's side of the tractor trailer pulls the big rig further to its left, sandwiching the car between the truck and the three-foot high concrete highway divider as the tortured mess continues its deadly ride, sparks spewing from the twisted steel sliding along the concrete and asphalt paving. As the vehicle nears a halt, gasoline spraying from the car's ruptured fuel tank ignites, and the horror for the trapped passengers reaches its zenith. The car and the rear of the truck are engulfed in an intense fireball.

The Bronx Communications Office receives the first of several reports relayed through the 9-1-1 dispatcher at 11:41 p.m. The callers' descriptions reporting a severe accident and fire prompts the decision dispatcher to "load up the box" for the 19th Battalion. The dispatcher assigns three engines, two ladders, and Rescue 3. The vehicles have come to rest only a few hundred feet from the quarters of Engine Company 81 and Ladder 46, but the Expressway is about 30 feet lower than the adjacent service road at this point, so those companies have to go up to the next entrance and proceed southbound to get at it. In the meantime, Engine 75 and Ladder 33 are trying to reach the site from the northbound lanes, but they are slowed by traffic in those lanes that has stopped because of the accident. Rescue 3 is also approaching the scene in the northbound lanes, but is several minutes behind 75 and 33, since we have to come across from Manhattan farther south. A quick-thinking officer in one of these units advises us to get off the highway and take the service road up, since the traffic northbound is now stopped completely as the companies work to knock down a heavy body of fire. We follow their advice and as we pull up even with the scene on the service road, the red glow below dims and then goes out. We stop at this point and call down to the 19th Battalion to see whether the chief wants us to come down.

From our vantage point at the railing 30 feet above the scene, all we see is a trailer fire that was centered at the rear wheels. From here it seems to be just a truck fire, probably the result of a hung-up brake shoe. There is no sign of any other vehicle. Immediately however, the battalion orders us to proceed up to the next exit ramp and come south in the northbound lanes, "We have a car sandwiched between the truck and the divider!" By the time we make our way down to the site, Engines 81 and 75 have extinguished the blaze except for some smoldering tires on the bottom of the car that they just can't hit because of the position of the vehicles. Most of the gas has been consumed, but the engine companies are applying a foam blanket, just in case. Ladder 33 has its Hurst tool out and operating from atop the highway divider, and has pried open the passenger's side door of the wreckage. When they peer in, they are greeted by a nightmarish tangle of twisted, fire-blackened corpses staring up at them. One man loses his dinner right there. What a mess.

Lt. Joe Spor climbs up onto the barrier with Gene Scherer, a Vietnam era ex-Navy Seal, and sizes up the operation. From the top it appears obvious that everyone is dead, but we still have to check. I am standing by with Pete Ferrante, the chauffeur, to bring over whatever tools the boss calls for.

Shortly he tells me to bring over the Hurst cutters; he wants to cut away part of the roof to gain better access to the bodies. We hook up our cutters and pass them up to Gene, who clips the top two posts. I then crawl in under the truck to get at the bottom posts. As I prepare to cut through the rear "C" post, my flashlight is fixed on a human back, wedged into what used to be the rear window. The skin is charred like a well done steak, but amazingly it is moving! I focus the light on the seared remains and come to the shocking conclusion that the slow, rhythmic movement is being caused by the expansion and contraction of the chest. "Hey, Lieu, we got a live one down here!" I call out. Everyone freezes at this revelation, not sure if I mean what I said, or if I know what the hell I'm talking about. "What did you say?" Lt. Spor asks. I repeat again, "There's at least one of these guys still alive down here at the bottom of the pile!"

Now the race is on. The top layer of corpses is unceremoniously hauled out, skin peeling off and limbs pulling out of the bodies as the crew at the top tries to separate the intertwined mass of human remains. We would never treat a body so callously under other circumstances, but now this is necessary to try to save a life.

To make access to the victim easier, we would like to move the tractor trailer away from the wreckage of the car. Since the tractor portion is not severely damaged, the obvious answer is to simply drive the big rig forward. As the driver is replaced by one of the ladder company chauffeurs, Ferrante and I survey the underside of the wreck to see what will happen if we do that. As I look in between the sets of wheels that are the dual axles, I realize there is a human head wedged in there! Working my way back around to the front of the tires where I could see the back that was breathing, I realize this head is attached to the back. "Shit, this guy's alive with his head between the wheels!" Driving the vehicle forward or backward is liable to crush the skull like an egg. Now what?

Lt. Spor comes down for a look. He concludes that we will have to move the rig sideways. What I want to know is how? Lt. Spor tells Pete to get our rig pulled up in line with the truck's axles, at a 90-degree angle to the truck. He has the Rescue rig's winch stretched over, and he and I wrap the trailer chassis with a chain: then the winch is secured to it. Ferrante engages the winch and draws in the ⅝-inch steel cable. The winch is rated for up to 20,000 pounds, but it doesn't budge the trailer. You may have seen television shows where human beings pull tractor trailers and airplanes and such in "World's Strongest Man" competitions, but that only works going forward or backward where the wheels can roll smoothly. Trying to pull tires sideways doesn't work.

We apply chain saw bar oil to the pavement, in hopes of lubricating it, but still no movement. Then Lt. Spor has another idea. "Get out the airbags and the cribbing we use for lifting subway cars," he tells me. Gene and I scramble out and return shortly with a pair of high pressure air bags rated at 25 tons each. We know from our experience lifting train cars (because people have fallen, jumped, or been pushed under moving trains) that you don't have to lift the whole weight, only part of it, since the other end of the vehicle is supporting the other half of the weight. We crib up under the rear axles' "banjo" and insert the pair of bags, one atop the other for extra height. Spor calls all of us out from under the truck and Gene inflates both bags, raising the truck up about six inches. "Stand clear!" yells Lt. Spor, then he radios to Pete to try pulling again. This time the bags teeter, and then shoot out to the side, as the truck falls off the center of the bags. The truck has shifted over a good four or five inches, though. Spor tells us to get back in there and build the bags up again the same way. We do and soon the rig is six inches off the ground again. Now I see the plan but I think this is going to take too long to move the behemoth over the five to six feet we want in order to fold the car's roof down. But Lt. Spor has another trick up his sleeve. "Splash some more oil down all along the path of the wheels now, so we can get under the tires." We do, and again he calls out the stand clear order. Now he tells Ferrante to lock the winch in place and get into the cab and drive forward, again pulling the big banana truck off the air bags. Pete puts the Rescue into drive, releases the brakes, and eases onto the throttle (this will be faster than reeling in the winch for the five feet that we want to move the truck). Again the blocks and bags fly out in various directions, but when the tires fall onto the oil soaked pavement the rear of the truck keeps going! Unfortunately, before we get the whole five feet we want, Pete runs out of room, the front bumper of the Rescue is nosed right up against the side of the highway cut.

Quickly he jumps out and engages the winch for the last foot. Even as the trailer is being winched the rest of the way, all hands are at work flapping the roof down. Ladder 46 has chained the car back to their rig on the other side of the barrier to keep it from falling toward us as the trailer is pulled away, so we now have safe access to the whole top of the car.

As the last few bodies are separated from the pile and we can reach the living victim, a soul-searing scream comes from the pile of charred remains. I'm looking right at the face of the guy we've been working to reach, and I know it's not him that's screaming. He is mercifully unconscious. That means there is somebody else alive in this charnel house! Frantically we lift our survivor onto a backboard, exposing a baby. "Oh my God," I think. How in the hell did this happen? One of the guys from 33 Truck carefully lifts the baby out of the debris and hands him out to EMS. He is also burned and bloodied, not to mention absolutely terrified, but he is alive. This is some kind of miracle. All I can think of is that if we had left the scene intact until the Police Accident Investigation Squad had finished with their investigation—like they wanted us to—a process that would take many hours, these two people would most likely have died. As it is, I'm not real sure of their chance of survival. They both have horrible injuries. In addition to the broken bones and burns, the young man's arm has been ground off above the elbow as the car slid hundreds of feet along the pavement. The same is true of much of his face and head.

A great team effort using all the skills of the six companies on the scene, including those of a master rigger, has helped save two lives. It tempers the sorrow of seeing those other five horribly burned corpses lined up under the sheets on the side of the highway.

CHAPTER 8

Railroad Pizza

It's late in 1984. I've been assigned as a firefighter in Rescue Co. 3 for almost a year now. The department has recently modified the response policy for our company. Until now, we responded to "all hands" alarms, where all companies assigned on the first alarm have already been put to work. At about this time, the department decided to assign the rescue companies to respond on the "10–75" signal, which is the initial report of a working fire. Rescue Co. 2 in Brooklyn had been responding this way for several years now, and their experiences had convinced the department's command of the benefits of having its heavy rescues start out early for any incident where there is a serious potential for a loss of life. Given the great distances that must be traveled by these units to reach the outlying edges of their response areas, the extra five to seven minutes between the receipt of the 10–75 signal and the transmission of many all hands signals makes great sense, especially since the majority of 10–75's result in the use of all hands. As the old saying goes, "It's better to have the help you need, than to wish you had it." Start them out. You can always send them back if you don't need them.

On this particular fall evening, I am performing housewatch duty in the "booth" of the firehouse on 181st Street where Rescue Co. 3 is housed at the time. "The Big House," as it is known, is unique in the FDNY in that it houses the most FDNY fire companies in service at the time: Engine 93, Ladder 45, Rescue 3, and the chief of the 13th Battalion: a total of 20 men (with the exception of the tours when one woman, FF. Susan Byrne of E93 was working) per tour. This is the largest assignment of personnel in the FDNY since mid-1975 when firehouses like Sheffield Avenue in Brooklyn and Intervale Avenue in the Bronx had four companies, each with six or seven men assigned. There is something else that is somewhat unusual about this housewatch as well. Since Rescue 3 is assigned as the rescue company for all of the Bronx, as well as a tremendously large section of Manhattan (everything north of 110th Street at the time), the housewatch has two radios to monitor fire traffic in both boroughs. On a busy evening, with both radios going, the volume of radio traffic makes keeping track of ongoing operations quite a challenge. To make matters worse, the Bronx radio shares its frequency with Staten Island, so you may hear a box transmitted that sounds like it is in your response area, when in fact it is 30 miles away.

At 6:30 p.m. this particular evening, Hugo Herold and I are sitting in the housewatch booth. Hugo is a very low-key, experienced firefighter who has transferred to Rescue 3 from Ladder Co. 14: a busy truck company in Harlem. Like me, Hugo was looking for the "unusual" incidents that the rescue companies respond to when he decided to transfer to "Big Blue." We're about to get one of those incidents, but neither of us knows how serious it's going to be.

The boss for this tour is Lt. Joe Bryant, newly assigned to Rescue 3. Lt. Bryant is not new to rescue company work, however, having been assigned as a firefighter in Rescue Co. 1 for many years, working alongside my mentor, Steve Casani. The chauffeur is John Mulroy, a senior man with nearly 30 years on the job. Nicky Giordano and John Hendries round out the rest of the crew for the 6x9 tour.

Hugo is regaling me and Joe Morstadt of Engine 93 with classic "Tommy McTigue stories." Tommy is one of the most colorful characters in the firehouse, if not the entire department. He also transferred to Rescue 3 from Ladder 14, and Hugo is filling us in on some of Tommy's antics while he was a member of Ladder 14. I'm sitting with my right ear just inches from the Bronx radio speaker when I hear the tail end of a conversation....

"Your fire is now reported to be an occupied commuter train in the tunnel at 154th and Morris Avenue. We've transmitted your box." Then, "Engine 41 to the Bronx, 10-75 at 154th and Morris. We got a fire on a train."

I spin around on the chair and bang out four bells on the alarm, then hit the loudspeaker switch and call over the speakers spread throughout all four floors of the massive firehouse. "10-75, Rescue goes, Rescue only, we're going to the Bronx, 154th Street and Morris Avenue. Fire on a train." Lt. Bryant slides the two poles from the third floor Rescue office and joins the rest of the crew, who are busily dressing for the response. As the house watchman, I am waiting for the teleprinter to begin printing its urgent message so I can give a printed copy with the details on it to the boss. I wait, and wait some more, but still no teleprinter message. Lt. Bryant has finished dressing now and walks back over to the housewatch booth. With no teleprinter or voice alarm message dispatching us, he wants to know what's going on. I explain that I thought that I had heard a 10-75 on the Bronx radio. "What do you mean you 'think' you heard a run for us?" he asks. I tell him about the message about the fire on the commuter train. He wants to know how I can be sure it's a run for us and not for Rescue 1, which we share the Manhattan radio channel with or Rescue 2, which would respond to 10-75's in Staten Island. I reply that the location is reported as 154th Street and Morris Avenue; I don't know of any 154th Street in Staten Island and 154th Street in either the Bronx or Harlem has to be ours. The lieutenant still seems skeptical. This whole business of responding on the 10-75 is still new to us, and he wants to be certain that we are actually assigned before we start pushing the big "R-Model Mack" through the traffic-snarled streets of Washington Heights and the Bronx.

Part of the problem lies with the fact that while Rescue 3 is responsible for fires in the Bronx, the firehouse is actually located in upper Manhattan. The alarm teleprinter circuits and voice alarm circuits are both routed through the Manhattan Central Office. The Bronx dispatcher has no direct contact with our quarters. Finally the chauffeur, John Mulroy, an old South Bronx firefighter, convinces the boss that 154th and Morris Avenue is definitely our box, since it is located just below Yankee Stadium. He knows the area well. Lt. Bryant gives the go-ahead, and out the door we go. As we leave the apron, the teleprinter finally starts typing the response message, the victim of a "glitch" in the still new system.

"Mullie" points the rig across the Washington Bridge, which connects Washington Heights and the Bronx. At the end he muscles the 45,000-pound behemoth off onto Edward L. Grant Highway, down the hill headed for River Avenue. On the way down the incline, the radio crackles with the preliminary report on the fire.

> "Battalion 14 to the Bronx, at 154th Street and Morris, we have a fire in a passenger train in the rail yards, right at the mouth of the tunnel. The extent of the fire is unknown at this time, but we have a heavy smoke condition in the tunnel and apparently at least one rail car fully involved. We are having difficulty getting water on the fire due to a lack of hydrants in the yards. Trucks are conducting searches for occupants. Notify Metro-North (the railroad that operates the trains in that area) to shut down power in the area."

Now looking through the pass-through opening to the cab, we can see a column of black smoke rolling skywards ahead of us on Morris Avenue. This sounds like it could turn into a real disaster, since a smoke-filled tunnel is no place for a train packed full of rush hour commuters. Silently we each conjure up what kind of chaos there will be inside that subterranean tube. Just as we make our descent down the ramp leading into the yard, the Bronx dispatcher notifies the Battalion that Metro North confirms power off on all tracks from 135th Street to 167th Street. Then we jerk to a stop and the men begin piling out of the rig, heading toward the mouth of the tunnel, which is spewing an inky black, plastic-laden smoke.

A glance around the rail yard produces a number of chilling sights. The first due engine has pulled down into the yard, in proximity to the blazing train. With no hydrants in the area, they have begun the attack with water from the 500 gallon booster tank on their apparatus. That tank is now empty, and only a tiny dent has been made in the fierce blaze. A relay operation, using 3½" hose from the second and third due engines, is now being organized, but will take a long time to get water, as the hose has to be carried by hand over multiple tracks.

The blaze itself now becomes visible for the first time as we thread our way through the tangle of apparatus, hoselines, piles of railroad ties, and rails. One and a half cars of a passenger train are protruding from the yawning mouth of a very wide, six- or eight-track tunnel. The train on fire is on the track nearest us, with heavy smoke pumping from the open door of the second car. As we arrive, the members of the engine company are backing out of the involved car, their limp hoseline offering them no protection against the searing heat inside the stainless steel car. No advance will be possible here until the relay is completed and establishes a reliable water supply. The members of Ladder 55 tell us this is not the main body of fire; it is only the adjoining car. The fire is extending its way down the length of this car while the engine waits for water, and in a short time it too will be fully involved.

The chief has called for several additional units to assist in the relay. Now he calls over his handie-talkie to inquire about the status of the search of the rest of the train. Ladder 17's officer replies that the front 1½ cars have been searched and are negative, but that access to the rest of the train is impossible through the interior. Ladder 55 radios that they are trying to get past the blazing cars along the tunnel wall, but are being stopped by flames venting from the windows and radiant heat from white hot steel along the length of the 80-foot cars. Hearing this, Hugo Herold and I look for a route around to the "outboard" side of the train. It is an old style train, not made for stopping at stations with raised platforms, but equipped with a set of fold down stairs built into the doorway for passengers to climb up or down on directly from grade level. We climb up into the second car, cross over to the other stair, and working by touch in the swirling smoke, manipulate the mechanism to open the far door and lower its stair. I descend the stair quickly, anxious to get past the fire and get a look at conditions deeper into the tunnel, past the blazing cars. As I am about to step out from the bottom step, I get a sudden premonition of impending doom. I am frozen in place, unable to move, despite Hugo's prodding, when suddenly a monstrous, screaming apparition flashes by, not a foot and a half from my face. It is another train, "highballing it" out of the tunnel at 70 miles per hour, its horn shrieking a tremendously loud and shrill warning that would be entirely useless to anyone unfortunate enough to be in its path. I have missed certain death by a split second. One more step would have placed me right on the right front bumper of the onrushing train. I scream into my handie-talkie, "Get off the tracks, the fuck-ing trains are still running!" along with a few choice words for the people who told us that the power was off to these tracks. The chief replies with strict orders for everyone to stay off all the tracks until he finds out what the hell is going on.

After several minutes and numerous calls between the dispatcher and the Metro-North power control office, the problem becomes clear-er. Metro-North had done exactly what they had been instructed to do; they removed all third rail power from their tracks in that area. The problem is that the tracks are also used by two other railroads, Amtrak, and ConRail, each of which has its own power source, separate and distinct from Metro-North's third rail. Amtrak uses an overhead wire to supply electricity at 11,000 volts down through the top of the train, while ConRail uses diesel locomotives on its freight trains. The train that nearly made railroad pizza out of me was an Amtrak express. The train engineer, seeing the blazing cars filling the tunnel with the poisonous soup they produce when plastics burn, knew he could not stop his train full of passengers in that deadly tunnel, so he must have decided to "put the pedal to the metal" and get his passengers out of that inferno as quickly as possible. Fortunately, I was the first firefighter to be approaching that side of the train. So much for being first! Thankfully "the Fire God" was looking out for me that night (like so many others) and stopped me in my tracks. As for the blazing train, it was an unoccupied leftover, having been laid up on a siding for storage overnight. There was no civilian life hazard on board, only the passengers of the passing trains, and (oh yeah!) the firefighters. Always the firefighters.

CHAPTER 9

I'll Make You a Detective!

The day tour of April 8, 1987, begins as a rather mundane tour for the members of Rescue Company 3. The annual inspection of the firehouse is coming up in a few days, so the pace of sprucing up and cleaning is picking up. Instead of a mere sweep and mop of the floors, today is scheduled for a major stripping and waxing job for all the second floor tiles and washing of the walls. It's a big job, given the size of "the Big House on W. 181st Street" in Washington Heights, but since there are three companies in the house, Engine 93, Ladder 45, and Rescue 3, plus the 13th Battalion, there are plenty of hands to "make light the work." That is, until about 9:45 a.m., when the teleprinter sends "Big Blue" heading south, relocating to the quarters of Rescue Company 1. The boys of Engine 93 and 45 act pissed off that we're getting relocated. It doesn't happen very often, and some of them think that one of the guys in the Rescue must have made a call to a friend in the dispatchers' office to somehow wangle this trip, just in time to get out of "committee work." In reality though, this is the real deal. Rescue 1 is operating at a four-alarm fire off Canal Street, which by itself is no big deal, but Rescue 2 is operating at a "man in machine" job that could tie them up for a good while, and Rescue 4 has just responded to an "all hands" in Queens, leaving a large portion of downtown and midtown high-rise areas with no heavy rescue coverage. With only one heavy rescue left in service, the dispatchers' protocol requires them to relocate that unit to the quarters of Rescue 1 to cover the high-rise, high-value district of Midtown Manhattan. This is a bit of a treat for the troops of Rescue 3 because they don't usually let us anywhere near mid-town—plus, we're getting out of some of the committee work on 181st Street (93 and 45 will try to save some work for us we know, but since the work has to get done anyway, they'll do whatever needs doing).

The morning spent in the quarters of Rescue 1 was routine: a small fire in a subway train, an odor of smoke at a department store, and a fire in a restaurant grease duct. Just after we back in to the firehouse on West 38th Street (Rescue 1's firehouse on W. 43rd Street had been destroyed by a collapsing wall during a spectacular 10-alarm fire in the piano factory next door in 1985, so they have been sharing quarters with Engine 34 and Ladder 21 while a new firehouse is constructed on 43rd Street), a group of firefighters from the city of Malmo, Sweden, who are in New York on vacation, stop by the firehouse. They're looking for Rescue 1, and seem disappointed when they find us there instead. We give them the "nickel tour" of the big rescue truck, and these foreign brothers are impressed with the number and variety of specialized rescue equipment on hand: three rotary saws, one with a steel cutting blade; two gas and one electric chain saws; hydraulic rescue tools; air bags; cribbing; three sizes of torches; trench jacks; two generators; pavement breakers; oxygen, carbon monoxide, and combustible gas meters; and hundreds of other tools. As the tour progresses, it is mentioned several times that many of the tools might be used only a few times a year, but when the need arises, nothing else will suffice. This statement will prove to be prophetic on this particular afternoon.

Immediately following lunch, Rescue 1 becomes available again, so Rescue 3 returns to our quarters, only to discover that we have missed "a job" in the Bronx. "Oh, well," the old-timers say, "There'll be others." The younger members don't want to hear it. We don't want to miss any. The members settle down to prepare the apparatus and equipment for the upcoming inspection. As the cleaning, checking, painting, and repairing are just winding down, a 10-75 signal is received for a cellar fire on 138th Street at Willis Avenue in the Bronx. A welder was using a cutting torch in the cellar when an explosion occurred, severely injuring him. As the first-alarm units darkened down this blaze and removed the badly burned welder, another explosion occurs approximately five blocks away, at Third Avenue and 143rd Street. Rescue 3 was just arriving at the first incident when the dispatcher requested its response to the scene of a major explosion and building collapse on Third Avenue. Lt. Paul McFadden verifies we're not needed at the first incident, and all members assist chauffeur Don Peterson in backing the big Mack out of 138th Street. This turns out to be a really great turn of good luck, for very shortly we're headed for 3rd Avenue and 143rd Street, our response distance now shortened from nearly four miles to just six blocks!

Several fire units have already arrived at the scene and sized up the extent of the damage. Chief Alfred Galdi of Battalion 26 has requested a full second alarm (six engines, four trucks, a rescue, another battalion chief, and a deputy chief) and reports victims trapped in the debris. On arrival, rescuers are informed of several totally or partially buried people under debris on the sidewalk. Immediate rescue of these surface victims is started by the first engines and ladders, and three survivors are being removed. Meanwhile, the members of Rescue 3 have begun our size-up and determined that the explosion and collapse resulted in two kinds of debris piles. On the right side, the second-floor joists are dangling from their wall sockets, while on the left side, the wall supporting these joists has been totally blown down. This creates a "lean-to"-type void from front to rear along the right-side wall (exposure 4). The remaining two floors, the roof and the left-side wall have blown down in a pancake fashion, on top of exposure 2: a one-story grocery store, flattening it.

Lt. McFadden confers with his members and then with Deputy Chief Neil McBride of Division 5 and Chief Galdi of Battalion 26. The officers agree the only people to be found alive in this pile of debris would be those in voids caused by heavy objects or building features. Chief McBride special calls Rescue Companies 1 and 4 to back us up, as well as the department's Hazardous Materials Company. Members of Ladder Companies 19 and 55 have begun searching for voids, hampered by sheet steel approximately ⅛-inch thick, which had covered the roof of the grocery building as a deterrent to burglary. Lieutenant McFadden, realizing the potential for survival was good in the lean-to area along the exposure 4 side, directs me to cut through the steel security gates covering this area to search for victims and access to other voids. Joined by firefighter Mike Loftus of Rescue Company 4, I cut a 3-foot × 3-foot triangle through the roll-up steel gate across the first floor of this building and enter the 12-foot-wide by 50-foot-deep void that lies beneath the sagging timbers of the second floor. Mike and I search quickly through this area, looking for people or access to the area under the main debris pile, without any success. I call Lt. McFadden on the handie-talkie with this report and he tells me to join him and Bobby Greene in the back of exposure 2 where they think they have found a likely sheltered area for victims to survive.

McFadden and the others have located a fairly large void in the pancake section and entered it to begin a search. Lieutenant McFadden and firefighters Harry Christensen, Nick Giordano, and Mike Milner of Rescue 4 begin exploring beneath a three- to six-foot-high pile of highly unstable debris. Calling for complete silence, McFadden calls out to possible victims and is greeted by a muffled reply from the front of the building. As the members move carefully through this black hole, they are constantly aware of a strong odor of natural gas. Orders have been given to shut off the gas supply to the building, but that will take a long time because both the meter wing cock and the curb valves are hidden under many tons of debris. (A leaking gas meter with three .38-caliber bullet holes through it was later determined to be the cause of the explosion. It seems two mutts were engaged in the illegal sale of the gun. The buyer wanted to try it out, so the pair went into the cellar of a vacant building for a test firing. The dopes drew a target on the wall and opened up. Unfortunately, they didn't bother to look on the other side of the plaster wall to see if there was a gas meter in the line of fire. There was! When they realized their mistake, the two fled, but didn't alert anyone else. Several minutes later, the gas odor has permeated nearby buildings, whose occupants did call the gas company, but not the fire department. Eventually the gas found a source of ignition and we got called to the resulting explosion.) Fortunately, no further explosions occur while the victims and rescuers are under the debris. Engine companies have stretched and charged hoselines as a precautionary measure and extinguished a small fire in the rubble. Additionally, hazmat has provided a combination combustible gas, oxygen, and carbon monoxide indicator to members operating under the debris.

As the rescuers move forward approximately 15 feet into the two-to three-foot-high void, they hear a desperate plea: "Get me out of here, please!" Harry Christensen descends eight feet further into the void, now in the cellar area, in search of the voice. As he is moving debris, a hand suddenly pushes its way up and claws at him. Shocked, he immediately calls for help to uncover this fully buried victim. Unfortunately, the space limits all movement and the number of members who can assist. Mike Milner and Fire Marshal Louis Rocco help Harry hand-lift the 60- to 125-pound fieldstones that are pinning the victim. These stones, measuring eight inches wide by 12 inches high and more than 2 feet long, were part of the foundation wall that now serves as Emilio Landa's tomb.

As the debris is removed, the extent of the victim's injuries can be assessed. Immediately apparent are multiple compound fractures of both legs, which begin to bleed profusely when the pressure of the debris is lifted off them. Direct pressure is applied to the most serious bleeding and speed becomes a factor, lest Landa bleed to death before he can be fully extricated. The victim, speaking incoherently in Spanish, is unable to inform his rescuers about other injuries. A stokes basket is lowered to the firefighters, who lash him in place. The opening to the cellar, approximately 24 by 20 inches, is just big enough for the basket and only one man to lift. At the next level, Bobby Greene and I each get one hand on the stokes and with the aid of the members below and a rope attached to the basket, manage to lift it up from the cellar into the section of the void we've entered. We then pass Landa out to members of Rescue 1, who carry him over the debris to a waiting ambulance. In the meantime, Nicky Giordano has made contact with another pinned victim—the person who had called out before rescuers found Landa. By crawling on his stomach into a tiny crevice approximately 12 inches high by two feet wide, Nicky can just make hand-to-hand contact with Norberto Luna, a grocery store employee. Nicky calms the frightened victim, questioning him about the extent of his injuries while assuring him the fire department would get him out alive.

Lieutenant McFadden, Battalion Chief Galdi, and Deputy Chief Philip Burns of the Sixth Division who has now assumed command, discuss the situation. Luna is pinned from the waist down by several feet of debris.

Tunnelling to him from inside the void seems impossible because rescuers would have to cut through a steel I-beam that is precariously preventing further collapse of the debris. Additionally, this approach would be agonizingly slow, because gasoline-powered tools could not be used in the small, confined space; their exhaust fumes would poison Luna, even if the firefighters were protected by their SCBA. We can't even see Luna's face to put a mask on him. Luna can be reached only by cutting from the top, through approximately six feet of mixed debris consisting of steel, brick, wood flooring, and timbers. This plan will allow most of the members to work from the relative safety of the top of the pile instead of the extremely dangerous position within the void.

The safety of the top area is by no means assured, however; a 60-foot-long by 30-foot-high portion of a 12-inch brick wall towers three stories above the rescuers, part of it leaning dangerously toward them. A member is positioned in a tower ladder basket to observe the wall for any indication of weakening or movement. McFadden orders Nicky "to stay in there with him until I tell you to get out." About a year prior to this collapse, a construction crane had toppled over in Midtown Manhattan, trapping a woman as it fell. A NYC police officer had crawled under the crane and stayed with the woman for the duration of her rescue, and as a result he was promoted to detective by the Police Commissioner, so McFadden adds as an attempt at some gallows humor, "Nicky, if you come out of there alive, I'll make you a detective!" Nicky will remain in this tiny hole with Luna for the next three hours to maintain contact, offer comfort and psychological support, and keep a constant watch on the victim's level of consciousness. He keeps up a running conversation with Luna, discussing everything from baseball to marriage, and frequently praying with him. He is not relieved until after we make contact with Luna from above.

Before this rescue operation commences, Chief Burns orders Don Peterson of Rescue 3 and the members of Rescue 1 to stabilize the conditions of the void as best as possible. Working against the rubble, which does not provide a secure base, the firefighters use trench jacks and timbers to shore the worst areas. Fortunately, though several minor shifts of debris occur and one jack is knocked out of place, no major secondary collapses occur.

Lieutenant McFadden has remained below with Giordano to observe conditions and watch for any dangerous shifting, so he orders me to take charge of the remaining members on top. The problem of how to locate and dig down to Luna has to be tackled. The ambient noise level of the overall rescue site makes attempts to locate the sound of Giordano pounding through six feet of rubble below impossible. Tommy Reichel of Rescue 1 provides the solution, a tape measure, one end of which is passed in to Nicky, who holds it out to Luna's fingers. Positioning himself at the entrance to the void with the other end of the tape, Reichel marks a distance of 22 feet. He then passes me the tape and we measure 22 feet out on top of the rubble, with Reichel at the mouth of the void directing me to match the angle created by himself and Nicky. This places us directly above the victim, but we are still separated by six feet of rubble.

Steel roof plating is the first obstacle to overcome. This consists of 4-foot × 8-foot sheets of ⅛-inch thick steel, welded together to form a solid sheet on what used to be the roof of a building. This extraordinary security measure is necessary to keep the mutts from breaking into a building simply by cutting a hole in the wooden roof. Using rotary saws with metal cutting blades, we make an 8-foot × 8-foot opening, gaining access to the next layer of debris. It takes multiple carborundum blades to complete this hole, and one ladder company is assigned to just the task of changing blades, to keep the two saws operating continuously. After the steel is cut off we confront the next layer: the original wooden roof, consisting of one-inch-thick roof boards and 3-inch × 12-inch roof joists. This layer is cut through in one operation using gasoline powered chain saws.

From his position beneath the rubble, Giordano can see Luna is pinned by the remains of a tin ceiling against an 8-inch steel I-beam. Since these ceilings often are nailed directly to the floor joists of the floor above, power tools cannot be used to cut any flooring until it is determined that the tools will not cut into Luna's back. For this reason, a relatively narrow exploratory shaft is sunk next, instead of clearing a large and time-consuming work area. Besides, the shaft will give faster, safer access to the distraught victim. Work progresses cautiously, with much of it done by hand. We use chain saws to cut through three collapsed floor assemblies, each separated by a layer of bricks. Beneath the third layer of flooring, we encounter a wire lath-and-plaster ceiling which, when pushed down, reveals a 2-foot-deep void. Bobby Greene and Don Peterson think we must be getting close, but Nicky says he can't see anything from his vantage point and that leaves us only one option: somebody is going to go down into the hole to see. Since I'm an EMT, and also the thinnest guy in the group, that means me. With Bobby, Don, and Mike Loftus holding my legs, I am lowered headfirst into the narrow shaft. Sure enough, we're right on target. As soon as my eyes clear the underside of the debris, I am looking right at Luna's belt buckle, one foot away (fig. 9–1).

After two and a half hours of frantic work, Luna is now accessible for medical evaluation. From my upside-down vantage point, I can see that from the waist up, the victim has relatively minor injuries and full movement of his body. From the waist down, however, Luna has no feeling in his legs and no movement. His left leg is wrapped around the I-beam, pinned to it by a partition made of wood studs, with metal lath-and-plaster on one side and one-and-a-quarter-inch-thick tongue-and-groove lumber on the other

Figure 9–1. Rescuers at center lower the author through a shaft they have cut through the debris to reach Norberto Luna.

side. In addition, his entire lower body is pressed to the floor beneath him by the tin ceiling, which is the last of six layers of debris. Ironically, the partition pinning Luna is what has saved his life, preventing the debris above it from continuing its deadly descent.

This first contact with the victim has lifted the spirits of both the victim and his rescuers. For the first time in nearly three hours, Norberto Luna feels he really is going to survive the ordeal. He isn't home free yet, however—the small matter of a building sitting on top of him has yet to be resolved. After Greene, Loftus, and Peterson pull me back up the shaft, I explain conditions to the gathered firefighters. The little hole in the rubble will become the focus of the firefighters' collective efforts for another three hours.

The plan is for rescuers to strip away the debris from the top, layer by layer, assisted by a man in the hole who will inspect each piece to be removed. This plan will allow the stability of the pile to be monitored constantly and permit the use of the various tools that best suit the conditions. For example, the guide man will be able to direct the depth of the cut with a chain saw, which would be rendered useless if it strikes the layer of bricks over which it is cutting. Mike Loftus of Rescue 4 is the next person to descend into the void. Since we now know Luna's condition is stable and where he is, with a bit of expansion of the edges of the hole, Loftus can enter the shaft feet first.

Debris is cut away from the mouth of the widening hole as quickly as possible. Tools have to be changed at each successive layer. A rotary saw is used to cut through the steel roof plates to the wood roof, where the 20-inch gas chain saw is then set in motion. When the roof is sectioned, brick and mortar have to be removed by hand, uncovering another floor assembly of asphalt tiles, wood floor, subfloor, and joists that are cut with an electric chain saw. The process slowly repeats itself at each successive layer of flooring until operations approach the victim's layer, making these tools too dangerous to use for fear of seriously injuring the immobile Luna. Further cutting in clear areas is then accomplished using a reciprocating saw. Tin snips are used to strip away the last layer of sheet metal or tin ceiling that cover Luna's body, leaving only the partition pinning his lower legs to be dealt with.

During the one-and-a-half hours that have now elapsed since my first contact with him, Luna's spirits have first soared and then ebbed. Mike

Loftus, who has replaced Giordano as the closest contact, now attempts to keep Luna's spirits and level of consciousness high. Mike jokes with, teases, and questions Luna. At one point, when Luna lapsed into several minutes of silence, Loftus issues the ultimate threat: "Norberto, if you don't start answering me, I'm going to have to start telling you my mother-in-law jokes!" This draws a quick reply from Luna!

By now, the night tours of Rescues 1, 3, and 4 are on the scene to relieve the day tour crews, but the bond between victim and rescuers has become a personal one. It is agreed that the crews that have worked so closely with Luna should remain to effect his removal. The partition pinning Luna is cut away board by board, with additional trench jacks and shoring used to replace the load it had supported. As this is being done, two more victims, both deceased, are discovered amid the last layers of debris on the other side of the wall. One of these bodies was actually part of the material that held Luna's left foot in place. The removal of these bodies before Luna's rescue would be time consuming, so they are left in place. A small, 6-inch by 12-inch, four-ton airbag is placed alongside Luna's left leg and inflated, lifting the last remaining beam pinning him. At 9:40 p.m., Norberto Luna is placed into a stokes stretcher and lifted out of the mountain of death that had been his prison for five hours and 50 minutes. He and Emilio Landa are among the extremely lucky survivors (fig. 9–2).

In all, six people died this day. Including the two near Luna's feet, five victims were in the store when the explosion occurred. A sixth victim was passing by the front of the store when a brick wall crushed the life out of him. The toll of injured reached 29, including a baby who, with its mother, was buried on the sidewalk. Many of the injured were police and firefighters who put their own safety aside while working desperately to free those whose lives depended on them. Out of the deaths and injuries of that day came several lessons. One point reaffirmed was the value of strong incident command. Control of the scene is absolutely essential, as has been proved at each collapse incident in recent New York history. The desire to help, by civilians and rescuers alike, is too strong to be stopped altogether. The knowledge that victims are buried can cause rescuers to take actions that have not been planned properly, further complicating the situation. At this incident,

Figure 9–2. A "Banner Year" for Rescue 3, when seven past and current members of the unit received Department Medals for Valor at City Hall: five as a result of the 3rd Avenue explosion. From left with red sashes and medals are David Riechel of Rescue 5, formerly of Rescue 3; Harry Christensen; Lt. John Norman, formerly of Rescue 3; Nicky Giordano; Lt. Paul McFadden; Vincent Rogers; and Fire Marshal Robert Greene, formerly of Rescue 3. (*Courtesy of Harvey Eisner*)

attempts to move or chop through debris caused the shifting and dropping of debris onto members working beneath the rubble. Chief Burns made the decision to halt all further work atop the debris until the removal of the first victim (Emilio Landa) was complete and shoring was in place to protect the second victim (Norberto Luna), as well as the firefighters beneath the debris.

A final lesson learned was the value of keeping the same people with the victim throughout the ordeal. In this incident, Nick Giordano established initial contact with the victim. He remained in place for several hours, even after he was officially off duty. This served several purposes. First, it allowed someone who had become familiar with the victim to monitor his condition, especially his state of consciousness. In addition, it provided reassurance to the victim. It does not instill confidence in the trapped victim to turn around suddenly and say, "I'm sorry, guy, but I have to go now, see you sometime!"

One caution, however, is that rescuers must be monitored for their fatigue. Members working in dangerous, tiring environments should be relieved at regular intervals to keep them from getting careless. The FDNY recommends rotating members working under gruelling conditions if operations exceed 30 minutes. In this incident, Giordano was allowed to remain with the victim because his part in the operations did not tax him physically; however, his mental condition was monitored closely by Lieutenant McFadden. In the end, he was unable to make him a detective, but Nicky was just happy to get back to the firehouse on 181st Street in Washington Heights, and then to his family up in Washingtonville, New York.

CHAPTER 10

The Southside

My first tour as a lieutenant saw me working a night tour in Engine Company 221, in a neighborhood known as the Southside. Many people think it is simply part of the much larger Williamsburg, but the unique topography of the area, being sandwiched between the Brooklyn-Queens Expressway and the East River, produced a tight knit, if gritty community. In 1987, the Southside is an area full of old loft-type factory buildings, 5- and 6-story tenements, and more heroin than you can imagine. The neighborhood should be thriving. It has great access to highways, bridges, and subways into Manhattan. Instead, the very life blood of the community (and many of its residents) is being wrung out of it by the heroin junkies. They break into apartments, cars, and factories and steal anything that isn't bolted down or welded in place (they have already stolen everything that was merely nailed down), forcing the owners to go to extreme security measures, spending every last dime of "disposable income" in an effort to remain safe. People literally seal themselves into apartments behind wrought iron bars, doors with steel bars across them like the gates of Fort Apache, and more locks than you can shake a stick at. The result is a forcible entry challenge for firefighters who have to fight fires in these buildings that is almost as severe as forcing your way into a bank vault.

The junkies bring other problems to the firefighters of the Southside as well. The tenement right next to the firehouse is one of the main distribution points for most of the dope sold in the neighborhood. The proximity of the drug den makes the firefighters' cars a prime target for the thieves desperate to score their next fix. You quickly learn not to leave anything of value in your car. Of course, the car itself is a target, as are the tires, battery, and any removable parts such as windshields, bumpers, catalytic converters, you name it. That is one reason most firefighters drive beat up old cars to work— "ghetto cruisers" in the vernacular. (Another reason is the low salary the city pays its firefighters in exchange for their blood, sweat, and tears.)

There is a steady stream of customers into and out of the tenement. When a shipment is late, the line extends out the door and down the block, right across the front of the firehouse apparatus doors. At times we have to physically move these walking dead out of the way so we can go to a fire. They go in nervous and desperate, afraid they won't be able to score the poppy powder that they live for. They come out less nervous but just as desperate, looking for a place to "cook" and nod off. Some of these junkies are "upstanding citizens" from the suburbs. They drive in to the Southside to score and then shoot up in their cars. Too often, in their desperation to get the drug into their system, they neglect to put the transmission into park before pushing the plunger on the hypodermic, sending the magic poison coursing through their veins. When the car rolls out into an intersection, or into a building, or someone notices the "dead" person behind the steering wheel, the firefighters are often the first on the scene.

The members of Engine 221 and Ladder 104 discover very quickly that the automobile accident was not the real cause of the patient's medical difficulties. They learn to quickly identify the telltale signs of the heroin overdose: the nodding semi-consciousness, the pin-point pupils, the track marks on veins of the legs and arms, between the toes, and even in the groin, depending on how far the junkie has descended on his or her trip to hell. The firefighter also learns to look carefully for the deadly needles, being careful not to be stuck by them. A shot of heroin would be bad enough, but we are more worried by the possible contact with the hepatitis C and AIDS that is so prevalent among the junkies.

The junkies accidentally start fires in vacant buildings as they cook their heroin, then nod off from the dope. Or they purposely start fires in an attempt to get the firefighters to rip open the walls as we search for hidden fire. After the firefighters leave, the junkies descend on the hapless premises like a pack of vultures, stripping the structure of all metal, copper, lead, and cast iron plumbing and electrical fixtures (we call it "urban mining") that we have exposed behind the plaster walls and ceilings as we chase the fire. They also start fires intentionally to burn the plastic insulation off the wiring they have ripped out, since they can't sell the wire to the scrap metal dealers with the plastic still encapsulating it. The blazing plastic insulation, full of PVC (polyvinyl chloride), creates a huge cloud of dense black smoke, full of hydrogen chloride gas that really clings to your clothing, and most importantly, eats at your lungs. The junkies light the wire on fire and stand back. The insulation is often burned off by the time the fire department arrives four or five minutes later, and then the obliging firefighters cool the red-hot metal off with their hoses as they put out the remaining green, blue, and purple flames so the junkie can walk right over and pick up his pile of cleaned copper and rush to the scrap dealer as quickly as possible. After going to two or three of these fires in an afternoon, the firefighters finally get pissed enough to confiscate the burnt debris, depriving the junkie of his fix. This usually sets off a cycle of retaliatory false alarms, vandalism to the firefighters' cars, and an occasional booby-trapped vacant building set on fire to injure the firefighters that so callously kept "his" ill-gotten profits. This cycle usually only ends due to the "self-cleaning oven effect"—when the junkie finally burns himself out, either with a hot shot of heroin or by nodding off into a fire. Unfortunately, the siren call of heroin is so strong it soon pulls in another poor soul who will begin the same flaming downward spiral, until they too crash and burn with a needle stuck in a vein.

At 6:00 p.m., I make my way into the kitchen at the back of the firehouse on South 2nd Street to conduct my first roll call assigning the members to their duties. I introduce myself, and give out the assignments—nozzle, back-up, door, and control, plus the all-important engine company chauffeur or ECC: the person that drives the pumper truck to the fire, connects it to a working hydrant and operates the pumps to supply water at the correct pressure to the hoses. As I finish, one of the guys says, "Hey Lieu, I saw on the [department] orders that you just got made [promoted] yesterday.

I guess you haven't given a 10-75 yet, huh?" I shake my head no, and start to say, "This is my first tour," when someone else chimes in "Well, it'll be a real long time before you give one around here." The whole kitchen breaks out in laughter. The Southside, despite its incredible drug-related problems, does not have a reputation as a "busy house": at least not in comparison with some of the other companies in the city, such as Engine 290 and Ladder 103, or Engine 82 and Ladder 31, who are still averaging around 5,500 runs a year per company. The units of the Southside, because of their isolation by the highway and the river, average "only" 3,500 runs a year apiece: still a good bit of work. The laughter in the kitchen hasn't even died down when the "tooo-dooo" tone of the voice alarm and teleprinter interrupts the mirth. The house watchman calls out from the desk at the front of the apparatus floor, "Everybody goes, South 2nd and Driggs, fire in a vacant building."

The rigs pull down the block to the corner of Driggs Avenue: the scene of the shooting of the famed NYPD detective Frank Serpico during a botched drug raid. In 1973, the story of the neighborhood's drugs, crime, and the corruption they inspire became a hit movie starring Al Pacino in the title role as Serpico. It hasn't changed much in the intervening 14 years. Smoke is pouring from the ground floor of a six-story vacant building. The windows are covered with sheets of tin, and the doorway is largely obstructed by the remnants of cinder blocks that the city had installed in a failed attempt to keep the junkies out of the vacant building. "Engine 221 to Brooklyn K," I radio, "10-75 for Box 217, fire on the first floor of a vacant six-story MD." Vito Oliva, the chauffeur, looks over with his broad grin: "A job on your first run—the guys are gonna love you, Lieu."

Vito stops the rig just past the front of the fire building, leaving enough room for Ladder 104 to pull their tiller aerial ladder into position to raise the 100-foot ladder to the roof. All the windows on the upper floors are covered with tin, meaning ventilation will be extremely slow, which in turn means anyone working inside is going to get beat up by smoke and heat. I step down from the pumper's cab, pulling my mask from its bracket as I step backward. The firefighters of Engine 221 are already pulling off three lengths of 1¾ inch hose from the rig when I reach the tailboard: the back step of the pumper where the various size hoses are stored. Seeing the crew has enough hose to accomplish our assignment, the control man climbs up onto the back step and waves to Vito, who drives down the street about 200 feet to a hydrant, trailing hose behind him.

The main entrance to the building consists of a hole about two feet square through a cement block wall. Harry Stefandel, the nozzle man, pulls his length of hose to the doorway and glares at the opening. His frame will barely make it through the hole, with much twisting and contorting. He's not a happy camper. He struggles through though, pulling the hoseline behind him, then flakes out the 50 feet of hose in a debris–littered public hall. As we wait at the door to the fire apartment for the hoseline to be filled with water, Ladder 104's crew starts working on improving the access and ventilation. One man swings a sledgehammer at the cement blocks, smashing them out one at a time. The original hole the junkies had made would be a serious bottleneck if we had to get out of the building in a hurry. None of us want to crawl through the inky darkness; unknown dangers might cause us all to suddenly have to retreat through that tiny hole. No one wants to be stuck behind Harry if that happens. The sledge solves that problem, but the exertion it requires in the process means that the firefighter swinging it will suck down the air in his mask cylinder at a much faster pace. The same is true of the firefighter outside who is struggling to pry the sheet metal coverings off the windows.

Vito has finished connecting the pumper to the hydrant, which unusually hasn't been vandalized, while the control man has broken the coupling of the hoseline and connected it to the pumper's discharge gate. When Vito opens the gate, the water flowing through the hose pushes air ahead of it toward the nozzle. Harry and I were in the same probie class together nearly eight years earlier. I don my mask facepiece as I watch him bleed the air from the nozzle until a steady stream of water appears; it makes no sense to direct more air at the fire without water to extinguish it. Turning to ensure the rest of the crew are properly masked up, with boots pulled up and gloves on, I pat Harry twice on his broad shoulder and shout through the mask facepiece, "Let's go, Harry." He simply nods, and pulls back on the brass handle of the controlling nozzle, sending 180 gallons of water per minute into the fire room in front of us. As the flames are driven backward, clouds of smoke and steam descend on us. With the tin coverings still on the windows, there is nowhere for the hot gases to go as the water turns to steam and expands hundreds of times in volume. The steam's only escape is back through the cinder-blocked door opening, right past us. It is lights out! We begin crawling into the next room, Harry swinging the nozzle from side to side along the ceiling. I am sweeping the floor in front of us with my flashlight, looking for booby-traps, holes cut in the floor and covered over with cardboard or linoleum to cause an unsuspecting firefighter to plunge into the cellar below, as well as looking for any junkie that may be lying unconscious on the floor, and most importantly, any evidence that there might be fire burning underneath us in the cellar.

As the beam of the flashlight moves from side to side, it has an odd glimmer to it. Initially, I can't tell what it is through the heavily scratched lens of the mask facepiece, but it seems there are little twinkles of light sparkling along the floor in our path. Fearing we may be operating above a cellar fire, I put my face and light right down on the floor for a better view. What I see stops me in my tracks. The floor is littered with discarded hypodermic needles! The flashlight beam glints off hundreds of steel slivers as Harry swings the nozzle overhead, causing the falling droplets of water to splatter the needles as they land on the floor. "STOP, Harry!" I yell, as I slap the big nozzleman heavily on the shoulder. "Get up off your hands and knees," I call to the rest of the crew, "the floor is covered in needles." "What are we gonna do, Lieu?" asks Harry. "Keep the nozzle going Harry, but get up off the floor; we're not going any farther than this." It is very, very difficult to advance a flowing line while "duck-walking," squatting on your heels in order to stay relatively low, and firefighters tend to have to put a knee or hand down to steady themselves. The needles make that risky. The alternative is to stand up and advance, but that puts us up in the heat and smoke, which severely hinders visibility, making it easier to fall into booby-traps or rotten sections of flooring. "But Lieu, there's still another room going back behind this one, and I can't hit it from here," Harry says. "I don't care," I reply. "We'll use the reach of the stream and try to bounce the water in there if we have to, or we'll wait until somebody else can hit it from outside, but I don't want anybody to get stuck." I radio the 35 Battalion about our situation, and Chief

John Devine radios back to "hold your position while I get another line around to the back." "You don't have to tell me twice," I think.

Holding your position at a fire is not a fun thing to do, especially in a sealed-up building. As long as you are moving forward, you tend to push the fire away from you, and the steam produced as the fire is knocked down picks up momentum and knocks down fire still further ahead. Once the momentum stops though, the fire, in this case just one room away from us, tends to re-group and intensify. We throw water at the doorway, trying to bounce some of it into the room, but that only knocks down the fire right in the doorway. The material off to each side continues to burn freely. The heat starts to soak through our gear. To make matters worse, we are up off the floor, kind of crouched to get away from the hypodermics. The nozzle reaction as the water leaves the hoseline pushes back at Harry. Thankfully, Harry has a few spare pounds to push back with, but this position is going to get real tiring, really, really fast.

We hear the truckies of Ladder 104 and Ladder 108 pounding on the sheet metal covering the windows. The crashing staccato sound makes me think of the way old-time sound effects men on early radio and television replicated thunder: exactly the way the truckies are doing now, by striking and waving a sheet of metal. One of the guys has gotten a bite into a piece of the galvanized steel with the wedge of a Halligan Hook and is starting to work it free. He is joined by another man with a hook, and they struggle to pull loose the roofing nails that hold the tin to the window frame. The opening soon allows smoke and heat to exit the room, and daylight and cooler air to enter.

With the extra lighting and improving visibility available, the rest of the company can now see the mass of potentially deadly needles that litter the floor of the drug den. "This place must have been a shooting gallery for years," one of them says. I tell Harry to sweep the floor with the hose stream to flush the needles away from us. Now that the outside team has also vented the window in the fire room, the flame is being pulled out that opening instead of being drawn toward the open doorway behind us or the window in our room. It makes our position a lot easier to take, since we can now stand up instead of having to remain in the crouched position we had been forced to assume.

As we are about to advance into the last room, Chief Devine calls over the handie-talkie "35 to Engine 221, we're about to hit it from the outside. Are you in a safe position?" I call back, "Give us 30 seconds to back out to the hall, chief." I tell Harry and the crew to shut down and back the line out to the apartment door. Once there, I call, "221 to the 35, we're OK now, chief, go ahead and hit it." Soon conditions in the apartment are visibly better as Engine 211 has knocked down two remaining rooms of fire from the rear yard. Ladder 108 reports that they have minor extension to the floor above, and Engine 216's officer calls 221's chauffeur, Vito, telling him to charge the hoseline that they have stretched to that floor. Chief Devine appears in the hallway and makes a quick survey of the original fire apartment.

Ladder 104's inside team, Lt. Gary Wendell and two firefighters, are already inside the burned-out rooms, poking holes in the walls and ceilings, looking for fire traveling in the void spaces. The crew works quickly and thoroughly, making preliminary openings near light fixtures, electrical switch plates, and pipes: all the man-made openings through the fire resistive barrier of the plaster walls and ceilings. If no fire appears there, they move on to the next opening, since if any fire is going to penetrate the plaster, it will do so around one of these openings first. Sure enough, in the ceiling of the middle room, right where the highest concentration of needles is located, there is fire in three bays of the ceiling right around the light fixture. Gary calls for us to bring the line in and to hit the fire after his members have finished pulling the plaster off the ceiling. While we wait for the truckies to open up some more, I tell Harry to flush the needles into the room's corners with the nozzle. Now that we are standing instead of crawling, the needles pose less of a threat, since they are not likely to penetrate the steel sole of the rubber boots. Here are yet two more reasons for that steel sole that probably weren't thought of by the designers when they added it to early fisherman's hip boots 50 years earlier: AIDS and hepatitis.

The temptation in a situation where fire is burning in the ceiling bays overhead is to open up the nozzle as soon as the fire is exposed, but that usually makes matters worse. The water won't hit fire that is farther down between the beams away from the opening, and the truckies will then have to work under scalding hot water to finish pulling the ceilings. It is better to wait until they have fully exposed the entire ceiling, and then hit the entire fire at once. The only exception to that rule is what we experience here at this fire—fire is traveling along the ceiling bay until it comes to a vertical shaft for the bathroom drain pipes that travels up through the building. Once fire gets into that large vertical opening, it is on an expressway to the upper floors, the roof, and the cockloft. Gary notices the shaft first and quickly orders his men out of the way. "Engine, you have to get water going up that shaft, it looks like it is starting to take off." Harry moves quickly into position below the shaft, where he can direct the nozzle stream almost straight up, taking care to stay out from under the cascades of steaming hot water. Wendell radios Ladder 108's officer about the location of the shaft, but we know almost instantly that they are already on it, when we hear the sound of an axe striking plaster and see debris falling down out of the ceiling opening. This was the extension they had previously reported. Harry shuts down and we move the line back out of the room so the truckies can finish their job of stripping the ceiling.

After the members of Ladder 104 have completely stripped the ceiling of the last room, they quickly move out of the reach of the hose stream. They really don't want to be in the way when Harry gets down to the final wash-down. In a vacant building like this, where there is no property to be salvaged, we will use the maximum amount of water to reduce the chance of a hidden fire escaping our attention, or of the junkies starting another fire in here tonight.

My next tour back on the Southside was a week later. I had received a call from the Division Commander, Deputy Chief John Regler, asking if I would be interested in working the 11th Division relief group. The relief group covers vacant shifts in six companies on a rotating basis, including the two companies on South 2nd Street, Engine 216 and Ladder 108 a short distance away in Williamsburg, and two single-engine companies over in the industrial neighborhood of East Williamsburg. A relief group is not as good as having a permanent assignment, or covering vacations, since you can't make mutual exchanges of tours, but I liked the prospect of working steady tours, so at least I could plan for when I was scheduled to be off. I would work with the same group of firefighters in each company each time I was there, and I also enjoyed the variety of assignments that this spot offered, from very active tenement work to factories, oil storage tank farms, and the Brooklyn Union Gas Company's Liquefied Natural Gas (LNG) facility, which I was familiar with from my time in Haz Mat Co. 1. I thanked Chief Regler for the opportunity and accepted gladly.

One of the tasks that firefighters conduct when not fighting fires, training, and maintaining the firehouse and apparatus, is building fire prevention inspection duty, or BI for short. To say that many firefighters view BI with something short of unbridled enthusiasm is an understatement. They hate it. BI means going out into the neighborhood and entering people's property, examining their buildings for fire or structural hazards that pose a threat to the lives of the occupants or firefighters, and issuing orders for corrective action if anything hazardous is detected. As one senior firefighter put it when I first got to Engine 290, "All we really do on BI is piss people off. They don't listen to what we tell them anyway, so why bother?" In reality, they do listen, but sometimes it is only because we have gotten serious and threatened them with legal enforcement of the violation order. This is done by issuing a criminal court summons for failure to comply with an order that was issued earlier. That's when people really get pissed off, though. The summons means they will have to take time to go before a judge in a municipal court who theoretically can fine them and make them take the needed remedial action to correct the hazard. One problem with this is the firefighters have to be "the bad guys" by issuing the summons, a process which in itself is tedious as well as distasteful to people who like to help the community, not hurt it. In the wake of the social unrest and civil disturbances of the 1960s and 70s, issuing a summons is viewed as a potential trigger for a confrontation. Sometimes it is.

One such confrontation occurred when I was working in Engine 237, re-inspecting buildings for compliance with orders that had previously been issued. One building we visited was an old commercial loft, housing a number of sewing factories producing women's and children's clothing. This was a five-story building, but the only area we had to visit this day was the top floor, where the company had previously inspected the place and found that the tenant had locked the exit doors to the two staircases with locks that required a key to open from the inside: keys that only the owner had possession of. That is a serious violation of the New York State Labor Law, and it is exactly what caused the deaths of 146 young immigrant women who were sewing ladies' clothing in a sweatshop known as the Triangle Shirtwaist Company in lower Manhattan in 1911. The conditions that Engine 237 had discovered were very similar to the Triangle fire, only these immigrants spoke Spanish and Haitian instead of German and Yiddish. The Labor Law, which was written in the aftermath of the Triangle fire, states that a locked exit in a factory during business hours is a very serious offense, requiring the immediate issuance of a criminal summons and an order to remove the lock. Apparently, Engine 237 had issued the order two weeks earlier to remove the lock, but had not issued the summons, in an attempt to convince the owner through an appeal to reasoning that it was the right thing to do. That appeal and the order had fallen on deaf ears. Now the firefighters had returned to find the situation unchanged. One of the members radios down to the rig where I am writing up the inspection activity log, "Hey Lieu, you better come up here, we've got a problem."

I arrive on the 5th floor to find the members being browbeaten by an old woman of Russian or Polish descent yelling at them in barely understandable English for "trying to scare a poor old lady." I see the men are upset at the way she is yelling at them, and that she is extremely agitated herself. My first priority is to calm things down. I also see that she has ignored the earlier order to remove the locks. Something has to give here, and it is not going to be the FDNY. I ask the men to step out of the office a minute to give her a chance to calm down. Using a soothing tone, I ask the woman to please calm down so we can try to resolve the problem. Immediately, she settles down as I ask her to show me what the problem is all about in her terms, since this is my first time here, and I really don't know everything that has gone on in the past, so I have no idea what she is yelling about. The woman starts to speak, but then stops when one of the men steps back into the room. Instead she takes my arm at the elbow and tugs me in the direction of a back office. In turn, I pull her back toward the factory floor, asking her to show me the problem. She leads the way into a hall, and as soon as we round the corner from the office she quickly glances both ways and pulls a wad of cash out of the sleeve of her sweater. Pushing it at me, she says "Here, take this. Make those men go away!" Initially, I don't realize what she is saying or doing, but when she tries to put the money in my shirt pocket, finally the light bulb over my head goes on as I realize what is happening. "I am being bribed!" suddenly dawns on me. The realization of that fact infuriates me. "What kind of whore do you think I am?" I shout at the startled woman. "Do you really think I am going to risk my career for whatever you have in your hand?" I shove her hand away roughly. There is no way that I, as a serious student of Fire Laws and the history of the fire service, including the penalty phase of the Triangle Shirtwaist fire where the fire inspector went to jail for murder for the death of the young seamstresses—there is just no way I am going to walk away from this problem until it is fixed. Now I am pissed off.

I call out to the members of Engine 237. The old lady was slick; she took me out of sight of the men so there would be no witnesses. "Get the summons book up here and write this wench up," I order. "Give her two summonses: one for the locked exit, and one for failure to comply with an order from the Commissioner." I can see the smiles on the men's faces as they realize what is going on. I then order the men, "Bring up the irons, too; we're going to remove the locks so that they can't be relocked. And go through this place with a fine tooth comb to see what other violations she has."

John Atwell, a probationary firefighter, arrives first with the summons book. "Do you know how to write a criminal summons?" I ask, thinking that since he's a probie, he may not have any experience at it. "No problem, Lieu. I used to be a cop," he replies. We issue several new violation orders for minor items like an electrical outlet box that is missing the cover and the two summonses, and leave the sweatshop, exchanging icy stares with the owner the entire time we are there. When I climb back into the cab of the pumper, the chauffeur, who has been monitoring the department radio all this time for any alarms, sees that I am shaking with rage. "What happened up there?" he asks. "The probie said you were really pissed when he came down for the summons book. Did everything go OK?" I give him the full run down. "Lieu, you better let the Battalion know about this. Chief Eberlien is working, and he's a stickler about notifications. Besides, that woman might call the IG [the inspector general] on you, saying that you asked for a bribe. Then you'll really be in trouble." We return to the firehouse so I can call the chief.

Now it's Chief Eberlien's turn to get pissed, which he promptly does, not so much at me in particular, as the whole situation in general, although he is far from happy with me. "She's got some nerve trying to buy a fire officer like that. You have to report her to the IG: she committed the crimes." He orders me to call the IG immediately and forward a written report to him as soon as I am finished with the IG.

The person I reach at the IG's office is very enthusiastic. He is a civilian investigator who spends most of his time trying to find firefighters doing something wrong, like drinking on duty. "A real cash bribe attempt, heh? How much was it?" he asks. "I didn't even touch it, never mind count it," I tell him. What was this guy thinking, that I counted it and turned it down when it wasn't enough? I tell him, "I wouldn't have taken it if it was a million bucks. I have a long career ahead of me and a wife and two kids at home that I want to see, not end up in jail." What I don't add is that I had only weeks before read an article on the Triangle Shirtwaist Fire that not only described the horrible details of the deaths of the 146 mostly young female factory workers, but also included the vilification of the fire department for their role in allowing the conditions that contributed to the deaths to go unabated. The IG then asks if I'd be willing to wear a tape recorder and go back to try to get the woman to repeat her bribe offer, on tape this time. Now I have to confess my anger to him as well: something I had avoided mentioning up to this point. "Well, I guess that shoots the shit out of that option." He too requests a full written report and advises me to "leave out all the gory details." I get home that night two hours late, after finishing all the paperwork for the inspections and the "full written reports" of the attempted bribe, a little wiser in the ways of the world.

On my very next set of day tours, I am back working in Engine 221 and Ladder 104. We are scheduled for BI on both days. I groan as I read the day-books that outline the week's activities. I go through the Building Record Cards in Ladder 104's "next inspection cycle" file. "No easy ones here," I say to myself. The plan calls for inspecting several large industrial buildings, all with outstanding violations. "Great," I say as I pick up the stack. "Another day of headaches." At 10:00 a.m. sharp, I drag the members of Ladder 104 out of the kitchen and get on the rig. We head west toward the river for our first inspection. As soon as we are fully onto the block, the radio calls us for a report of a building fire, a block behind us on South 2nd Street. If we had waited 30 seconds more to leave the firehouse, we could have turned left out of quarters and been right at the scene. Now we will have to go all the way around the block, through the traffic-clogged streets.

We pull up nose to nose with Engine 221—a situation that should be avoided if at all possible, but in this case it works out OK. There is a working hydrant right in front of the fire building that 221 has connected to, and from where we are located, the aerial ladder will reach all but one line of windows. If needed, Ladder 108 will be able to cover them when they arrive off Driggs Avenue. I get on the radio to tell 108 to make sure they come in off Driggs.

The fire turns out to be pretty minor, except in the eyes of the apartment's tenants. An extension cord had been run into the kitchen from an adjoining bedroom, and was being used to power the refrigerator, a toaster, and a microwave; it had overloaded when all three of those appliances called for power at the same time. The wire ignited the back of a bookcase, and fire was just starting to roll along the ceiling when we arrive. Ladder 104's can man knocks down most of the flames rolling across the ceiling using the 2½ gallon water can, and I call out to Ladder 108 to have their can man join us as well. Being the second ladder to arrive, Ladder 108 normally would proceed directly to the floor above the fire to search it for any trapped or overcome occupants, as well as to check it for fire traveling upward in any hidden void spaces. I estimate, however, that if we have another 2½ gallons of water right away, we will be able to control the fire without having to use Engine 221's hoseline. 108's can man crawls in just as our extinguisher sputters empty. The second can just about finishes all the visible fire, but we have a good deal of smoldering material on the book shelves. Engine 221 charges their hoseline to complete the extinguishment, and they are not happy, as the two extinguishers knocked down all the flame and they are only getting to spray water on smoldering embers. They now will have extra work draining and repacking the wet hose without getting a real piece of the action. All work and no play.

As we start to pull the charred books off the open shelves, it becomes apparent that many of them are family albums filled with photographs, newspaper clippings, and other memorabilia. One is a wedding album, shot many years ago in a place a lot more appealing than this squalid apartment on the Southside. It is clear that many of the items are heavily damaged, and most likely will never be salvaged or replaced, but I tell the members to be very careful with them anyway: they are a family's history and are truly irreplaceable. We take the smoldering albums into the kitchen sink, for a judicious spray from the faucet, and then lay them out on the counter top and table to begin drying out. The whole thing only takes 20 minutes, but the family will be able to pick up the pieces of many years of memories.

After packing hose and washing up, we return to our intended inspection duties. This time, we no sooner introduce ourselves to the manager of a large feather factory on Kent Avenue when the radio sends us to a stuck elevator on Wythe Avenue about 10 blocks away. The feather factory manager, a Hasidic fellow, probably breathed a sigh of relief and thanked God that the *"goyim"* were gone. He also probably alerted all his employees to clean the place up because the damned firefighters are here again—for when we return 20 minutes later the place is in decent shape, better than I would have expected. The exit doors are all unlocked, the aisles are not full of carts or debris (but the sidewalk is) and the manager is now suddenly "gone for the day" even though it is barely 11:00 a.m.

I tell his replacement, "That's OK. We don't really need him, just show us around the place." Suddenly, this fellow only speaks Yiddish. I tell the firefighters to follow me and we start walking up the stairs heading for the roof to check on a previously reported problem with the fire escape, to see if the repairs that had been ordered have been completed. At the roof bulkhead, we find the door is locked with an approved type panic hardware which will allow escape from the inside, but it has a built-in alarm that will sound if the door is opened, unless the owner turns the alarm off with a key. I ask our escort for the key. "No key." He holds both hands palm up and shrugs his shoulders, giving me a "fuck you" sneer. I push open the roof door and the alarm sends an ear piercing shriek throughout the stairwell. We hurry out onto the roof to get away from the damned thing and begin inspecting the attachment of the fire escape to the top of the building. It has not been repaired. Almost immediately, the noise stops. I see our escort has suddenly remembered that he did indeed have a key, right there on his key ring.

Sam Giamo calls me over to the exposure 4 side of the building where he points down a flight of stairs on an exterior stairway-type of fire escape, which are generally very good, safe ways of getting out of a burning building if they have been properly maintained. That is a big "if" in Brooklyn. Several of the steel treads are dangling from one flight and others are missing entirely. This could be a real life-threatening condition to anyone who tried to use that stair in an emergency, especially at night or if there is a lot of smoke hiding the broken and missing treads. I call for our escort, but now he is among the missing. I tell Sam to go get the NOV (Notice of Violation) and summons books. We continue down through the building, noting several less serious violations such as fire extinguishers that haven't been inspected in three years instead of annually as required, several of them totally empty, and a sprinkler system that hasn't been inspected by a licensed inspector in three years as well. The good news is that having been a sprinkler system engineer and inspector for seven years before becoming a firefighter, I could tell that the system at least appeared to be operational. We go down to the office to serve our written notice of violation and a summons for failure to comply with the previous order.

A notice of violation does not mean a penalty or fine, or anything like that, except for the most grievous offenses, the locked exit door in a factory, or smoking in certain dangerous locations. The notice of violation is just a written way to explain to the owner or tenant what is wrong, and how to fix it. We have written orders to fill all the empty fire extinguishers and have them and the sprinkler system inspected, and to have the fire escape "scraped, painted, and repaired." Because the firefighters had previously issued an NOV for the fire escape and it is still not repaired a month later, I intend to issue a criminal court summons to the person in charge for "failure to comply with an order from the Fire Commissioner." That could result in a fine, if a judge feels it is warranted. (Too often they don't, and the owner just thumbs his nose at the order.) The owner has 14 days to get the fire extinguishers and sprinkler systems serviced: and because it is a much bigger job, 30 days to get the fire escapes fixed. After the designated dates, Ladder 104 may or may not return to see if it is fixed. When we get back to the factory office to deliver the N.O.V and summons, only a janitor is there. Everyone else has left, he tells me. We leave a copy of the N.O.V and summons tacked to the office door and leave feeling thoroughly frustrated. (Pissed off is a better description.)

One of the drawbacks of the building inspection process in place at the time was that the same officer is rarely working on the day the re-inspection comes due. As in my case, as a covering or overtime officer not normally assigned to the same unit, I may not be back in that company for months, if ever. Under the NOV system, only 10% of the violation orders are ever re-inspected until the next time the building is due for its routine inspection, which can be anywhere from one to three years later. The system relies on the building owner or tenant to "self-certify" that the violation is corrected, by signing and sending back a copy of the order within the allotted time frame. The 10% that are re-inspected are randomly done to "keep the system honest." It is ripe for abuse. I left the Southside after one year, working in 221 and 104 every three weeks, without ever going back to the feather factory again—for three years.

One warm, sunny afternoon I am working in Rescue Company 2, returning from a fire in East New York, when I hear the dispatcher send out a Class 3 alarm (automatic fire alarm) for an address on Kent Avenue that sounds vaguely familiar. The dispatcher almost immediately calls the 35 Battalion and tells him the original alarm was for a sprinkler water flow, but now they are receiving reports of a fire in a feather factory on Kent Avenue and South 2nd Street. I tell Richie Evers to start heading for Williamsburg.

Battalion 35 gives a 10-75 while we are still at the Broadway Junction: a good 10-minute ride away. When we report in to the chief, we are told that there was a flash fire that had traveled almost instantly across the whole 4th floor, spreading on the surface of the chest-deep piles of highly flammable, processed goose and duck feathers that covered the entire 150-foot wide, 100-foot long area of the top floor. Piles of eiderdown pillows that had already been stuffed with feathers in other parts of the floor had also ignited, but the sprinkler system had operated and was keeping the fire in check, although the water was rolling off the surface of the feathers like rain off a duck's back, while the fire burrowed underneath the surface. The chief ordered us to the top floor with our thermal imaging camera to try to locate the deepest-seated fires so the hoselines could have access.

When we get to the fire floor, visibility is very bad. The steel shutters that cover all the windows are sealed shut, and the smoldering piles of pillows and feathers produce tremendous quantities of dense, carbon monoxide-laden smoke. Fortunately, the sprinkler system has cooled the atmosphere to the point where flashover is no longer an immediate danger, and a person can stand, which is at times, necessary to find your way around the maze of feather-laden carts that now seemingly block every aisle between the plywood and chicken wire enclosures that form bins for different grades of feathers. After spending just 30 seconds under the soaking spray from the sprinklers, and passing through the piles of feathers, everyone that has entered the fire floor is covered from head to toe in the damned feathers. We look like a bunch of alien roosters from some kids' cartoon.

Suddenly there is an urgent message on the handie-talkie. Ladder 104 is totally out of air and lost in the maze of feathers and carts. The officer, Lt. Wicker Kobes, has removed his plastic mask facepiece in an attempt to see past the feathers clinging to it in order to locate an exit, and he is nearly instantly overcome by the extremely high carbon monoxide (CO) levels. Luckily for him, one of his members was holding onto his coat when he lost consciousness from the CO, and knows exactly where the fallen officer lies. That member immediately gives a mayday message over the radio, and begins to share his own very limited air supply with the unconscious Lieutenant. Rescue 2 immediately goes into the downed firefighter recovery mode, adrenaline starts pumping and all non-emergency tasks are dropped. Now all we have to do is locate the man giving the mayday, without losing our own reference point to a safe escape route, and carry the unconscious member out while leading the rest of the firefighters with depleted air cylinders to safety. Easier said than done.

I order Timmy Higgins to tie off a search rope in the stairwell that we had passed on our way in. Tim tells me he had done that before we even entered; he recognized the deadly potential as we walked up to the building. Tim knows exactly how to get back to the staircase. Locating the downed lieutenant and his now desperately low-on-air rescuer is made possible by the thermal imaging camera (TIC) that we carry. I simply tell the firefighter who gave the mayday to stand and wave his arms over his head, to distinguish himself from the 20 or so other firefighters operating on the huge floor. We advance along an aisle, trailing our search rope behind us, looking down each side aisle, searching for the waving arms. Because I had been in the building three years earlier on the inspection, and thus was somewhat familiar with the layout, I was able to navigate more readily through the large central room, guided by the TIC. Fortunately, we reach the downed Lieutenant and his team on the first attempt, but now we face the daunting task of crawling back nearly 200 feet through the maze, dragging the soaking wet, feather-encrusted officer back to the stair we had entered from. Just as we begin this trek, a cool breeze and an increase in outside noise signals that the members working from Tower Ladder 119's basket have managed to force open one of the steel shutters, right at the end of the aisle we are working in. We immediately change direction, and drag the unconscious officer to the window and manhandle him out into the basket and fresh air, right into the arms of two astonished firefighters who are trying to make sense out of the ball of feathers that has just been dropped on their feet. Quickly, they figure out which end is which, clear the officer's nose and mouth of the ubiquitous feathers, and begin rescue breathing. Three or four breaths are all it takes to get spontaneous respiration started, and the Lieutenant resumes breathing on his own, a very good sign. The other members of Ladder 104 join their officer in the basket, and Ladder 119 delivers the whole crew to safety on the ground where an ambulance transports the still unconscious Lieutenant to Bellevue Hospital where he will make a full recovery.

Rescue 2 is not out of the woods yet, though. The exertion of searching the large floor and carrying the unconscious officer to the Tower Ladder has depleted *our* air supply. The low air alarms are sounding on all our masks. Now we might not make it back the 200 feet to the staircase. I recall the fire escape from my inspection visit; it should only be 20–30 feet off to our right. I wish I had not laid the thermal imager down in the feathers as we struggled to carry the overcome officer, but that is water over the dam now. "Come with me," I tell the group. We will have to navigate by feel and memory. Fortunately, there have been no new alterations to hide the exit door, and we soon find the large steel door, but when I push on it, it doesn't budge. I call for the Rabbit Tool, but it is lost somewhere in the pile of feathers. "Mel, hit this door hard," I tell Mike Esposito. Mikey comes closer, feels the door and jamb and lock mechanism, then lays down on the floor, drawing his knees up into the fetal position. I hold the panic hardware on the lock open, while Mike kicks out with both feet, smashing against the lock side of the door with all his considerable strength. It creaks open about six inches. Mike creeps closer to the door and lashes out again. The door springs open another six inches. A third blow from Mike's powerful leg kick launches the door back against the end of its hinges, allowing a wave of mercifully fresh cool air to flow in over us all. "Watch out, this fire escape was in really bad shape the last time I was here," I say to the group, then I gingerly step out onto the metal strips of the landing, testing its remaining strength. My mask sucks hard against my face as I do so, totally out of air. Talk about perfect timing!

Four stories below, in the yard, are my two "friends," the owners. "No, don't come down!" they yell up at us. "The steps are very bad." I look. The stairs above and below are as bad or even worse than they were three years earlier when we gave them the notice of violation. I yell back at them, "I told you to fix them three years ago and you didn't. Now I am going to come down them, and if I fall I am going to sue the living shit out of you! I will bankrupt both of you! I will own this entire block!" This draws screams of protest from below, and chuckles from the men behind me, but the stairs are just too dangerous to even attempt to use. So much for N.O.V's that are not re-inspected by the same people that issue them, or more importantly, enforced by the courts. Thankfully, there is no fire chasing us, so we don't have to jump. We have the luxury of waiting at the doorway for Ladder 119 to return to the fire floor with the bucket. It takes several trips to get all of us to the ground, but at least we make it all in one piece—looking like we had been tarred and feathered, but in one piece. It takes another hour to change air cylinders, retrieve the search rope, and find the thermal imager and other tools dropped during the rescue effort, and hours more to remove all the damned feathers from everything they touched, but it is worth it: another fine job by all hands. But then, that is just another day of business on the Southside.

CHAPTER 11

A Tree Grows in Brooklyn

On August 8, 1987, the City of New York decided that it needed to promote a group of firefighters to lieutenants to make up for vacancies that had occurred due to retirements and promotions to captain. I and 99 others were to be promoted from the civil service list that had resulted from an examination we had taken the previous December. Among that group of the top 100 were four members of Rescue Co. 3: John Salka, Don Peterson, Joe Jove, and me. All of us had studied exceptionally hard for that exam, spending eight to 10 hours a day locked away in spare bedrooms converted into studies at home, on the beach while the kids played in the water, or any place else where we could find some peace and quiet. Many of us had taken to listening to tape recordings of lessons as we commuted to work—either recordings of promotion class lectures, or "books on tape," where we read the many volumes of department publications aloud into a tape recorder. These tapes then served to convert the hour and a half commute to and from work to additional study time. As a result of all this diligent effort, and aided by the fact that several of my actions during the previous seven and a half years had been deemed to be worthy of official recognition, I was promoted to the rank of lieutenant.

The system that the FDNY uses to select members for promotion is a combination of written examination, seniority credit, rewarding courage under fire, educational requirements, and job performance evaluations. It may not be a perfect system, but I believe it is a very good one and it is above all, fair. There is no way to inject favoritism into the system, as is possible when subjective evaluations are made, such as during an oral interview. This helps to eliminate discrimination in the promotion process which, unlike in other cities, New York has not had to deal with in the fire department ranks. (The FDNY promoted its first black battalion chief in 1938 at a time when the United States, including the military services, was still very heavily segregated.) If you pass the tests, do your job, and keep your nose clean, then you get the job. If you screw up, there is no one to blame but yourself.

The written examination—usually 100 multiple choice questions about firefighting tactics, duties, department regulations, and procedures—is designed to determine if the candidate knows the job he or she is aspiring to. There is no room for mistakes, since an officer will be issuing orders that put others' lives at risk. The person giving the orders must be technically competent. Book knowledge alone, of course, does not make a good officer. Unlike the military, there is no Annapolis or West Point to graduate fire officers in the United States. Nearly every one of them is a "mustang," having risen up through the ranks. There are training classes for the new officers, a month at the First Line Supervisors Training Program (FLSTP or "flips" for short) for new lieutenants, two weeks of captains' management training and two weeks of chief officers' development for new battalion chiefs: all run in-house by the FDNY. The National Fire Academy (NFA) runs the Executive Fire Officer training program at its campus in Emmitsburg, Maryland, that some members apply for, and many take training at the Naval Postgraduate Institute in Monterey, California and the Response to Terrorism program that is presented in conjunction with the U.S. Military Academy at West Point. You have to earn your way into those programs by earning your promotion first, however.

Realizing that a newly promoted ensign out of Annapolis does not know a fraction of what a seasoned old petty officer does about the real world, the FDNY seeks to take seniority into account in the officer selection process. Typically, the written examination accounts for 50% of a candidate's final mark. Seniority and awards make up the other 50%. That's because experience is a spellbinding teacher. If a person has read about how to do something, he or she may or may not be able to do that task correctly the first time they attempt it. After having done it several times, they will get good at it, and if they are good at it long enough, they should be able to teach someone else how to do it. A firefighter with less than three years on the job has not been around long enough to know all the tasks that one will have to perform, to say nothing of how to teach someone else how to do them, which is a large part of a lieutenant's job. Therefore, those with less than three years on the job are not eligible to take the exam. With three years on, you can take the test but your seniority mark is only a 70. For each additional year in grade, you earn two additional points for the first five years, then one point a year for the next five, up to 10 years for a maximum of 85 points in seniority. Thus, if a member with three years on wrote a 90 on the written exam, he or she would have a final score of 80, 70 for seniority and 90 written, averaged to 80. That member would be lower on the civil service promotion list than a 10-year member (85 seniority) who wrote an 80 for an 82.5 average. In the time the members are waiting for the list to reach them, they are gaining additional experience, helping to improve their skill set. A civil service list is usually valid for four years, after which time if the examiners haven't reached your number on the list, you have to take the test all over again.

"Meritorious acts" are rewarded with additional seniority credit. If a firefighter or officer performs an act of heroism that the department feels is worthy of official recognition, that act might earn the person who performed it anywhere from one-quarter of a point up to three full points on top of any seniority credit. The intent of this is to promote officers who have acted decisively in the past to resolve life-threatening situations, which is something that indicates the ability to make quick, correct decisions under pressure: one of a fire officer's primary functions. (In 1997, the United States Marine Corps was studying the way that they select combat officers in a desire to improve that process. During this time, they came to study the FDNY to determine how we select and train fire officers, recognizing the obvious similarities that require split-second decision making.) Additional points are awarded under state civil service laws to veterans of the U.S. military.

Job performance evaluations are filled out annually or as required by the member's immediate superior. There are four broad categories of performance that members are rated on, ranging from aggressiveness at fires, to performance of maintenance duties, to the public image the member presents. Each category is rated from one to 10, with one being the poorest rating and 10 being the highest. Getting all 10s on an evaluation (something I've never heard of—we all have room for improvement!) would not get anyone promoted any sooner, but getting any twos, threes, or a significant number of fours (below average) could get a promotion delayed or even, in theory, stopped. To prevent a situation where a personality clash between a member and their rater might result in an unfairly poor rating (or an unwarranted high rating), each evaluation must be reviewed by the next level of command. The member being rated is asked to make a written comment on the evaluation and has the right to appeal what might be perceived as an unfair rating.

Then, finally, there is an education requirement. Each rank from lieutenant to deputy chief has a minimum number of college credits that the person to be promoted must have successfully completed in order to be granted tenure in that rank. The member has one year after promotion to achieve this status. As I said, all in all I believe the system is a very efficient and fair way to select leaders in an 11,000-member organization where the top decision makers simply cannot know everybody. Thus, I find myself leaving Rescue Company 3.

Prior to promotion, each candidate is given the opportunity to submit a "wish list" of assignments they would like: as to what division they work in upon promotion. My first choice, since I had to leave the 5th Division, was the 15th, in eastern Brooklyn. It was where I started in 1979, in Engine 290—at the time, the busiest engine company in the world, and still in the top 10 every year in New York. The 15th Division was close to home for me, while some companies were only 20 minutes away and had a large number of very busy companies to work in. At a very large explosion and collapse in July 1987, I met Chief Michael Cronin who was working in the Fire Commissioner's office at the time. Chief Cronin had been the captain of Ladder 103 while I was in 290 and Ladder 103. When he made battalion chief, he was assigned to the 13th Battalion, which shared quarters with Rescue 3 on 181st Street in Washington Heights. When the collapse scene had stabilized, he approached and we exchanged pleasantries. Anxious to know about the yet to be released list, I asked about upcoming promotions. He assured me that I had done very well, and inquired about where I might want to be assigned. When I mentioned the 15th Division, he stated that no one would be going to the 15th because the overtime average for lieutenants there was very low, and the new promotions would be assigned to divisions where the lieutenants were working a lot of overtime; but he said he would see what he could do. Assignment to the 15th was not to work out for me, but in the long run, I could never have asked for anything better.

When the promotion order was published early the next month, I was assigned to the 11th Division. After a year of working the Relief Group, I was given a somewhat more permanent assignment, covering "until further orders" (UFO) at Ladder 111: a very busy tower ladder in the heart of Bedford-Stuyvesant. Ladder 111 was a really great place to work, with a lot of fire duty, and I had put in my transfer request to be permanently assigned there but was "bumped" out of the spot by a good friend of mine, Donald Hayde, who had more time as a Lieutenant than I did. Fortunately, another officer in 111 retired at the same time, and I went directly into his group, again UFO, pending the next transfer order. After spending a year at Ladder 111, and still not being permanently assigned, I got a call from Captain Ray Downey of Rescue 2 telling me there was going to be a vacancy in the Rescue and asking if I'd be interested. Interested? Hell! I'd be thrilled at the chance to get back into a rescue company. Now I had to tell Captain Frank Pampalone of Ladder 111 that I was going to pull my transfer request, after he had endorsed me for the spot. He rightfully wanted to know why—after all, 111 is one of the premier companies in the FDNY. I was forced to rationalize my decision to myself before I could begin to explain it to Frank.

The 11th Division, and especially Ladder 111, had been enjoyable, but I missed being in a rescue company. Ladder 111 was a great spot. They go to a lot of fires, including many of the same fires that Rescue 2 goes to, but they don't go to all of the fires in the rest of Brooklyn, and they don't go to all the unusual incidents—building collapses, plane crashes, ship fires, cave-ins, and subway wrecks—that the rescues are called to. No other unit faces the vast variety of assignments that a rescue company does. In addition, the nature of the job often means there is no game plan written for how to deal with what may be a once in a lifetime event. I had been to so many Brownstone fires while working in 111 that it seemed I could fight one in my sleep. There is a well-defined bulletin on how to fight Brownstone fires. The jobs that a rescue company must deal with, however, are diverse and in many cases rare events, which means that the rescue officer is quite literally thinking on his or her feet, developing the strategy and tactics for handling a specific event as it unfolds. In short, it's a huge challenge. If I got the chance, I would leave a great company for the opportunity of a lifetime. In mid-May, 1989, I got the call from Rescue Operations. I would be brought in to cover in Rescue Operations, and I would be detailed UFO to Rescue 2 after August 1st, when Lt. Craig Shelley would be promoted to captain.

Things went well in this new assignment. The guys were really great about accepting a new boss, even one that did not yet have 10 years on the job. I had gotten to know some of them during my year in Ladder 111, which is quartered only about a dozen blocks north of Rescue 2's old brick firehouse on Bergen Street. More importantly, I had three big boosters in my corner: Paul McFadden, a lieutenant in Rescue 2; Pete Lund, who had been a firefighter in Rescue 2; and Marty McTigue, who had also been a firefighter in Rescue 2 and who was now a captain in Rescue Operations. All three men had been my lieutenants at Rescue 3 and spoke up for my abilities. Finally, there was Ray Downey. The members believed that if Ray thought that I was capable, it was good enough for them. I wasn't Ray's first choice. That was Jack Kleehaus, a legendary firefighter who had spent years in Rescue 2 before getting promoted at the same time as me to lieutenant. The Chief of Department, Bill Feehan, believed strongly, however, that it was not a good practice to have officers supervising firefighters they had worked with when they were firefighters and refused to allow Jack to return to Rescue 2. That is how, on the night of August 18, 1989, I was working with a group of firefighters that were about to save a dozen peoples' lives.

Box 907 came in at 4:56 a.m.: a late night run to "fill out the alarm… second source." Engine 214 had been sent to investigate an ERS—No

Contact at Box 907 at Sumner Avenue and Halsey Street. Fires at this time of night are often very deadly, since most occupants are asleep and there are fewer people on the streets to spot a small fire and call for help, so the fire grows exponentially. As we were turning out, the dispatcher was advising, "Engine 214, your box has been transmitted, the second source [neighbor's telephone calls] reports a fire at 383 Halsey Street, Lewis to Sumner Avenues, in the cellar. Engine 214, receive?" As we turn onto Troy Avenue, which runs directly into Lewis Avenue, 214 pulls up to the address to find fire venting out of a cellar grating in front of the building, and radios the "10-75, fire in an occupied brownstone." We pull into the block just behind Ladder 111 and we can see there is a lot of work to be done. There is heavy fire blowing out of the cellar grate and fire is visible at the basement windows as well as the entrance door under the stoop. (Basements in a brownstone are several feet below the street level; the cellar is located directly under the basement, completely below grade.) Heavy smoke is pumping from all the openings on the upper floors. Many windows are open due to the hot weather, and there are panicky faces showing at nearly every one. We're off the rig in a flash. Everyone knows their assignments, and they sprint for the door.

At the stoop, I meet Captain Frank Pampalone of Ladder 111. Frank tells me that his inside team will take the basement and cellar, and he has his chauffeur and OV putting the bucket up to get the occupants in the front windows, but a neighbor just told him that there are jumpers at the upper floor rear windows. I take four of the Rescue guys inside with me, to try to get to the upper floors, while Bobby LaRocco, a new guy from Ladder 176, heads for the roof. We know there are no fire escapes on most of these four-story buildings, a condition that was legal when they were built, since they were only supposed to house one or two families. That makes the open wooden staircase a vital escape route. Engine 214 is stretching their hoseline to this stairway via the stoop and is preparing to descend it to begin knocking down fire in the basement and cellar. We can't wait, however. I leave Pete Bondy and Denis (one "N", or Murph for short) Murphy to search the parlor floor (second floor) and take Patty Fox and Tony Errico up above with me.

At the third floor landing I think I hear a human noise. I tell Tony and Pat to continue on up to the top floor, that I want to search this level. The top floor is a critical location that must be reached ASAP, but I've got a hunch about this sound. I kick open a door at the top of the stairs and crawl forward blindly, toward where I think I heard the sound coming from, sweeping my arms in front of me and calling out. Suddenly my hand lands on a soft, yielding mass. I grope in both directions with my hands, coming across long, thin arms and legs and realize it is the gangly form of a child. I can't make out any features, or tell if this child is breathing or not. I give a "10-45" on the radio, scoop the child up, dropping my hand tool so I have both hands free, and retrace my steps back to the stairs. As I descend to the parlor floor, the heat rising around me is tremendous. The doors to the basement and cellar are open and fire is venting right up this chimney that I'm trying to get down with this lifeless child. I do my best to wrap him up in my arms and chest to shield him from the tongues of flame shooting up past us. At the front door, the swirling smoke parts, revealing huge tongues of flame belching out of the basement windows and snaking out from underneath the stoop. I hear Patty Fox's voice calling over the handie-talkie as I scramble down the stoop, "Rescue irons to Battalion 37, 10-45." Pat and Tony have another victim on the top floor. Then it's Bobby (Rocky) La Rocca's turn, "Rescue roof to Battalion, we've got a jumper in a tree outside the top floor rear windows. We're trying to get to her. We need at least a 35 foot ladder in the rear." That's not going to happen anytime soon. It would take at least two full companies to haul such a large ladder up to the roof in front of the building, and then lower it into place in the rear yard, as well as a considerable amount of time to achieve. There are no spare firefighting units on scene to do it. Rocky and Ladder 111's roof man, Teddy Krowl, are on their own for a while.

I run out across the street with my victim, a handsome-faced boy, 8-year-old Dwayne Gill, right to Engine 214's chauffeur, Jimmy DiMeo. I tell him to get the resuscitator out as I begin giving the kid mouth–to–mouth. I give three or four breaths, then check for a pulse in the kid's neck. I think I feel one, so I give two more breaths, then check again. Yes, that's definitely a pulse: faint and very rapid, but definitely a pulse. Now Joe Borst, Ladder 111's OV, is giving another 10-45. I turn to DiMeo, who has just turned the resuscitator on and tell him, "You got him, Jimmy. I gotta get back inside." He tells me, "Not to worry, Lieu. EMS is on the way. I can handle it 'til then." I pull my gloves back on as I bound back up the stairs, two at a time.

Meanwhile, up on the roof, Rocky has his hands full, literally. Twenty-five-year-old Celestine McIntosh, *very* pregnant and deathly afraid of fire, is screaming hysterically for rescue, with good reason. She and her 67-year-old grandmother have attempted to escape via the stairs, only to find it blocked by fire. When they opened the apartment door, the roaring inferno below knocks them back, and they retreat to the rear window. Panic stricken, and taking a terrible beating from the heat and smoke entering through the open apartment door, the pregnant McIntosh jumps several feet to the limbs of a skinny "ghetto palm tree": actually an *Ailanthus glandulosa,* or Chinese Sumac, *the* "tree that grows in Brooklyn." These willowy little weeds are everywhere in the borough, even growing out of the third-floor window of at least one vacant building on Saratoga Avenue. They are not, however, a mighty oak or elm tree that you would want to climb onto if your life depended on it, which it now does for Celestine and her grandmother. The grandmother, Viola Whittaker, visiting from the isle of Barbados, has tried to climb out onto the tree also, but lost her grip, and has fallen four floors to the rear yard. Miraculously, this tree is somehow holding the other woman, with the bulging belly, but there is no way in hell she can climb down four stories along its bamboo-like shoots.

Rocky and Teddy lay flat on their stomachs at the roof edge and lean way out into space, but can barely touch her fingertips, which are desperately clutching the thin stalks. Rocky tells Teddy to lay across his legs while Rocky leans still further down and out over the roof edge. With his entire torso dangling fully out over the smoke-filled abyss below, Rocky is able to reach the woman's wrists. Taking a firm hold, Rocky wrenches himself around and screams at Teddy to pull with everything he's got. They swing the helpless mother-to-be back over to the building, and drag the terrified woman to safety on the roof.

I pull the 4.5 mask's hairnet over my head as I race back up the stairs. At the third floor, I meet Pat and Tony carrying the unconscious form of 66-year-old Idela Norgrove. I grab one arm from Tony as he backs down the stairs. She is not a big woman, but the stairs are very tight and cluttered with debris. Once we reach the parlor floor, the wide stairs of the front stoop make my assistance unnecessary, and I drop off and again head back upstairs to continue the search. Ladder Companies 123 and 176 are in now, Engine 214 has beaten the fire into submission with their line, and conditions are starting to improve. I meet Bondy and Murphy at the third floor, and they tell me the primary search of the parlor floor and the third floor are negative. I relay their report to Chief Kotz of the 37 Battalion out in the street, then call Rocky to see how he's making out. He replies that he and Teddy have got the one jumper and are bringing her down through the adjoining building, but he sees another jumper down in the rear yard, and there are guys (meaning firefighters) working on her. Chief Kotz tells me he's got Ladder 132 back there working on the jumper, but she's a "Code 1": dead on arrival (DOA). There are a bunch of firefighters on the top floor now, the fire is knocked down, and there is more than enough help at hand. I call the guys to meet out in the street.

I think aloud, "It's amazing what a few minutes mean in terms of getting enough manpower on scene." The "bean counters" down at city hall, and the "efficiency experts" hired to study fire department staffing around the country don't get this part. They look at computer printouts that show that the late night and early morning hours are the slowest in terms of alarms per hour, and decide that (like when scheduling buses or trains), since there is not as much demand in the late hours, you should be able to make do with less service, meaning fewer firefighters or even fewer fire companies at night. That is craziness! The late hours are when the life hazard in dwellings is at its zenith, with people sleeping. I look back at this fire that we've just beaten, and try to imagine it with fewer firefighters, or fewer companies, or companies that were spaced just another two minutes farther apart, and I am horrified at what the consequences might have been. Yes, we have one DOA, and three others who are unconscious, burned, and might die (all eventually lived), but there are over a dozen people alive right now who would be dead if the firefighters of Bed-Stuy hadn't gotten there when they did. What a great job! To be a part of a group of people like this gives me an absolutely magnificent feeling.

CHAPTER 12

Flight 5050: Who Likes Those Odds?

The night tour of September 20, 1989 promised to be a very pleasant, enjoyable fall evening. The weather was warm and skies clear. I arrived at the squat, red brick fire house on Bergen Street early, since it seemed that even the traffic that Wednesday night was cooperating. I relieved Lt. Paul McFadden at around 5:00 p.m., so he decided that my report about light traffic would make leaving at the height of the evening rush hour more bearable than usual. In fact, he might even make it home in time to have dinner with his family instead of dining alone. Paul handed over his riding list from the day tour, and I proceeded to update our "tracking system," placing my name at the top of the 3×5 sheet of paper and crossing out his on the line marked "Officer." A few minutes later, Mike Pena knocks on the office door as I'm reading the Department orders: catching up on changes in Department personnel, policy, and procedures. He is letting me know he has relieved a member and I write his name in on the riding list as well.

At 5:50 p.m., the Department radio has a fairly steady chatter. Somehow you start to separate the routine, unimportant radio traffic from the messages that could mean "business." On this evening, the dispatcher is sending out box after box, most rather routine, when he announces, "Brooklyn box 2786, the location: 86th Street and 13th Avenue, reported to be a vacant building at that location." Almost subconsciously I say to myself, "All right, that one's got potential." I know the location well. There is a Knights of Columbus hall on the northeast corner where I have both taught and taken many promotion study classes. I can't picture a vacant building on that block, since the area is very well-to-do. A report of a "vacant," therefore, is probably the real thing, since a prankster would be unlikely to think up such a report to send out such a false alarm. I start down the stairs, preparing for roll call, when the dispatcher informs the 42 Battalion,

> "We're giving you three, two, and Rescue, 42. It sounds like a job; we're getting calls reporting the first and second floor of a vacant on 86th Street between 12th and 13th Avenues."

The 42 acknowledges just as the teleprinter in quarters starts sounding "the charge." As I sprint for the cab, I hear the chatter in the back as members of the day tour who have taken early relief bitch and moan about how the incoming night tour guys have "stolen" their fire. Most of the guys in Rescue 2 ride right up to the last minute, rather than risk missing "a job" due to early relief. Today is an exception though, and two unhappy firefighters are left on the apparatus floor as the rig blasts out the door.

Bensonhurst is about seven and one-half miles from Rescue 2's quarters on Bergen Street in Weeksville, right between Bed-Stuy and Crown Heights—a pretty long ride given the dense traffic so often encountered, so it can take upwards of 25 minutes even with flashing emergency lights and siren. Many times on such runs we are "10-2'd": told to return to service before we even get there. Not tonight, though. Engine 284 gives the "10-75" as we pull out the door. As we cross Coney Island Avenue, the 42 Battalion aide gives his first preliminary report.

> "We're using all hands for a fire in a three-story frame, 30×60, Queen Anne. We have heavy fire on the first and second floors, with possible extension to the third floor and attic. Exposure 1 is a street, 2 is a similar detached, separated by a 15-foot driveway, 3 is a rear yard, and 4 is a two-story brick 20×50. The situation is doubtful, request an additional engine and truck."

From this initial size-up, I can tell we're probably going to go to work. "Doubtful" indicates that the chief in command doubts that there are enough resources yet on scene to control the blaze. Queen Annes are notorious as tough, snotty fires. They are large, 100-plus-year-old frame houses, with many hidden voids in the walls and floors for fire to burrow into. The stairways above the ground floor are narrow and winding, making the advance of hoselines difficult at best. Fires in the first or second floor are problem enough but when fire is in the cellar or attic, you can count on getting beat up. The access is worse and so is the ventilation, with only a few very small windows at best.

As we swing the rig left onto 86th Street from Fort Hamilton Parkway, we get an idea of what we're in for—heavy gray-yellow smoke (I call it puke yellow, not so much for its resemblance to vomit as for its ability to make you do so) is blanketing 86th Street, billowing from every nook and cranny on the front and sides of the building. A slight breeze from the north is making life miserable for everyone in front of the building and blowing the smoke across the wide street into the Dyker Beach Golf Course, ruining a lot of people's games on a beautiful late summer evening. As we get closer, Tower Ladder 149's boom appears through the swirling gray cloud. The firefighters in the tower's basket have just finished operating the master stream, knocking down the heavy fire on the first and second floor. From the quantity of smoke remaining, though, it's very plain that there is still a lot of fire in the walls and floors that the tower's stream has not been able to extinguish.

As we report into the 12th Division, which has assumed command, we are directed to assist Engine 330 in advancing a line to the attic. Mike Pena, Lee Ielpi, Denis Murphy, and I are the inside team, while Al Washington and Bruce Howard are the outside. As we climb the smoking, charred remains of the staircase to the attic, one of the treads lets go beneath Mike Pena. He catches himself on the riser below as he goes through, and we are able to pull him up and out without major injury, but he has seriously bruised and pulled some muscles in his back. The line is behind us, coming up the narrow attic stair and the attic is red hot. There is fire burning freely behind the knee walls, half-height walls that rise from the floor to the underside of the sloping roof beams, forming a large open space behind it. I would love to send Mike down for a blow, but we need the line up there right now, or else this attic is going to light up. The engine has to get up here immediately with water, and Mike will have to wait on the side until they clear the stairs. Lee Ielpi is encouraging 330 to pick up their pace, as it's getting a little bit warm now. His exact choice of words escapes me but its delivery was something like, "Get the fuck up here with that line, guys, or somebody's gonna be toast!"

A lot of hot, fast, and dirty punching of holes and pulling of ceilings exposes the hidden fire to the hoseline, and soon the beast is tamed. After a tough fight, we emerge from this smoldering pile of embers that once was a proud Victorian home, satisfied that the situation is under control. We regroup at the rig and change air cylinders, getting ready for the next job. Mike tells me that his back is starting to tighten up and that he wants to "take a mark," to file an injury report. I let the Battalion know all of the particulars before we leave the scene. Everyone is hot and sweaty, and the first order of business is to find a bodega (a Spanish grocery) and get some cool drinks. As we head back toward Bergen Street, Lee notes, "It's a great way to start a tour, huh Lieu?" Being out in Bensonhurst, we don't see any of the familiar yellow and red aluminum awnings with the flashing yellow lights that seem to mark every bodega in New York City. Instead, we settle for a green canvas roll-up awning that marks Vinnie's Deli and Meat Market. As we gulp down our Gatorades and iced teas, we hold an informal critique. The building was set back from the street about 40 feet and was built on a small hill, which prevented the aerials from reaching the peak of the roof. Al and Bruce are describing the difficulties they faced trying to vent the roof, and wonder if maybe they should have joined us and tried pushing the roof open from inside, since they never did get a really good hole cut before we knocked the fire down inside. The whole inside team is unanimous in saying that no, there was no room for them inside, and that their efforts really were appreciated. On the way back to Crown Heights we pick up the meal for the night tour.

It's 7:30 p.m. by the time we back into quarters. I go right upstairs after finishing the night tour roll call and entering the particulars of Mike's injury in the Company journal. I call the notification desk and ask them to have the on-duty doctor come to take a look at Mike's back. He has a large bruise and is walking very stiffly. A few minutes later, the Department phone rings. It's Dr. Kelly, one of the Department Medical Officers. She asks whether Mike needs to go to the emergency room, or if he can wait about an hour or so until she can get to us. She has to visit several other injured firefighters from the first alarm assignment. Mike says he can wait and takes a couple of aspirin to hold him over. Work has begun on the evening meal, and the men settle into their comfortable routine. The bantering back and forth is typical of all firehouse kitchens, with many of the remarks centering on Mike's bad back. The comments range from the sympathetic, "You're walking like you got a broom stuck up your ass," to the profane, "If you get sent home, remember to call and let your wife and her boyfriend know you're coming. Spend a quarter and save a marriage, you know." Mike's wife, Diane, is a really sweet girl who would just die if she knew what the guys were implying about her—all totally untrue. But nothing is sacred in the firehouse once you owe your life to the guy next to you. The one exception is the guys killed in the line of duty and their families.

At 11:30 p.m. the voice alarm announces:

> "In the borough of Queens, a third alarm has been transmitted for Box 37, LaGuardia Airport Control Tower. Repeating... in the borough of Queens, a third alarm has been transmitted for Box 37 at LaGuardia Airport."

Denis immediately says, "Gee, I didn't hear the second alarm come in, I wonder what they got going over there?" Lee explains that "Box 37 is the crash box located in the tower. Just like at JFK where we go, it's an automatic third alarm as soon as the tower transmits it. It means they got a plane down!" Now everyone is glued to the scanner, which has been set on manual to the Queens radio channel. Rescue 2 is assigned as the second due Rescue Company for the crash box, Box 269 at J.F. Kennedy Airport, and we attend annual drills at that location. Rescue 4 and Rescue 3 are the two assigned Rescues at LaGuardia. We all know though that a serious crash with trapped survivors could possibly bring all five rescue companies to the scene. Getting a radio report from the scene will give us an idea of whether we'll be going or not.

This preliminary size-up is critical to us in Rescue 2 this night, for an even more unusual reason than normal. Our regularly assigned apparatus is out of service for mechanical reasons. We had been using the Department spare until two days ago, when it too was taken out of service because it was contaminated with asbestos fibers at the scene of a major steam pipe rupture at Gramercy Park in Manhattan. As a result, there are no more spare rescue trucks. Rescue 2 is temporarily riding around in a spare pumper and a spare chief's Chevy Suburban, which carries a total of maybe one-quarter of the equipment and tools that we would normally have onboard the rescue rig. The remainder is scattered all over the apparatus floor. The dispatchers have been instructed to notify us of the nature of any special calls, so that we can hopefully select the needed equipment from the pile on the floor. Now is one such time.

The Queens dispatcher is relaying information to the 14th Division as it is received from the tower at LaGuardia. Initial reports indicate a commercial jetliner has failed to complete its takeoff and has gone down in the East River. Immediately, I order the men to start loading all the extra water rescue gear aboard the two pieces. Rescue 2 is a SCUBA trained rescue, as are Rescues 1 and 5. I know that we can beat these units into LaGuardia and as I am about to call the Queens dispatcher to ask if they want us to start out for the box, I hear Car 13D, the rescue liaison duty captain, call on the radio, asking if that initial report of a commercial jet down in the river has been confirmed. The dispatcher replies that no fire department units have reached the crash site yet, but that the tower now confirms it is a US Air 737, Flight 5050, with at least 50 people aboard, and it is in fact down in the river at the west end of Runway 31. Captain John Cerato in Car 13D immediately replies, "Start out all three SCUBA Rescues on this box. We can always return them if they're not needed."

The members of Rescue 2 have finished loading extra SCUBA set-ups, spare bottles, an inflatable boat and motor, thermal recovery capsules that help warm people exposed to cold water, ring buoys, cold water exposure/flotation suits, and extra lifelines. I grab one of the flotation suits myself as I climb up into the cab of the Mack CF Pumper. Mike Pena climbs up into the back of the pumper behind me, carrying his SCUBA gear. "Where do you think you're going? The doctor is coming to see you!" I ask. "Hey, Lieu, it sounds like you're going to need divers," he replies. "My back can wait till later if this is as bad as it sounds." I know he's right, judging from the urgent chatter coming over the Queens radio.

As we career through the streets of Bed-Stuy, heading for the Brooklyn-Queens Expressway that will take us right to LaGuardia, I am kept busy monitoring the Queens radio, sounding the sirens and air horn and trying to communicate with the three members riding in the Suburban, which is following the pumper. Radio reports now from the crash site indicate a big jet is indeed down in the river just off the runway, which is actually built on a pier sticking out into the river. There are numerous survivors in the water, but worse, there are reports of persons trapped in the wreckage, which while partially supported on a large timber lighting tower that projects up from the river bottom, is slowly sinking. The chiefs' aides sound frantic as they request all available fireboats and scuba divers to the scene. I silently thank God that Pena had climbed aboard.

In Rescue 2 this night I have four Fire Department divers working who are trained in black water rescue diving: Mike Pena (with his bad back), Denis Murphy, Bruce Howard, and Lee Ielpi. Al Washington is not a diver. As the chauffeur, Lee Ielpi will probably be quite busy bringing tools and equipment from the rig. At this time, I am only a sport diver, not yet Fire Department black water qualified. That's okay, though, since the way we dive, each diver requires a tender on the surface to handle their communications rope. Fire Department dive procedures call for all our divers to be tethered to shore or boat at all times for the diver's safety. We treat all the water in New York Harbor as polluted water, and dive with Viking Pro full dry suits and AGA Mark II full facepiece masks. In theory, water never contacts the skin. In practice, given the always tight fire department budget, things are slightly different.

The masks are equipped with microphones and speakers, which are connected to the tender's mic and speaker via hardwire, which is woven inside a rope that is attached to a harness around the diver's chest. The system we are using in 1989 is not totally reliable and, as a backup, we use a series of tugs on the rope to send messages from diver to tender. As we approach the airport, I don the cold water flotation suit, thankful that I am in the relatively luxurious space of the CF cab instead of in the severely constricted space I would normally have in the front of our regular American LaFrance rescue truck cab. I do not plan on having to enter the water, but having been around piers and docks since I was a child, I know that a lot of people who never intended to go in the water end up getting wet. The flotation suit is a precaution.

Pulling into the airport property, we have been informed by radio that a staging area and entry control point has been established at the 94th Street gate. The plane has not caught fire, and there is little need for the second and third alarm pumpers. As a result, all the engine companies are being diverted at the gate by the battalion chief who is acting as the staging area commander. And along comes Rescue 2 all dressed up to go diving, but riding in a pumper! The chief jumps out in front of the rig as we go to proceed through the gate, screaming at us, "What the hell's the matter with you? Don't you listen to your radio? We said all engine companies are to stage over there," gesturing with both arms flailing to the area where eight other pumpers exactly like ours are lined up, if not exactly neatly, in some semblance of order. I start to try to explain that we're not an engine company but a rescue company. That drew about the most disbelieving look I've seen in quite a while, but it did not slow that chief down one bit. He kept on ranting and raving about how we're screwing up the whole works and for us to pull immediately to the right and let the other units get by. Meanwhile, the department radio and the handie-talkie are both alive with reports of people in the water and calls for an ETA on the dive rescues. I turn to Lee as the chief keeps on jumping up and down at my window demanding immediate compliance, and I say, "Fuck him, Leo, let's go! Follow that ambulance!" Lee nails the throttle and pulls left, leaving the chief sputtering in a cloud of diesel exhaust. The chief's Suburban follows us through unimpeded, since God knows you can never get enough chiefs on the scene of a major problem.

We pull up to the pier edge behind a big yellow Port Authority crash truck. I climb down from the cab in my orange "Gumby Suit"—so named because its large, loose fit makes you walk just like that children's animated TV character—and report into Chief Neville of the 14th Division, the Fire Department's incident commander. He tells me that there are at least 40 people in the water and several people trapped in the sinking plane, the back emergency door of which is right at the water line, about 50 feet off the end of the pier. The plane had aborted its takeoff and had nearly come to a safe stop when it ran out of runway and plunged off the end of the pier, careening off this approach light tower about 100 feet out from the pier. The passengers were unaware that the plane was in the water as the emergency evacuation began. The doors were opened onto a pitch black night and they began jumping out of the plane, landing in Flushing Bay. Fortunately, the impact of the plane knocked loose large pieces of timber from the wooden light tower where it came to rest. These timbers, along with other floating debris, served to keep many of the survivors afloat as they did not take their seat bottom cushions with them initially.

First-arriving Aircraft Fire Rescue personnel find themselves stymied. The runway at this end of the airport is actually built out as a huge pier over the bay. The edge of the runway is about 18 feet above water level. The survivors in the water are being swept in under the pier by the tide, right past the rescuers atop the pier. Boats are urgently needed to pick up survivors, but Rescue 2's mission is to get divers out to the plane's sinking fuselage.

Al Washington and Bruce Howard grab a 35-foot extension ladder off the rear of Tower Ladder 117, and prepare to lower it into place beneath the pier. It's nearly 20 feet down just to the water level, so they manually extend the "fly" section and lash the two sections together with the halyard. Then they are joined by Denis Murphy and Mike Pena and the four of them lower the ladder off of the pier, hopefully to rest the end on the bottom of the riverbed. Naturally, there is more than one "Murphy" at work here and the one with his own "law" takes over. The fully extended 35-foot ladder does not touch bottom and is swaying in the current. Lee Ielpi grabs a life-saving rope, 150 feet of $^9/_{16}$-inch nylon normally used to lower firefighters from the roof of a burning building to rescue trapped occupants who are out of the reach of ladders, and now uses it to tie the ladder in place, securing it to a ship mooring bollard on the edge of the pier. With the ladder thus "secured," the three divers, Mike, Denis, and Bruce, now are able to clamber down it in full SCUBA gear and they don the fins they carried, then swim out the 70 feet to the door of the plane. The swaying ladder and strong current pushing against their swim worked against their efforts, and in most situations would have been cause to abort the dive. But this, of course, is no ordinary situation. All three divers make their way to the plane, but due to the amount of wreckage and victims in the water, must disconnect their safety tethers. They connect one each to the doorway of the plane and to the right wing surface. This will allow other swimmers to make their way back and forth to the plane with less danger and fatigue. Besides, with all the stuff in the water (wires, cables, people, etc.) swimming around with this rope on would be even more dangerous. Mike Pena assists a number of people from the wing of the plane into some small boats and rafts that have now arrived or been deployed. He is to be on standby at the wing exit as crews from Rescue 3 work inside the sinking plane, extricating a woman who is trapped in her seat, partially submerged (fig. 12–1).

Bruce and Denis join as dive buddies and begin an underwater search of the perimeter of the plane, looking for any victims who may have been swept into the wreckage and trapped beneath it by the current. Finding none, they then descend 20 feet to the river bottom and search the area beneath the sinking hull for victims. As soon as the divers have disconnected their tethers, Lee Ielpi starts descending the dangling 35-foot ladder. There are still a lot of victims around the plane: some in the water, some sitting on the wings. Rescue companies all carry inflatable four-person life rafts known as "Switliks" for use in such situations. Lee has brought Rescue 2's Switlik to the pier edge and now asks for permission to deploy it. I am at first hesitant about letting him climb down the free-hanging ladder with the 30-pound package, which is stored uninflated in a case about the size of a small suitcase. I know the divers just climbed the same ladder with more than twice that weight on, but at least they were wearing most of it fairly evenly distributed and had both hands free. Lee will have to carry the package down by hand and then stabilize himself on the precarious perch in order to activate the

Figure 12–1. The wreckage of Flight 5050 at rest in Flushing Bay. FDNY units saved numerous lives in the sinking plane and surrounding waters. (*Courtesy of FDNY photo unit*)

inflation mechanism. Then he will somehow have to keep control of the rapidly expanding raft as the CO_2 cartridge blows it out of the suitcase; otherwise, the current will merely carry it away from the plane. Reluctantly, I give my OK to Lee's plan after telling him that if there is any kind of a problem, to abandon the raft. Lee starts down the shaky ladder with supreme confidence. He is Lee Ielpi, one of the best damn rescuers in the City of New York, if not the world!

Of course, there is a problem. Remember, "Murphy's Law" is still in full force and effect. As Lee is trying to maintain his grip on the ladder, which is actually canted in under the pier, pushed by his weight and the current, and also to manipulate the operating mechanism to inflate the raft, he hears people screaming for help from deep within the pitch blackness under the pier. He now yells up, "Hey, Lieu, we got people screaming under the pier here! What should I do?" Quickly, I consider the options. All of my divers are committed to the plane, two more dive rescues are en route but are not on the scene yet, and Lee is having trouble getting our raft inflated. I think, "motorized craft cannot travel under a pier due to the maze of criss-cross bracing typically found." I am very much aware of this last piece of information, thanks to an under-pier firefighting operation I took part in with Rescue 3 about four years earlier. The only way to maneuver under the average pier is as a swimmer—even the raft will not be very useful, I'm thinking. My options are suddenly very clear. There is no option. I am going to have to swim in under the pier and go after these people, since I am the only one immediately available with any kind of flotation equipment. I yell to Lee that I'm coming down and to hang on, then I hand my handie-talkie to Al to hold. It's now my turn to climb this ladder that pitches away from my feet, being the opposite of any ladder that I've ever climbed before. Fortunately for Lee and me, I didn't have to get right down alongside him and work. I just wanted to get close enough to the surface to drop in without striking any submerged obstacles too hard. Been there, and done that! I plop in alongside Lee and tell him to send reinforcements in after me as soon as Rescue 1 or Rescue 5 arrives, or our divers get free.

I swim easily in under the pier, buoyed by the exposure suit and carried by the current. This is not what I had expected. There's no maze of timbers and cross ties. This pier is built on massive concrete pilings, each about 48 inches in diameter. The pilings are spaced about 10 feet apart in every direction. I can hear the people yelling for help from well ahead of me. I wish that I had brought along a SCUBA flashlight. I can't see anything in the pitch dark, but I can hear the voices calling from off to my right, slightly. I try to swim in that direction, but without fins I find it nearly impossible to make any real progress against the current. I make my way to the nearest piling and pushing off from it, make my way sideways to the next row of pilings, but the current has pushed me in, another piling in depth. The voices still seem to be to my right and I call out to them to keep shouting to guide me. I must judge their position accurately. If I pass them, or even line up at the right depth, but on the wrong line of pilings sideways, I will miss them. My best guess is that I either am on the line with them, or they are one line to my right and several columns still ahead. I start swimming down the middle and now, after my eyes have adjusted to the darkness, I can see several forms clinging to a piling ahead of me to the left. I was on the correct column line.

I dog paddle my way up to the shapes calling out, "Hold on! Help is on the way!" As I get closer, I see that these are not plane passengers. These people are in some kind of raft, *and* in uniform. When the victims finally see my head bobbing toward them, they are very thankful. They're still screaming like a bunch of school girls that just saw their first flasher, but now they're yelling at each other. It seems that these three would-be-rescuers arrived at the crash site in a police truck, similar in some respects to a rescue truck. Their truck carried a small inflatable raft similar to Rescue 2's and the heroes decided to deploy it and paddle out to the plane. So, inflate it they did, and all three jumped in, but none of them brought along any type of paddle. Naturally, as soon as the current got hold of them, they found themselves "up the creek." As I reach them, the reason for their excitement is evident. The current has carried the rubberized fabric raft in under the pier, bouncing it off the barnacle encrusted pilings like the winning ball in a pinball game. The raft quite naturally has taken offense at such treatment and has decided to spring several leaks. The raft is now taking on water and the little darlings are getting their tootsies wet, much to their dismay. To make matters worse, the current keeps pushing them deeper and deeper into the inky darkness under the pier, and farther from help.

I try to gain control of the situation, reassuring them that with the inflatable suit I'm wearing, I can easily keep all three of them afloat. "What, are you shittin' me?" one of them screams, "I'm wearing three guns and a heavy (bulletproof) vest! I'll go to the bottom like a fuckin' anchor!" I suggest that the fellow might want to ditch the "weight belts," the guns, ammo, handcuffs, etc. The reply was immediate and unanimous, "Are you fucking nuts, ditch our guns?" Then suspicion, "Hey, what truck are you with anyway?" My reply only worsens their mood—"Rescue 2," draws groans of "Oh no... not a firefighter!"

The sinking raft and the incessant push of the current prevent pride from interfering and the three cops readily accept my plan for remaining afloat. They'll all link arms in the raft while I wrap myself around the next piling. That way, we can hold onto each other, and prevent us from going further into the unknown. I assure them that between the raft and myself, we'll stay afloat until help arrives, which doesn't take long. In a matter of minutes we are joined by Paul Hashagen from Rescue 1 in SCUBA gear who swims in with a tether line that could be used to pull ourselves back against the current to the pier face. Then Denis Murphy joins us in Rescue 2's raft, following the rope in with paddles as well. The raft is small when it carries a big guy like Denis, also in SCUBA gear, in addition to victims. Only two of the cops will fit inside the raft: one will have to get wet, hanging onto the outside between Paul and me, as Denis guides the boat out. This causes a lively debate between the three officers to see which one of them is going to be so unlucky as to be immersed in the polluted, jet-fuel-covered bay. At one point, I think they might come to blows, but cooler heads prevail and the two senior men make it into the inflatable. As we are about halfway out from under the pier, we are joined by another inflatable raft, this one being operated by a police officer. It too is small, so again all three cops cannot fit in it. Now another argument ensues as to which ones are going to get in it, and which poor sucker is going to have to be seen being rescued by the fire department. This argument is nearly as intense as the first, with the two guys in our raft fighting just as fervently to get out of it as they were to get into it just two minutes ago. Eventually, the junior man ends up in the fire department raft, the two senior men in the police department's, and Paul and I hang onto the outside, hitching a ride back to the now flood-lit pier edge.

After unloading our "passengers," I scramble back up one of the now numerous hanging ladders, and make my way to the command post. I report the conditions under the pier, the need to make a thorough secondary search for persons who may have been swept underneath by the very strong current, as well as the feasibility of using an outboard motor powered rubber boat. I then rejoin Lee and Al at the edge of the pier. Numerous victims have been plucked from the water or from the wings of the plane. Members of Rescue 3, Rescue 4, and Haz Mat, as well as divers from Rescue 1, Rescue 5, and Rescue 2 have made their way out to the plane and have encountered several people trapped within the sinking wreckage. Chris Blackwell and Jerry Murtha of Rescue 3 are working to free one of the passengers who is trapped in her seat by the overhead luggage compartments, which have fallen on top of her. The plane has sheared along its circumference in two places, just ahead of, and just behind the wings. The front portion is being supported by the lighting pier it crashed into, but the rear half is unsupported and the rear has dropped down about three feet, pinning Ms. Alethia Crews in the wreckage.

An airplane's wreckage is rather substantial. The aluminum and plastic that constitutes the bulk of the construction is no match for the power of a Hurst or similar hydraulic rescue tools. But when a plane is partially submerged 60 to 70 feet off the edge of the pier, getting the 76-pound Hurst tool jaws and their 70-pound motor into the plane is something of a challenge! Murtha and Blackwell immediately start prying the material away from Ms. Crews with their bare hands, while assisting her in keeping her head above the rising water. A 100-foot tower ladder is brought into position along the pier edge and a Hurst tool is placed in the basket. The jaws and cutters are passed in through the back door of the plane while the power unit remains in the basket. In the meantime, Rescue Co. 5 has backed their rig up to the edge of the pier in an effort to keep the tail of the plane from sinking any lower. Their winch cable is unwound and Denis Murphy and Bruce Howard, working with NYPD divers, now swim the ⅝-inch steel cable through the murky blackness under the tail of the plane and bring it back up through the rear doors, where it is hooked back on itself with a shackle, forming a loop under the sagging tail: anything to buy time for the rescue effort.

With the Hurst tool in operation, Murtha and Blackwell, assisted by Mike Milner of Rescue 4, make good progress. They are working in chest-deep water inside the plane, with only a few SCUBA lights for visibility inside the fuselage. Every time a police launch approaches the plane, it rocks back and forth in the wake of the boat, and sinks a little deeper. The shriek of tearing aluminum is continuous as the rear settles deeper and deeper. Soon Ms. Crews is freed from what promised to be a watery grave.

Unfortunately, two women sitting in the row in front of her are not so lucky. They are in the row right in line with the break in the plane's fuselage, and were crushed by the top of the front half of the plane when it split. Their bodies are underwater, and the rising water level makes the situation too dangerous to risk trying to extricate them at this time. A last search of the cabin proves all survivors, 52 total, are out of the plane and ashore now. The two women in front of Ms. Crews are the only fatalities, thanks to the FDNY and other emergency responders. Soon, fire department units are being returned to service. The scene will be handled by the

National Transportation Safety Board (NTSB), the Port Authority, and the airline's salvage crew. Rescue 2 gathers up our gear and prepares to return to Brooklyn, having had enough excitement for the evening. Somebody comments about the number designation of the flight: "Who the hell would want to get on a flight called 5050? I want much better odds than that, like 10,000 to 1." That gets a chuckle from everybody. As we prepare to leave, the command post calls: "Come back by before you leave."

At the command post, Medical Officer Dr. Kerry Kelly wants to let us know she hasn't forgotten about us. She is automatically assigned on the third alarm though, so she has not gotten out to see Firefighter Pena yet. I reply, "That's okay, he's here with us. I'll send him over." Imagine the look on her face when this wet, bedraggled figure, dressed in SCUBA underwear, shows up to tell her about a back injury suffered at a fire at the opposite end of Brooklyn eight hours earlier. A quick examination reveals a severe strain of the lower back with multiple contusions. "That's it, mister, you're done. You're on medical leave as of this moment. And be glad you're not getting in trouble for not remaining in quarters to await the doctor, like you are supposed to," Dr. Kelly says with a furtive smile. She knows the deal here. Her father was an old-school firefighter. Mike did what needed doing. He wasn't like some guys who might have stayed in quarters. He went where he was needed, like the old cavalry saying: "Ride to the sound of the guns."

Back on Bergen Street, all the SCUBA gear, rafts, ropes, and everything else has to be cleaned of the jet fuel and made ready for service before there's any rest for the weary. You never know when the next call is about to knock on the door. We'll find out soon enough.

At 4:30 a.m., I climb the squeaking old wooden staircase for the first time since just before 6:00 p.m. the night before. At 4:32 a.m., precisely as my head hits the pillow, the tone alert goes off, sending us all back down the stairs again. (I swear there's a switch under that pillow that senses pressure and automatically sends us out if a head makes contact!)

Lee is reading out the teleprinter ticket as we come down the stairs. "Box 957, Nostrand and Herkimer, reported to be a store!" "Oh, great," I think. That's a bad box, one with a whole lot of potential trouble. A couple of blocks of stores with dwellings above surround the intersection and all are heavily fortified due to the high crime rate in the area. And to make matters worse, we're down a member, since Pena is now officially on medical leave. The rigs start out the door and I remind Lee to make sure to stay away from the scene, since we don't want our pumper, with no hose or other equipment, to be mistaken for an engine company.

At this late hour Atlantic Avenue is empty, and we fly down the thoroughfare hitting all green lights. Lee stops us just short of Nostrand Avenue so we can look up the block toward Herkimer Street. We are first to arrive. From the corner we can see smoke coming from a store on the east side of the block, between Atlantic and Herkimer. I tell Lee to leave the rig where it is and transmit the 10-75 signal.

We hustle down the block to find heavy smoke pumping out from behind the roll-down gates of a religious goods supply store. The store is on the ground floor of a four-story wood frame building, but its entrance is actually a few steps down. There is an exterior stoop leading up a few steps to the second floor, where there is an entrance to another store, but no entrance to the upper floors. Ladder 111 is pulling in as we arrive. I tell Denis and Bruce to help with forcing the store entrances, while Lee and I look for a way into the apartments above. We look all over but don't find a staircase. From the look of the smoke at the upper floors, there is fire upstairs already. Now Lee makes a discovery, a hatch within the upper store, that leads up to the third floor. Pushing the cover over the hatch upward with his hook, we can see a very heavy smoke and fire condition on the third floor. It seems somebody has been selling dope out of the second floor store. There was a rope ladder hanging over the lip of the scuttle that the dealers would climb up to make their getaway if raided.

I tell Lee to forget this joint, it's too much of a maze, with steel-plated walls and doors, trap doors, and the fire is too advanced. It's a crack den, and not worth getting trapped over. We'll have to try to cut the fire off before it gets into the attached exposures. I give the news to Chief Harry Rogers, who tells us to get into exposure 2; he already has people in exposure 4. As we make our way up the stairs in this old wooden building, we find a heavy smoke condition pushing through a large hole in the wall between the buildings. The dealers' escape route included a hole smashed through the wall into the next building, confusing people looking to head them off. I call for a handline as we continue upwards. This building has occupied apartments and needs to be searched. I call Bruce and Denis up to help us. After a quick primary, we realize the occupants have all escaped safely. The rising heat and the crackling in the ceiling tells us there is fire in the common cockloft. Al Washington has our saw on the roof, working with 111's and 132's roof firefighters. Now I let him know to concentrate his efforts over our heads in exposure 2. Denis Murphy, Lee, and Bruce are pulling ceilings for all they're worth but, after a very short time, it's obvious the fire is past us already and is getting into exposure 2A. I call down to the command post with this latest bit of good news. Chief Rogers informs me he has sent the third alarm, but the companies are not in yet. "Can you get over into 2A?" he asks.

Wearily now, we make our way down from the top floor, past the hole in the wall where an engine company is trying to keep the fire back inside the original fire building. I tell them the cockloft overhead is roaring away. Just then, the chief calls them to tell them to back down, he'll be putting 111's tower stream into operation. We trudge up the four flights to the top floor of exposure 2A, where conditions are only slightly better than they were in exposure 2. Engine 235 has a line up there already though, and is looking for the ceilings to be pulled. Lee, Bruce, and Denis are pulling ceilings all along the wall between exposure 2A and 2: the building we just left. Again there is a lot of fire overhead in the cockloft. The first holes in the ceiling release a hot jet of stinging brown smoke that feels like somebody is injecting needles full of hot acid under the skin around the edges of my mask facepiece: real hot, stinging, cockloft smoke. Engine 235 has their line there, ready to go, but they don't want to open it up yet. Shooting water through the small holes we've made so far will only piss the fire off, extinguishing only a very small area but injecting lots of air into the cockloft. Like a venturi, as the stream passes through the small openings, it draws air in alongside the water. They wait as we pull larger and larger sections; soon we are standing under the largest toaster oven you've ever seen. Fire blasts along horizontally overhead, until 235 opens their nozzle. Almost as soon as they do, the powerful stream starts pushing the fire back. Conditions on the top floor are still really nasty, but the steam is better than the hot smoke—at least steam doesn't ignite! While the boys have been busy pulling ceilings, I've been in the other rooms, taking windows and searching beds. The primary search is negative in here also, and I report this to the command post.

With the fire knocked down here, I pull the troops together and tell them to get downstairs. At the street, I take a look up at exposure 2B. I head over to the command post for further orders, hoping that the staff chief thinks exposure 2B looks as good as I think it does. When I get back over to the troops, the boys look whipped. I report back our latest orders, "Take a blow." Murph looks like the weight of the whole world has been lifted from his shoulders. "Lieu," he says, "if you had come back here and told us we had to go pull the ceilings in the next building, I think my arms would have fallen off." I know exactly how he feels. This has been a hell of a night tour!

We go back over to our rig to change mask cylinders. The command post wants us to hang around for a little while, just to be sure conditions don't change, but from the looks of it, this one is done. We hang out at the corner for about 25 minutes without any further orders. Finally, we are told to take up. I punch the transmit button on the radio to tell the dispatcher we are 10-8 and available for Box 920 which he has just transmitted. Mercifully, he tells us to stay in service, the relocators have got it covered, and it sounds like it might be just food on the stove. Of course, it's only 6:30 in the morning. There are still two and a half hours left to the tour. Ya never know!

CHAPTER 13

Some Really Important Steps

Many areas of New York City have seen a wholesale exodus from neighborhoods as wave after wave of successive immigrants has flooded through the city. At times, it seems the whole neighborhood has moved en masse from one location to another. The occupants of the Grand Concourse in the west Bronx moved to the then new Co-op City in the northeast Bronx. Brownsville residents in Brooklyn flooded into North Woodmere on Long Island. Another of the neighborhoods that witnessed such an exodus is Crown Heights. At one time, the buildings of Crown Heights were among the most magnificent the city had to offer. Huge marble and granite lined lobbies, broad staircases of steel and marble, and huge apartments, often with three, four, or more bedrooms. By the late 1960s and early 1970s, the downward spiral of crime, neglect, lack of services, decreasing revenues, increasing costs, abandonment, and arson had left many of these once magnificent buildings little more than hulks of their former grandeur.

By 1989, many thousands of these buildings have been vacant like this for 20 to 30 years or more: the scenes of many, many fires. In the Crown Heights section of the borough, the north and south avenues are the primary commercial strips, generally two or three lanes wide, plus two parking lanes, all headed one way; the tendency to double and even triple-park in front of the more popular stores, social clubs, etc., can bring traffic to a crawl. The east and west side streets are usually narrower, also one-way streets, allowing only one lane of traffic through, if you're lucky. *Maybe* there's room to pass the gypsy cab that is defiantly stopped in the middle of the block, waiting for its fare to come out of one of the rows of apartment houses, despite your sirens, lights, and blasting air horn. Many times there's not. Rescue 2 crisscrosses this network of streets daily, often passing the same spot eight or 10 times a day en route to various alarms or drills, to pick up a meal, or for a wide variety of other functions. As a result, the officer and chauffeur, surrounded by windows, may see a building so often that they can visualize it in their minds if given the address or intersection. The remainder of the Rescue members, riding in the back with only two small windows to peer out of (like a bunch of watermelons as John Barbagallo calls them) do not get that view. Instead they listen for information from the department radio and shouted descriptions from the officer or chauffeur as the rig pulls into the block. ("Chauffeur's side, it's out three windows on the second floor!" or "Looks like they've got it knocked down.")

One of the street boxes that Rescue 2 responds to on the first alarm is Box 955, located at the intersection of Nostrand Avenue and Pacific Street. Pacific Street runs parallel to Bergen Street, two blocks to its north. Pacific Street and Nostrand Avenue is the location of a number of social clubs and liquor stores: two occupancies that tend to lead to a large number of false alarms from the ERS box on the corner. The companies in the area are more than familiar with the intersection and the corner stores. Largely ignored, however, is a vacant, six-story 150-foot × 150-foot former apartment house located 200 feet west of Nostrand Avenue. It's a building that I passed numerous times in my five previous months in Rescue 2, but paid little attention to—until the night it tried to kill me, and then later did kill a demolition worker.

Around 10:30 p.m. on December 3, 1989, Rescue 2 is assigned to a reported person trapped in a multi-car accident at Ocean Parkway and Avenue R in the Midwood section. This is a long run for Rescue 2, usually about 10 to 12 minutes, but this evening there is an added twist. There's a storm blanketing the area, with gusty winds and pelting rain. The trip to Ocean Parkway takes more than 15 minutes. As we cross Cortelyou Road on Ocean Parkway, the dispatcher announces the box is being transmitted for "Box 955 at Pacific Street at Nostrand Avenue." Normally, this is one of our assigned boxes, but we are already assigned to the auto accident, and must still continue in to it. In 1989, only about 40% of the ladder companies in the city carry a full complement of Hurst tools: extremely powerful hydraulic spreaders, cutters and rams commonly referred to by the media as "the jaws of life." The rescue company is assigned to all reported entrapments to provide an additional Hurst tool, as well as all of the other myriad tools in its arsenal. The five rescue trucks (in 1985, Rescue 5 is re-established in Staten Island) each bring with them the largest quantity and variety of rescue equipment imaginable. I call them "the world's largest moving toolboxes."

As we cross Avenue J on Ocean Parkway, Engine 235 transmits a 10-75 signal for Box 955. Now I'm a little upset because we could have been first due at that, but instead we are out on an auto accident. We haven't heard much from the scene of the accident, other than the 10-84, and I'm envisioning the carnage we could face. Ocean Parkway is a broad thoroughfare—seven lanes in the main roadway, with a service road on each side—but it is only a boulevard with grade level intersections, not a limited access highway like most of the other parkways in New York State. I've been to numerous accidents along here where a driver on one of the side streets attempts to "make the light" and fails to make it all the way across the nine traffic lanes, two parking lanes, and two intervening pedestrian malls (complete with park benches and checker or chess tables) before the sequentially controlled traffic lights release their torrent of cars into his or her path.

As we cross Quentin Road, the 57 Battalion Chief Harry Rogers, transmits an "All-hands," with a preliminary description of, *"Heavy fire on several floors of a six-story, H-type, 150-foot by 150-foot vacant."* "Son of a bitch," Larry Gray mutters, as he pulls the big rig into the northbound lanes, around the stopped traffic of the southbound lanes. We are a half-block away from the auto accident scene as I radio the 33 Battalion via handie-talkie to request instructions. "Rescue 2 to the 33, where do you want us, Chief?" The reply is casual, and infuriating—"Oh, Rescue, I don't need you. There was only one injury; I forgot you guys were still coming. You can take up." Forgot we were coming? What the hell is that? There has been a rescue company assigned to every person trapped in Brooklyn for at least 15 years that I know of: how do you forget that? Now I'm pissed. We've been out chasing a bullshit run and missing a job because of someone's forgetfulness? Damn!

Without so much as a word, the crew is off and holding traffic so Larry can make a quick U-turn in the middle of the Parkway. As soon as he has the rig turned around and pointed north, we hear the two urgent buzzes from the guys on the back step. "Go!" they're signaling. I call the Brooklyn Communications Office (CO) telling them we're up from Ocean Parkway and available for the all-hands. Before the dispatcher can reply, we're interrupted by the 15th Division, who transmits a second alarm for Box 955. The dispatcher acknowledges the second alarm, transmits the two shrill "beep... beeps," and announces, *"In Brooklyn, a second alarm has been transmitted for Box 955, Nostrand Avenue and Pacific Street. Fire in a six-story 150' by 150' vacant multiple dwelling."* Then she calls, *"Brooklyn to the Rescue, take in the second alarm."* Both my feet and Larry's are already pressing hard on the pedals. In my case it's the siren and air horn switches: in Larry's it's the throttle.

When we pull up to the corner of Rogers Avenue and Pacific about 12 minutes later, the third alarm is now in, at a box where we should have been first due. The streets are jammed with rigs, the fire being on one of the narrow side streets. I glance down the block and see the fire is closer to the far end of the block towards Nostrand, but I tell Larry to leave the rig where it is and let's walk down, since Nostrand is likely worse with traffic and rigs trying to get close.

We report to the deputy chief in front of the building. The block is a swirling sea of activity: a dazzling array of flashing lights, blanketing, choking smoke, throbbing diesel engines, snaking hoses, towering aerial and tower ladders. A heavy fire is present somewhere in this building, but from the street only clouds of heavy grayish-brown smoke, that nasty grey-yellow smoke peculiar to those water-rotted timbers of long-vacant buildings, are an indication of the fire. The chief tells us that the fire seems to be concentrated in the "B" wing, but it has reached the cockloft and is threatening to spread to the "A" wing. "Get your guys up to the roof and see what you can do to cut it off," he tells us. So off we go to Ladder 123's aerial for the 80-foot climb to roof level.

When we arrive on the roof, the fire is blowing through the rear and middle of the "B" wing roof. There are already about four or five saws operating, and there are three other officers up there supervising, as well as a covering chief whom I don't know. The truck companies have just completed a trench cut across the throat between the A and B wings to isolate the fire. I report to the roof chief, but we both know he doesn't need us here; he already has plenty of help. Instead, the top floor ceilings need opening up below the trench, so that's where he sends us. We make our way over to the roof bulkhead that encloses the top of the staircase that serves the "A" wing. The smoke is belching out of it like it's coming out of an old wood burning locomotive going up a 10% grade, but there doesn't seem to be so much heat present to indicate that it's about to light up. So down we go, carefully picking our way, mostly by feel, down the steel and marble staircase.

Normally, the stairs in vacant H-type buildings are suspect, due to the effects of weather as well as vandalism, but this staircase seems fine. The marble treads are all present, some cracked, but they are there. As we reach the top floor, we are met by Engine Co. 249, which is stretching a handline in from the front of the building, having stretched up an aerial ladder—a rare event in the day of the tower ladders. Usually, if a hoseline can't be brought up the staircase, the tower ladder stream is quickly put into operation.

This building is a true, classic H design, with four apartments in each of the vertical lines of the "H" and a transverse hallway running through the horizontal line of the "H" called the throat. There is also one apartment on each side of the hallway in the throat, for a total of 10 apartments per floor. The two staircases are located right where the throat joins each vertical line. The staircase and the public hall are required by law to be enclosed in two-hour fire-rated material to provide an area of refuge for fleeing occupants in the event of an apartment fire, and to keep fire in one apartment from spreading across the hall to other apartments. In a vacant building, however, the apartment doors are often missing, or at the very least, have holes piercing them where the doorknobs, locks, and "peep-holes" used to be until they were scavenged for scrap metal.

Engine 249 has their line at the staircase now, but it is still not charged. Timmy Stackpole has gone down the hall in the throat, and is calling for a line over there. Fire is coming across from the "B" wing. I look to the rear, though, and see that fire is already here in this wing in the rear. There is at least one room going real good back there, and there is a strong breeze blowing from the rear toward the front, right towards us! Engine 249's officer is trying to get through on the handie-talkie to his chauffeur to get their line charged, but there is just too much radio traffic. The fire is moving very quickly down the hall at us now. The temperature in the hallway is rising quickly, it's about 35 degrees outside, but here inside it's now 110, 115, 120 degrees. I tell Larry to shut the door to the involved apartment, to isolate it from the hall, but even as I say it I can see there is no door on the apartment.

Larry instinctively moves toward me to get to the door to the adjoining apartment. With his Halligan tool, he can rip it off its hinges and then we can place it in the door opening of the fire apartment. But the inferno is moving toward us like a juggernaut now, moving inexorably down the hall, covering its 30 feet length in about 30 seconds. I crouch down as I survey our predicament. I see the members of Engine 249 are still standing up, as they are now trying to pull their still uncharged line back toward the front apartments. Then suddenly there is fire out in the public hall, rolling along the ceiling, actually singeing the tops of some of the still oblivious engine men's helmets. I shout a warning to "Get down!" at the same time as they become aware of the seriousness of their plight and dive for the floor. I back up a step, towards the staircase to make room for everybody, just as a blast of flame fills the doorway with fire. Another step backward, and then all of a sudden, I'm falling backwards—falling, falling. Then just as suddenly, my left leg yanks me agonizingly to a halt. I'm lying upside down on my back, suspended on the skeleton of the rusted steel staircase, held only by my calf, which is wedged between two jagged steel stair risers. On top of me, adding to my discomfort, lie Larry Gray and 249's officer. Over our heads, flame boils angrily along the ceiling, driven by the wind blowing up the stairs, seeking the escape route offered by the open bulkhead door. The heat radiating down from the swirling ball of flame is impressive, but it's what is below us that has captured our attention. I am looking straight down at the cellar floor, seven stories of nearly uninterrupted flight below, where the stairs used to be. I see the hand light that I had dropped go out as it smashes into the cellar floor, followed by a distant clang as my hand tool also smashes into the concrete.

Engine 249 has gotten water in its line and is driving the fire back into the rear apartment. Now, hands are pulling us up from the rickety remains of the stairs. I borrow a light and look back down the stairwell, and what has just happened is now obvious: while the staircase to the roof is intact, scavengers stripping all the metal out of the building to sell for scrap have destroyed the steps leading down to the ground. They have smashed out the marble half-landings in the staircase, so they can drop the brass and lead pipes, and the cast iron bathtubs. The tubs going down have cracked the marble treads on the steps, so as a result, all that remains are the vertical steel risers leading down from the top floor landing. Nearly every other piece of the staircase is gone. Those that remain are some really important steps however, for if they were not still intact, at least three firefighters would have plunged over 70 feet to our deaths.

I get on the handie-talkie and call the Deputy, "Rescue 2 to the One-Five, Kay." After two or three tries, I finally get through. "Go ahead, Rescue, Division One Five." I tell him right out, "Listen, Chief, this building's a piece of shit. There are holes in the floor all over the place. The stairs are missing, and we just had three guys, including me, almost get killed in here. I strongly recommend we pull everybody out, and go to the tower ladders." The reply that comes back almost knocks me back into the staircase. "Listen, Rescue, just be careful up there. With the main body of fire in the rear, we can't get a good shot at it with the towers. We'll have to keep at it with the handlines. Just be careful!" I almost scream into the mic, "Careful? Chief, nobody's more careful than me in a vacant building, but this one is bad! We have very heavy smoke and can't see half of the dangers!" His reply is just as infuriating, "Well, move slowly and probe in front of you." No shit, Sherlock.

Other radio traffic floods back in now, washing over our conversation like waves washing away footprints on a beach. Now it's Timmy Stackpole calling me. "Hey, Lieu, I got a lot of fire over here and I need a line!" Mindful of my most recent encounter with the "Devil," I tell him to keep the apartment door closed until I can get 249 and their line over to him. As I crawl down the hallway in the throat toward him, I'm vaguely aware of some low rumblings: partial collapses of floors or roof sections, most likely. I find Timmy and Pete Bondy crouched by a door on the north rear side of the hall. I tell Timmy to open it carefully so I can get a look, but be ready to pull it closed if fire blows out. As the door swings open, there is plenty of fire, but it doesn't come out at us. It is going straight up through the roof. I look down and see that the fire is coming up at us from about three floors below us. The whole top three or four floors and roof in this area have collapsed into the lower floors! If anybody had taken a step through that door, which was quite possible in the heavy smoke we are operating in, the first step would have been an express trip to hell. I pull the door closed and tell everybody around me that this is bullshit. "This whole fucking building is falling down around us. Let's get back to the front apartments where 249 came in from the ladder."

As we make our way back to this area, we meet Chief Rogers of the 57th Battalion: one of the senior and most respected chiefs in Brooklyn. I tell him of the situation and lead him back to show him what I mean. He immediately gets on the radio and tells the command post that it's past time to get everybody out and have a roll call. This pile of shit is beyond saving. Chief Rogers' message brings a change of heart in the deputy chief. When a guy like Harry Rogers tells you a building or situation is bad, it's *really* bad. Everybody is ordered out of the building and off the roof. Finally! (See fig. 13–1.)

I gather up the troops and we make our way down Ladder 132's aerial. I report to the command post that all my members are accounted for, and that we are located out behind the command post. The staff chief is on the scene, and he is looking to start getting rid of some units, now that this will be strictly an outside attack. We are told to "take up." We don't have to be told twice, since it seems the temperature has dropped well below freezing. Warm, dry clothes will feel good after getting soaked by 249's line while lying in the stair. As we pull up in front of the old squat, red firehouse, Al Washington comes walking out the front door. Al lives in Brooklyn, and he has come by quarters to pick up his dress uniform from his locker so he can go to headquarters in the morning for a meeting. It's a good thing he did.

Figure 13–1. The hazards of vacant buildings are not worth the risk to firefighters' lives. Three firefighters nearly fell seven stories down the remains of this stairwell when a sudden change in fire conditions drove a scramble for survival.

As Al pulled up in front of quarters about 20 minutes before, a thin guy was climbing through the top of the window on the side of the apparatus door. The window has to be 12 feet off the ground and is covered by a heavy wire mesh. This guy had pried the mesh off with a piece of pipe, and was about to make his way inside when Al pulled up and scared him away. It seems that the presence of a four-alarm fire only five blocks away has given some of the more enterprising criminal elements some ideas. While most of the other firehouses are covered by relocating companies, Rescue 2's quarters are almost never covered since there are only four other rescue companies in the city. An empty firehouse seemed like an easy target. There are TVs and VCRs to be fenced, and there might be cash lying around in some of the guys' lockers. If only this guy knew what he missed in this particular firehouse. The Rescue firehouse is also loaded with lots of extra goodies: power tools, SCUBA equipment, welding equipment: all of which are easily carried and easily sold. Well, we scared him off without a loss and got lucky twice tonight. Or so we thought.

About an hour later, we responded to a "Class 3" alarm from St. Mary's Hospital. Since all the normal first alarm units are still over at the fourth alarm, or on "R&R" (rest and recuperation), we are responding with all relocated companies. Since we are the closest normally assigned unit, we get in before anybody else. The alarm is from a pull station up on the fourth floor, so we all go up to investigate. Finding no cause for alarm, I radio down the 10-92 signal to the Battalion. As we pull back up in front of quarters, it's now almost 4:00 a.m. and we are all looking for a few hours in the rack. Timmy Stackpole unlocks the door and goes inside to open the overhead apparatus door, as Larry backs the rig up to the door, waiting for it to go up. When it doesn't go up after about a minute, Pete Bondy goes in and hits the button, but then he doesn't come out either. Now we hear Timmy hollering from the backyard. Everyone runs in through the apparatus floor and the kitchen, and out the back door, leaving Larry sitting with the rig stopped out across Bergen Street.

We find Timmy standing over in the corner of the parking lot with a shovel, looking up at the top of the fence nearly 18 feet above. It seems our "friend" has returned. This time, he pried the heavy wire screen off one of the second floor windows, after climbing a fence topped with razor wire, to reach the roof of the one-story kitchen behind the apparatus floor. He must have gone in as soon as we left, because he not only had time to complete the break-in through the heavy wire mesh and enter (a feat that even one of our OV's would do well to accomplish in this short a time) but also to scout out the firehouse, remove the kitchen TV from its shelf, and rifle through our workbench. When Timmy walked in to open the door, this skell was walking the TV and a toolbox down the apparatus floor on one of our own hand trucks! When he saw Timmy come through the door, the thief bolted out through the backdoor with Timmy in hot pursuit. On the way, Timmy stopped to grab anything he could find that might be useful as a weapon. The first thing he saw was the five-foot-long shovel. When he got out the back door, the thief was already well up the fence, but the razor wire was between him and freedom so Timmy thought he had him cornered. He took one swing with the shovel while jumping up, which caught the perpetrator on the right ankle. That didn't slow him down in the least. Instead, it was like igniting a rocket in his ass. He slithered right through the razor ribbon, and jumped down from the top of the fence to the ball field below. By the time any of us ran back through the firehouse and out to the street, the mope was gone.

Our newfound friend wasn't through with us yet, though. Over the next few days he made several more attempted break-ins, most of which were thwarted by our increased security. After the first few break-ins, we had reinforced every window and door, welding drop-in bars in place like the gates of Fort Apache on the back door, and stringing so much additional razor wire and regular barbed wire that you'd have thought we were expecting infiltration by a regiment of North Vietnamese sappers (elite units). Eventually, things got so bad that the department began assigning light-duty firefighters to our quarters to remain behind as security while we went out on a run. After a few weeks, we got a break. One of the residents of the nearby Albany Houses Projects came by quarters to tell us he had bought an oxy-acetylene cutting torch off a guy in his building. Upon closer examination, he noticed the "Rescue 2" engraved in it, however, and realized it was probably stolen. He sold it back to us for the $20.00 he had paid. Now we had an ID and an address. A call to the 77th Precinct and the fire marshals resulted in swift apprehension. End of story, right? Not a chance. The guy does 30 days in Rikers Island and is back out on the street.

Everything starts all over within a few days of his release. Now we know where to find our prime suspect, however, right in a taxpayer subsidized, first-floor apartment on Bergen Street. It's a long fly ball away from our front door, and we go there all the time for fires, false alarms, water leaks, food on the stoves, and stuck elevators. The nine buildings, each 14 stories, hold many "titles." For at least four years in a row, at least one department medal was awarded to the members of the first alarm units for actions taken to rescue citizens trapped by fires in this complex in each of those years. In short, the sight of fully equipped firefighters traipsing into these buildings was not enough to even cause a second glance. Shortly after the latest break-in, Rescue 2 receives a "verbal" alarm for a gas leak in our friend's apartment. His apartment door is at the rear of the elevator bank, in the lobby. I knocked heavily on the door with my mini Halligan while standing off to the side and bellowing out, "Fire Department, open up!" After a minute or two with no reply, we insert the Rabbit Tool and with six quick strokes on the hydraulic pump, find our entrance cleared of the offending door. We give the place a quick toss, retrieving our stolen items, but there is no sign of our friend. I tell one of the guys to turn off the gas to the stove, and call the dispatcher to request police for security of the apartment, since when you force a steel door with a Rabbit Tool, it can almost never be relocked and will have to be replaced.

When the cops show up, I tell him we had a gas leak that was secured, but in the process, we happened to find what appeared to be stolen property. The place is littered with dozens of TVs, stereos, VCRs, car radios, leather coats, you name it. If it can be easily stolen and fenced, it was there. The cops thanked us for the info, but told us to be careful, "The guy that lives here is a real skell: an AIDS-infected heroin junkie." Well, that was not an altogether new revelation to us. We already knew he was a skell, and one look at his gaunt skin stretched over a bony skeleton told us that he had "the thins": ghetto slang for the withering effect of AIDS. We should know—we treated enough of them who couldn't quite make the walk to St. Mary's for their methadone fix and decided the firefighters next door to their favorite shooting up spot would have to do. The situation went on like that for another year. The mope would be pinched for one offense or another, serve his 15 or 30 days, and be back out on the streets, doing more burglaries and car break-ins as soon as he was released. I don't know if he ever went through detox, but if he did, I know it didn't work. Finally, either a hot shot of heroin or the thins got to him, and he was no more. Nobody in the neighborhood minded in the least. It was just part of the "self-cleaning oven effect," where the baked-on crud finally gets burned out.

About the time that we realize that we haven't been broken into for a few days, Rescue 2 gets called back to the scene where it seemingly all started: Box 955. This time there is no fire, but a major collapse has occurred while demolition workers were in the process of tearing down the remaining brick walls. Several workers are trapped in the debris, and one is killed.

It seems the ghost of the building would not be denied its blood sacrifice. It may have lost the three firefighters who got away, but it got one life in the end. The worker was reportedly killed in the remains of the staircase—the same staircase that nearly got us, except for some really important steps that remained in place.

CHAPTER 14

The Winter Offensive

"The Winter Offensive" is a natural fire phenomenon in most urban areas. As soon as the first prolonged cold snap occurs, the fire duty starts picking up. It is a time to make sure you are working as often as possible, if you want to go to fires. The reason is simple; numerous alternative heating sources are used to try to stay ahead of the icy grip of winter. Overworked heating plants struggle to keep up with the demand and soon malfunction, portable electric heaters overload ancient wiring that was never designed to carry that large a demand, illegal kerosene space heaters are placed too close to bedding, squatters move out of the weather and force their way into vacant buildings with neither heat nor electricity, and start "keep-warm fires" within the building they are occupying. The list goes on and on. December of 1989 sees all of these and more. The month is one of the coldest in many years. The temperature dives into the single digits overnight on the eighth, and never gets above 20 degrees for three weeks. By the tenth, the frozen sprinkler pipes placed many fire protection systems out of service, and the fires began, but with an added wrinkle.

That fall, a revolution in the manufacturing sector had seen a brand-new process introduced in the knitting industry: one that would make knitted clothing in New York profitable again. It was a very expensive technology that would take several years to pay for itself, but from then on it would be profitable. Many of the knitting mills in the city invested heavily in this new equipment, and everything from sweaters to hats was soon being produced. Three weeks later it was all obsolete. Another advance in computer-aided design made the entire process second rate, to the point where it looked shabby compared to the newest technology. The "going out of business fires" began almost immediately. By the end of the month, Rescue Co. 2 would respond to three five-alarm fires, numerous 2-, 3-, and 4-alarmers, and over 50 all-hands fires. Thirty-one multiples in a month, with most of it in only 21 days, is a lot of work. The knitting industry should have been so busy!

Five-alarm fires are so rare that many fire companies have established "5th alarm pools," where all the members contribute a dollar every two weeks at payday. The unit may go for a year or more without responding to a 5th alarm. The pool is won by the crew that is working on the day that they respond to their next one. Each of the winning firefighters can collect $150 or more. In Rescue 2 that month, the winning members would have barely gotten even money back on their investment: a return about as bad as what the knitting industry was going through.

The first real standout job that my groups go to was in a large old mill type building off Knickerbocker Avenue in East Williamsburg. In the 1960s, this area had seen many large fires in the lumber yards that dotted the area, but by the late 80s they were nearly all gone: burned out or moved to the suburbs. This night would give the fire companies of Brooklyn a taste of the bad old days. The low temperature chases most of the derelicts off the streets, which causes a drop in the number of minor fires in the city during times like this. It's too cold for the "mongo thieves" or "urban miners" to spend much time stealing copper wires from the abandoned buildings, the punks aren't hanging out on street corners pulling the handle on the fire alarm boxes, and even the dope dealers prefer to work indoors, avoiding turf conflicts. The Brooklyn fire radio is so quiet that several times guys have gone over to it and adjusted the squelch, just to see if it is working. That makes the dispatcher's voice sound even louder than normal when he announces, "Box 705 is in, reporting a fire at Porter Avenue and Thames Street. Repeating, Box 705, Porter Avenue and Thames Street for a structural fire. Brooklyn to the 28, we're loading you up, receiving reports of a fire in a factory." (The dispatcher is adding extra companies to the initial response, based on numerous telephone calls indicating a working fire. Typically that means sending three engine companies, two ladder companies, a squad company, and a Rescue company.) Everybody in the kitchen pushes away from the dinner table, even before the teleprinter starts to print. We know we'll be going to work on this one.

Dennis Mojica is the chauffeur tonight—a quiet, devilishly handsome guy who is soon to be promoted to lieutenant. Dennis is also an alumnus of Sheffield Avenue, where I started, and several of our mutual friends have spoken to Dennis about the "new lieutenant," letting him know that I'm OK. This is one of our first tours working together. By the end of the month we'll be lifelong friends, but as Dennis noses the LaFrance Rescue truck up Lewis Avenue, he's still trying to feel me out when it sounds like we might be going to something a lot more dangerous than the average tenement fire. "So, Lieu, you want me to stay with the line when we get in, or do you want something special?" he asks. Dennis has almost twice my time on the job, and I know he's quite capable of handling any situation he finds himself in. "Let's see what the chief wants when we get in, but bring a search rope," I reply. Company policy gives the chauffeur a great deal of latitude in his assignment, recognizing that this is usually the senior firefighter working. With that, the aide to the 28th Battalion comes on the radio as if on cue. "Battalion 28 to Brooklyn, transmit a second alarm for this box. We have fire in a four-story brick factory, 200 feet by 150 feet, unknown what floor the fire is on. Have the Rescue report in with both saws. We're having trouble gaining entry. The building is very heavily sealed." "OK, Dennis, you heard him, give us a hand with forcible entry," I say.

As we come across Bushwick Avenue, there is a huge cloud of dark smoke, dotted with winking red and orange embers, visible against the night sky. I yell through the window to the guys in the rear cab, "Take both search ropes," as Dennis parks us out on Flushing Avenue, around the corner from the actual fire building. We do not want to block out the many tower ladders that we know will likely be required. Ladder 124 is already on the block, and is setting up as Dennis locks up the rig, but we'll need more towers for this. We report in to Dennis Cross who is the ABC (acting battalion chief) in Battalion 28 this frigid night. I'm sure he'd rather be in his usual assignment as the captain of Ladder 102, where he can get inside the building and be warm, nice and closer to the fire than he will be tonight, out in the street as the battalion chief. "Rescue, give 108 a hand forcing those fire doors inside that loading dock over there, and make sure to get somebody to the roof, OK?" "10-4, Chief," is all the reply needed. The handie-talkie is chattering with reports from all sides of the building, all reporting heavy smoke that looks about to light up. This is not a good sign.

We make our way through a hole that has already been cut in a rollup steel door, expanding the opening as we do to make escape easier. We also open other doors along the left side of the building, looking for alternate approaches. Dennis says, "Hey, Lieu, I'm going to deploy my search rope." It's not a question seeking permission. It's more a cautionary note that this could be getting really bad. I nod and wink my understanding. The loading dock bays are relatively clear of smoke, clear enough to see across, but we all know how quickly that can change. The search ropes are our insurance policy, allowing us to quickly retrace our steps to the outside when we can't see our hands six inches from our faces. Engine 237 has a 2½-inch line in ahead of us, and we follow it through a doorway to a narrow, old brick staircase. The smoke here is so thick it's hard to tell which direction it's coming from, banking down from up above or rising from down below. The last thing any of us want to do is to get caught above a fire without warning. I take Dennis and his search rope with me down into the cellar, and tell Larry Weston, another senior firefighter, to take the rest of the company to check the ground floor on either side of our entry.

As soon as I hit the bottom of the staircase, I know that I made the wrong choice. The heat level has diminished, a sign we're below the main body of fire, but a strange sound draws me further into the basement. There is a tremendous, ghostly "whooshing" noise down here. We need to check it out. Down a narrow corridor, after opening several doors, we find a chilling sight: an opening into the base of a square shaft which rises up through the middle of the building. The shaft was built to let sunlight and fresh air into the inner reaches of the sweatshop. It has two large windows on each of its four sides. Every one of them, from the first floor through the fourth, is full of fire. I am about to get on the handie-talkie to give my report just as Ladder 108's roof man does the same. "108 Roof to the 28, there's a shaft about 100 feet back from the Grattan Street side that just blew fire out every window. We've got heavy fire on all floors." ABC Cross does not need either of our reports, though—"28 Battalion to all units. Back out. Back out. I want everyone out now!" When Dennis and I reach the staircase, we are met by a cascade of hot water. Engines 237 and 218 are both operating their 2½-inch lines into areas on both sides of the staircase. The fire is roaring around their hose streams, and velvety black smoke pulses out over their heads. The companies begin withdrawing the hoselines, but they must retreat with the hoselines flowing for protection. The rescue and ladder company members back out of their way, but the retreat takes time to accomplish in order to be certain that no one is left behind in the large floor areas.

When we reach the street, the companies are greeted by an impressive sight: the entire building is puffing fire and black smoke. Flames are blowing 20 feet horizontally out of about 50 windows on the south side of the building, threatening to jump the gap to ignite an even larger building that faces Flushing Avenue. Chief Cross sends us around the corner to see if we can delay that from happening. He says the Deputy has sent the fourth alarm, but the companies aren't in yet. "See what you can do," he says.

As we go back around the way we came, I tell Dennis to move our rig down the block even farther away. There are already multiple 3½" hose lines being stretched across our path, and more are likely, as well as the 4½" hose from the satellite units of the Maxi Water System, which will be needed to supply the many large caliber streams it will take to knock this fire down. I don't want our rig to get blocked in here, unable to get out. It will likely be a very long night for the guys in the tower ladder baskets spraying tons of icy water at the flames (fig. 14–1).

We make our way out of the fire stair on the second floor of our new building and I am heartened by what I see. Engine 217 has a 2½" handline stretched out onto the huge floor. The windows facing the fire building are wire-glass windows, which are meant to resist a fire exposure like this without failing like ordinary glass, and the floor is not very heavily loaded. Mostly it's row after row of tables full of sewing machines, with cardboard boxes full of fabric beneath each work station. There is only a little fire on

Figure 14-1. Tower ladders open up on the Flushing Avenue side of a knitting mill as members descend an aerial ladder from the roof.

this floor, where radiant heat has ignited some of the cardboard boxes over against the windows. We ought to be able to handle this.

The third floor is a different story altogether. It is a storage area, piled to the ceiling with boxes of clothing. The fire has already heated a good number of the cardboard boxes nearest the windows to the point of ignition. The smoke is just starting to roll along the ceiling, followed by darting tongues of flame. I radio Division 11 to "Get a couple of lines up to the third floor, quick!" The acknowledgment is less than promising, "When I get them, you'll get them." We stretch the "house line" off the standpipe system. It is so-called "first aid hose," not the quality of fire department hose, but better than nothing, but we get no water when we open the valve. "What the hell is that about?" Now there's a good bit of fire rolling along the ceiling. I watch and wait for the sprinkler heads to operate. I know that with the stock piled this high, and with fire igniting in several places at once from the exposure fire, the sprinklers will have a hard time getting complete extinguishment, but they should control it long enough for a hoseline to get here and finish off the fire. If only the sprinklers would go off!

I radio Dennis, who has yet to rejoin us, to see where he is. "Boss, I'm just coming in the Flushing Avenue door now," he replies. "Dennis, see if you can find the sprinkler and standpipe control valves in the cellar, and see if they're open." Five minutes later, it's too late. The entire third, fourth, and fifth floors in this second building have flashed over. We are driven out of the building by the spreading flames. With no water in the standpipe, 217 will have to go back to the street and hand-stretch hose from their rig up to the fire floor, but with three massive floor areas fully involved, there is no point. This will require several tower ladders to apply enough water to knock down. The 5th-alarm units will still have a lot of work to do when they arrive. I call Dennis and tell him to meet us in the street. The rest of the fire for us consists of checking exposures around the perimeter of the block, as this is now a tower ladder festival. We report periodically to the command post, hoping to be told to "take up." The command post however, apparently wants to keep us on the scene, just in case something bad happens, like a wall collapsing and trapping firefighters, but that should not happen in this situation. We have lost. Just keep everybody away from the buildings; they will burn until they are damned good and ready to go out. This is what we call a "no impact" fire. Nothing we have done here has made a bit of difference. It is disheartening. There is nothing here for a rescue company to do; it is up to the engine company chauffeurs to keep pumping the tens of thousands of gallons of water per minute to the tower ladder nozzles and deck guns surrounding the block to drown the beast. It's daylight before that happens. As we sit in the cab of the apparatus trying to ward off the cold, Dennis mentions that he found what he thought was the sprinkler valve—it was closed.

At 6:00 that evening, I am back in to work again, with Dennis Mojica and Larry Weston, but with three different members on their first tour. At about 10:00 p.m. we are assigned to Third Avenue and Ninth Street for another factory fire. Coming down the hill on Ninth Street we see a large cloud of smoke. It's "déjà vu all over again." We report in to Battalion Chief George Gander of the 32nd Battalion, a friend of mine with whom I teach firefighting on our days off at the Nassau County Fire Service Academy. Chief Gander directs me to go around the block to Eighth Street to get a size-up of the building and to see if we can get a shot at the fire from there. Ladder 122's roofman is reporting the building runs through the block to Eighth Street, but he can't tell from the roof level if we can get in over there or not. Dennis follows Squad 1's 2½-inch line in through another overhead door, while I trot around the corner. Again, he ties off the search rope before entering.

While the building does go through the block, the back is very heavily sealed, with steel plating over the doors and heavy wire mesh over the few windows. It is a single-story building, about 100 feet wide and 200 feet deep. I radio that information back to Chief Gander, and make my way back around the corner. I pass a water motor alarm gong and a fire department Siamese connection for the sprinkler system. The gong is silent. If the sprinklers were working, it should be ringing. I mention this to Chief Gander, who orders Engine 220 to stretch a 3½-inch line to the Siamese connection. I go back over to the left front corner of this place, to see if the sprinkler valve is there and if it's open. Not only is the valve closed, but a large section of a 6" tee is blown out, apparently from ice in the pipe, and now the water being pumped in through the Siamese connection is just blowing out onto the floor. The fire ends up a fourth alarm, and the tower ladder boys again learn to regret not having been assigned to an aerial ladder. It sure gets cold standing on that metal basket, spraying tons of water for hours at a time! On the way back to quarters, Dennis is convinced this is a plot. Two sewing factory conflagrations in two nights, with the sprinklers shut off at both of them. Somebody ought to tip off the fire marshals to this one. I tell him I'll do exactly that first thing in the morning. First let's get back and get some warm, dry clothes.

I have just stripped off my still frozen pants, when the teleprinter and the "tooo-dooo" of the voice alarm announce our next encounter with the millinery industry. "Box 2566, Thirteenth Avenue between Fiftieth and Fifty-First Street. Smoke from a store." As we round the traffic circle on Parkside Avenue, Battalion 40 orders the second alarm be transmitted. He also reports that Thirteenth Avenue is totally blocked off for repaving; no rigs can get in front of the fire building. Down Fort Hamilton Parkway we go, then left under the el on New Utrecht Avenue, and a left onto Fifty-First Street. It probably takes us 15 minutes to get there, time that the first alarm companies are getting beaten up.

The fire building is a two-story affair. It looks like it used to be a taxpayer on the ground floor with apartments above, but the picture windows in the store fronts are all bricked up with only one small window and one door in the fire store. The chief says he needs help moving the hoseline in to the seat of the fire, and also the searches of the apartments above need help.

Larry Weston has the "can" tonight. He and Dennis will help on the hoseline. I take Mikey Esposito and Timmy Higgins to the floor above. Ladder 114 has forced the door to the second floor, and Engine 247 has a handline at the top of the stair, but they are pinned down and unable to advance. It is hot as hell up here. I mean really bad. The hoseline is operating, throwing over 250 gallons of water a minute at the flames that we can see, but it makes no difference: the fire just roars along around the hosestream. The Rescue guys all have our hoods on (which the other companies don't have) and our earflaps are down on top of the hoods, and it's still hot. This is really not a good place to be. Timmy, Espo, and I try to get out into the apartments to search, out of the public hall, but it's too hot. This place is going to light up any second. Fire burns right up through the floor boards under us. We can hear deep crackling coming from somewhere to the rear, as well as over our heads, but we can't see a damned thing. Even with the hood on, my face feels like it's being stuck with hundreds of tiny needles, each pumping hot acid under my skin. This is bad. We can't advance until the companies downstairs put the fire out underneath us. I order everyone back behind the safety of the nozzle at the apartment doorway. I tell Espo and Timmy to "stand fast, I'm going down to see what the hell is taking them so long to put that store out."

In the meantime, Dennis and Larry have hooked up with the crew of the first due engine. They have been pinned down as well, by an unusual building alteration that causes fire to push at them with a vengeance every time they try to advance. What appears to be a typical 20-foot wide storefront from the Thirteenth Avenue side, is really another large knitting mill, once you get back about 50 feet. There is a loading dock-size doorway on the left wall, which opens into the large area that is full of fire. There is another opening into this same area about 20 feet past the first one. As the hoseline tries to advance, it meets fire from both doors; if it pushes water in one, more fire comes out of the other one, endangering their position: a Catch 22. A second line is positioned so that it can deal with the other door, but by now the air cylinders in the first alarm engine's masks are running low. To make any headway, the units will have to hump that 2½" line around the wall, making a U-turn with the line. That's a lot of work. The company is content to sit and throw water at the doorway until fresh troops, in the form of third alarm engine companies, arrive to relieve them. That is until Dennis and Larry arrive. They see the predicament, but they know they have to make a move, because there are units up above counting on them. The engine officer tells them he wants to wait, but Dennis is adamant. "We gotta go now." The officer, an old-time captain, asks if Dennis is trying to get them all killed. "Hell, no. I don't want to die, but we all got to go sometime. We got to move now, or else people are gonna die upstairs." The captain considers this for a moment, and then agrees, "Hell, you're right. We all gotta die sometime. Let's go!" The two remaining members of his unit hear that, and decide that their mask cylinders are empty, and skedaddle for the front door. They aren't going to die with these three lunatics. Larry, Dennis, and the captain start humping the line forward and around the bend.

I start to make my way in on the first floor, following the hoselines, groping blindly, keeping one knee on each side of the line as I crawl so I don't lose it. In several places the lines are intertwined, which causes me to slow down to keep from losing my guide back to safety. Suddenly, when I'm about 20 feet in, a major collapse occurs right in front of me. Fire lights up the entire area and all I see is a tangle of timbers and debris where the line goes. I give a "Mayday! Mayday! Mayday! Collapse on the first floor," over the handie-talkie. Division 12 calls, "Who gave the mayday?" "Rescue 2 to Division. There's a major collapse in the fire store. I think we have members trapped. I need two lines in here, now!" Then, "Rescue to chauffeur. Dennis, where are you?" No reply. "Rescue to chauffeur. Dennis, answer me, are you all right?" No reply. "Rescue to Can. Come in, Larry!" "Rescue 2 to any unit on the first floor of the fire store." Nothing. No reply. Silence is the scariest sound on the fireground.

I begin to crawl back to the front door to get help. Mike Esposito bumps headfirst into me, going the other direction. "Who's this?" he asks in the darkness. "Rescue 2," I tell him, "Who're you?" "Lieu, it's Mikey. What's going on in front of us?" "Dennis, Larry, and a couple of companies should be up there. Go out to the deputy and tell him we need two lines and two trucks in here right now, and tell him to get at least two more rescue companies coming here right away!" "No, Lieu, you go get the lines. The chiefs won't listen to me, but if an officer comes out, they'll listen to you." Then he says, "I'm going to get the guys." Then, just like that, he crawled right past me, right into a wall of fire—over and through a pile of burning debris. It was an unbelievable sight, worthy of a Hollywood special effects crew. I start to follow him, but I hear him call over the radio, "Go get the lines, Lieu! I'll be all right." Bravest thing I've ever seen.

Out in front, the scene is chaotic. The deputy and two battalion chiefs are standing on the sidewalk looking at the building as I emerge. From the street it doesn't look so bad, the wind is pushing the smoke and fire toward the back of the building. I rip off my mask facepiece and ask where the lines are.

"Lines? What lines?"

"Chief, for Christ's sake! I told you I need two lines right now, we got guys trapped!"

"What guys trapped? There's no one trapped!"

"Rescue to chauffeur. Where are you?"

Silence.

"Rescue to Can. Larry, answer me!"

Silence.

"Rescue to floor above Irons. Mikey, where are you?"

Silence.

"Chief, I'm telling you, there was a major collapse right in front of me, about two minutes ago and now I have three missing members of my unit, and God knows who from other units. I need a full assignment, right here, right now. And we need to get at least two more rescue companies rolling." The three chiefs look at each other, and then back at me with a look like I've got three heads.

"There's been no collapse, son, you're mistaken. Look around you, everything's all right."

"All right? Hell, I'm telling you we got guys trapped and they need help."

These guys are in total denial. They can't accept this news. Their minds are refusing to let it register. They stare at me for a full 30 seconds.

"Rescue Irons to Rescue 2, K?"

"Rescue to Irons!" I scream into the handie-talkie. "Where are you, Mike?"

"I'm out in the rear yard, Lieu. I got Dennis and Larry, and the engines are back here too, but we are in a fenced-in area. We need somebody with a saw to come around on Fiftieth Street and let us out."

My heart finally came out of my throat. Thank you, Lord! I take the rest of the crew around back, where Timmy and a metal cutting saw make quick work of the wrought iron fence, allowing the trapped firefighters to breathe easier for the first time in 10 minutes. Dennis says that if the nozzle team had stayed where they were when he and Larry got there, the entire collapse would have landed on their heads. Another one where "the Big Boss" let us off easy.

We spend several more hours on the scene, using the apparatus winch and two Grip-Hoists to pull concrete highway dividers out of the way, so that tower ladders can finally get into the front of the fire building, and we watch them get their workout. By the time we get back to quarters it's almost time to go home. The two fire reports will wait until I come back for my day tours. The call to the fire marshals won't wait. "Yeah, we noticed the pattern too," is the reply. "We'll look into it." Before the New Year, though, all of the entrepreneurs have created their tax write-offs. It sure kept us busy.

About two years later, I get called by the Brooklyn District Attorney's Office to come down and give a deposition about the fire on Thames Street. The marshals have made an arrest on that fire. A number of firefighters and officers are called to testify. The case goes to trial: the only one that I know of out of dozens of these obvious "torches." It seems like a pretty good case to me; the entire place was locked up tight. The fire was set with accelerants in an area that only a key holder could have entered. All the fire protection systems in the complex had been shut down just before the fire, something that would again require a key and knowledge of the plant. But because no one saw the guy actually light the match, he walked. I wish they'd let Dennis, Timmy, Espo, and Larry have a go at him. I would have just stood back and watched!

CHAPTER 15

My Turn to Be Schooled

December of 1989 had been an unusually cold month in the New York area and consequently an exceptionally busy month for Rescue Co. 2. In those 31 days and nights we did 31 multiple alarm fires, including many fourth and fifth alarms, plus countless all-hands. The extreme cold temperatures of December had abated somewhat by the middle of January, and so went the extreme fire duty. By the 15th of January we had done just eight multiples for the New Year.

The night tour of January 15, 1990 had been fairly slow—one all-hands at Dumont Avenue and Howard Avenue, where we spent a lot of time removing HUD sealed windows (windows covered over with plywood and 2×4s, to keep vagrants out) to vent the place, and a few minor alarms.

By 1:30 a.m., I had yet to put my head on the pillow. Around 2:00 a.m., I finally finished the last of the paperwork and decided to hit my crib, hoping to get a solid six hours in before morning. At 5:36 a.m., my sweet dreams of warm beaches and beautiful women were abruptly shattered by the *tooo-dooo* tone of the voice alarm. As we turned out, Lee Ielpi was calling out the information on the teleprinter ticket—"Fill out the alarm... Box 551 - 120 Smith Street near Bergen Street." (Fill out the alarm means that extra companies are being added to the initial response, usually due to numerous phone calls reporting the same fire.)

Rescue 2's quarters are located on Bergen Street, between Troy Avenue and Schenectady Avenue about three miles east of Smith Street. Bergen Street is a one-way, westbound street which is the direction we would have to travel to get to the box, so I was somewhat surprised when Lee made a right turn at Troy Avenue, heading north. I was about to say something about our going the wrong way, when I caught myself. Ielpi is one of the best chauffeurs in the job: very fast, yet very safe, with a good knowledge of Brooklyn, and a superior firefighter. He was about to teach me one of the many lessons I would learn from the men of Rescue 2. And in a little while longer, he would teach me yet another: one that has stayed with me to this day. The dispatcher is now telling the 31 Battalion, *"Your assignment is 3 (engines), 2 (ladders), squad, and the rescue."*

After three blocks on Troy Avenue, Lee swung the big American LaFrance westbound onto Atlantic Avenue, which runs parallel to Bergen. Lee knew from his years of cruising around the borough that at this time of night Atlantic Avenue is a wider, faster, and safer route to the west side of Brooklyn than Bergen Street. Up until that point most, if not all, of my runs in that direction had been either close by, where the difference would be negated by the extra few blocks up to Atlantic and then back down, or else they had occurred in day or evening hours when Atlantic Avenue, a busy commercial street, was severely congested. At 5:30 a.m., though, its four lanes were nearly deserted, and the wider intersections provided clear views of oncoming traffic. We picked up speed as the rig hit several green lights in a row. The department radio again crackled in my left ear, as the dispatcher started relaying messages to the 31 Battalion. The communications office (CO) was now getting many calls reporting a serious fire in a multiple dwelling. Engine 226 now turned into the block and radioed, *"10-75, fire in an occupied multiple dwelling. We got jumpers!"* (People jumping out of the windows to escape the flames.)

Instinctively each man's throat tightens a little, now knowing that other human beings are in the process of dying. My feet are pressing hard on the three foot-switches that control the air horn and the mechanical and electronic sirens. As we approached Flatbush Avenue, the 31 Battalion gives its preliminary report.

"31 to Brooklyn. Transmit a second alarm. We've got fire on all three floors of an occupied, three-story brick multiple dwelling. Address unknown at this time. Exposure 1 is a street (in front of the fire building). Exposure 2 is a similar, attached (the building to the left of the fire building). Three (the rear of the fire building) is unknown. Four (the building to the right of the fire building) is a two-story brick, attached. Further particulars to follow."

This terse initial description paints a very bleak picture, especially coupled with the reported jumpers.

As we swing south off Atlantic into Smith Street, we can see no visible fire, but the street is nearly obscured by biting smoke. We rapidly dismount the apparatus at the corner of Bergen Street and report in on foot, in order to allow the second alarm ladder company into the block in case it is needed. As we report in to the 31 Battalion, I quickly scan the building, sizing it up. What I see sends chills down my spine. A firefighter from Squad Company 1 (who I later find out is John Cullen) has placed a portable ladder to a second-floor window for VES (vent, enter, and search), and has just taken out the glass and sash. The window is pushing out such hot, black, boiling smoke, under such pressure, that it is obvious that the room he is entering is about to flash over. I know instantly that Cullen must believe there is someone trapped in there.

Simultaneously, the chief is giving me our assignment. "Rescue, see if you can get up above. We have a report there might still be someone trapped in there. Try to get above. And make sure somebody gets the roof!"

Turning to the company, I need few orders. They all know their assignments. Tony Errico is already on his way to the roof with the saw. I am about to say, let's get up the portable ladder, when John Cullen suddenly appears at the second floor window, in trouble. Fire is rolling along the ceiling and out the top of the window he had entered. The room has flashed! John dives out the window onto the ladder below, as the room and the window fills with fire. In seconds, two more windows on that floor also light up, as do two of the third floor windows. One window on each floor has not vented fire yet: the window of "the dead man's room." It is called this because it sometimes has its only entrance through the hallway—a hallway that at times like this can be blocked by fire. I call to Billy Lake to take the aerial of Ladder 110 into the top floor window. I take Mike Esposito and John Kiernan with me, heading for the interior stairs. As we approach the front door, Engine 226 and 204 have two hoselines operating just inside the door, and the interior teams of Ladders 110 and 105 are jammed up behind the engine companies, all milling around trying to get up the stairs.

Amid all the confusion, I can discern the source of the problem. Some sick, twisted son of a bitch has poured gasoline in the wooden ground floor hall and up the wooden stairs and lit it. That is how the fire has reached such great intensity so quickly on all three floors. In the process it has destroyed all the treads and most of the risers of the interior stairs, leaving only the 2×12 stringers that run from floor to floor and carry the weight of the staircase barely intact. One of the truck companies has brought in a 20-foot ladder, and is trying to place it over the burned-out stair, but in the tight space of the hallway, with hoselines struggling to knock down the fire in the stairwell and hallway, it is a confused mess.

I decide to bypass the logjam and find another way into the upper floors. We head through the area of the ground floor hallway that the engines have knocked down, toward the back of the first floor and the rear fire escape that I know I will find on each three-story multiple dwelling. What I can't know until I get there, though, is that every window in the back of the building is blowing fire 10 feet in the air. That sight nearly knocks the wind out of me. I feel helpless. We return to the front of the building looking to commandeer an aerial, but Ladder 105's outside team is already in their bucket enroute to the third floor and Ladder 110's aerial is being used to put firefighters on the roof of exposure 4. We're joined by Billy Lake and Lee Ielpi. Lake tells me that 110's chauffeur wouldn't put him into the third floor window with the aerial, saying it was too dangerous, and besides the troops on the roof needed the aerial for their escape route, and the chief has ordered him to use the aerial to vent the windows. We have no choice now. We have to go up the burned out staircase. The men of Rescue 2 are about to show me how it's done when "the excrement has hit the ventilator."

The two engine companies have made little progress, just barely reaching the second floor landing, after placing a portable ladder over the burned out stairs. I begin to urge them onward, "Come on guys, we've got to get up those stairs now!" But their reply is that it's too hot to advance, and there are holes burned in the floor. Just then, I can barely make out the form of Lee Ielpi, as he scoots past me up the remains of the stair stringer. The ladder is jammed with people who won't or can't move, so Lee doesn't even try to use it. Instead he "walks the tightrope" along the charred edge of the 2×12 stair stringer, up to the second floor. Once there, he can't go as far as the head of the stair, so he takes a Halligan and knocks out all the balusters (the upright posts in the stair railing). Now he rolls up onto the second floor, ahead of the nozzle team. The entire stairway to the third floor is ablaze, as are most of the apartments on the second and third floors. Calling to the engine companies at the head of the stair, Lee tries to coax them forward. Conditions are severe, and there is not time to worry about hurt feelings, these lines have to move. But the companies are starting to run out of air now. They've been here for seven or eight minutes before us, and have been working that much longer. The nozzle team of the first line has to get out now, but the second alarm companies are not in yet. They reluctantly pass the nozzle over to Lee. Again, along the stringer goes a Rescue 2 man. This time it's Mikey Esposito. I try to follow, but I'm getting bumped and bounced by desperate men trying to get down the ladder and out to fresh air, as we are still trying to get up. Finally, I make it to the landing and try to organize the attack.

The first line is now moving through the second floor apartment, with Espo and John Kiernan on the line. Mike is like a bulldog on the line, moving steadily ahead like he's strolling through his backyard instead of crawling through a fire-filled apartment. I move to the third floor stairway, and find it is in worse shape than the previous stair. I am about to call to have another ladder brought to this location, when Lee steps right by me. He and Billy Lake have the second line now, and nothing is going to stop them from making that top floor, not even something like a missing stairway. Espo and Kiernan are still in the second floor apartment, knocking down fire and working on the primary search.

Lee is stepping on the remains of the stair riser and stringer where it goes into the wall, since the outer stringer cracked when he first tried ascending it. It is impossible to operate the line as they climb, so out of necessity they have to take a terrific pounding from the remaining fire until they can get to the top floor landing. Ladder 105's tower has operated its master stream in through the front windows, knocking down fire in the front two rooms and the cockloft over them, but the rear four rooms are still roaring. I'm still at the second floor, trying to get some help for our push into the top floor. Some guys with hooks will certainly be needed to open the top floor ceiling to expose the cockloft fire and to ventilate the roof, but my main concern is the burned out stair. We have to get a ladder up here. I'm not so much worried about getting up there as I am about getting down, especially if things start to turn to shit, or if we do find a victim whom we have to bring down to safety.

I follow Lake and Ielpi up to the top floor. They push through two rooms of fire, but now the fire in the cockloft and in the middle two rooms is a problem. The cockloft is radiating a tremendous amount of heat down at us, as well as large quantities of highly irritating smoke. "We need hooks up here!" I call down the stairs and repeat over the radio, but no further assistance is forthcoming. The door from the public hall to the front of the apartment has burned through. As the hose stream is directed from the rear door of the apartment toward the remaining fire in the front, the fire blows out into the hall and comes around behind us, through the rear door, like a dog chasing its tail. We have to then turn the line and drive it back away from our only escape route, but that only pushes the heat and smoke back the other way. The problem is solved without a word being spoken, when Espo and Kiernan's heads pop up over the landing. They have brought the second line with them, and seeing our predicament, begin to drive the fire back through the front door.

Now Ielpi is up with the line, pushing forward into the remaining rooms, and then climbing up on kitchen furniture, first the table, then the counter, then finally up onto the top of the refrigerator. I can't see him clearly. The smoke is still that bad, especially up where he is at ceiling level, but I now know what he's doing. We are still operating alone on the top floor at this point, and we only have one hook that John Kiernan somehow dragged up with him. Lee doesn't want to wait for the rest of the ceilings to be pulled before he starts hitting the cockloft fire. The conditions are bad, and now our air is running low. Lee knows if he can get the line up into the cockloft, he may be able to finish it off before we're totally out of air. That way, we will be able to continue to operate without masks. In order to do this though, he must expose himself to tremendous punishment, literally climbing up into the cockloft with his head and upper body. I see now why Lee always "turtles up" at a fire, Nomex hood on, helmet earflaps down, and coat collar up. He, as well as Billy, Espo, and Kiernan are prepared to do *whatever* it takes, under *any* circumstances to beat the red devil. And now it is done.

The cockloft is finally knocked down, and conditions start to improve. The ladder is in place now over the stair, and other companies start making their way up to the top floor. I call the chief on the handie-talkie and ask for relief. I tell him we need two engines to man the lines we have with us, as well as trucks to complete the overhauling. I pause to reflect on what I'd just witnessed. The lesson that Lee and the others showed me was the value and the meaning of *tenacity*. These four men had just knocked down two floors of fire under very severe conditions, doing the work of four companies, improvising when others less well-motivated were content to sit back and wait for things that are not forthcoming. "We don't need no stinking hooks to get at a cockloft fire," I think, or for that matter, ladders, or even stairs. I'm proud of these guys; I feel honored to be permitted to work with them. And even though I don't know it yet, I really love them.

I'm brought out of my momentary "rush" by the chief's voice on the handie-talkie, "31 Battalion to Rescue 2." I acknowledge, "Rescue 2, Kay." "How is the search of the top floor?" he asks. I want to scream, "Chief, we just finished the work of four companies, and only now can we even begin the search!" but instead I give the standard reply, "The primary (search) is still in progress, Chief." Just then, Mikey Esposito calls me from the adjoining room. "Hey, Lieu, you better get in here! I got a 10-45! I got a bunch of 10-45s." (A 10-45 is the FDNY designation of a fire victim. It is accompanied by one of three sub-codes. Code 1: victim is definitely deceased. Code 2: victim is very severe, probably deceased. Code 3: victim has a serious injury and could die.) I immediately get back on the handie-talkie and radio, "Chief, we got at least two 10-45s up here, and we'll need a couple of resuscitators!" We start moving debris out of the way, clearing an area to work on the first two victims, when two more bodies are uncovered. By now the smoke has lifted somewhat, and we can make out the forms of a mother and three of her children. Her fourth child was the jumper, as the first engine pulled up. The boy apparently had told the first firefighters to arrive that his family was in the room he had just jumped out of, but nobody relayed that to the rest of the incoming units. Instead, we got a vague message that "there might still be people in there." This kid gave us an exact location, critical intelligence and absolutely reliable; he had just jumped three floors to deliver it, and it was wasted. It has been a long, hard fight to make this third floor, and now it is for nothing. Our company helps load the bodies into Ladder 105's basket for removal to the street. All four are pronounced dead—smoked, not roasted, so if somebody had gotten to them earlier, they might have survived. The bodies were located just under the sill of the only window on the top floor that did not have fire blowing out of it—the window that Billy Lake had wanted to climb into.

CHAPTER 16

The What If's?

The firehouse that Rescue 2 occupies sits right in the middle of Bedford-Stuyvesant and Crown Heights. The housing stock here is old, heavily over-crowded, and under-maintained. The fire companies that serve this area are located in a spread-out pattern, with many single company houses, like Engines 217, 222, 227, 230, and 235, as well as Rescue 2, with a few double and triple houses, Engine 214/Ladder 111, Engine 280/Ladder 132, and Engine 234/Ladder 123/Battalion 38. The practice of maintaining so many single units is foreign to many city planners, who would prefer to consolidate a number of fire companies in one house to save the cost of buildings and maintenance, as well as put more land on the tax rolls. That concept works fine in less crowded areas, but in most of New York City, this would result in excessively long response times due to traffic. The idea is to get someone to the scene as quickly as possible, to begin rescue, make a size-up of conditions, and call for extra help or return any unneeded units as quickly as possible.

One advantage of being in a single house (the only fire company in the building), is that as soon as the teleprinter's *tooo-dooo* tone goes off, every-body in the house knows that their unit has a run and heads for the rig. That gives your unit a several-second head start on the fire that units in double and triple houses don't get. They have to wait for the ever alert house watch-man to drag himself up out his chair, read the ticket or the monitor screen, see which units in the house are assigned, and then announce it with either a coded bell system, an intercom, or both. Those few extra seconds can make the difference between life and death!

On March 10, 1990, as I was holding roll call behind Rescue 2's rig, giving each of the five firefighters their tool and position assignments for the night tour, including primary and secondary SCUBA divers, the tone alert sent us all scrambling for the rig. Sterling Place and Schenectady Avenue, fire in a multiple dwelling, with children trapped! Richie Evers already has the air brakes released as I slam the cab door on my side, and the low-slung American LaFrance noses out into Bergen Street. Turning right out of quarters, going the wrong way against the light, oncoming traffic of the one-way street, we head for the corner of Schenectady Avenue. My two feet are alternating between the three foot pedals for the air horn and the electronic and mechanical sirens, lest someone speeding up Schenectady Avenue be caught off guard by the behemoth entering their roadway from the wrong direction. It's a bit risky, but it probably saves us 60 seconds compared to going with traffic, and circling the block. Richie swings right onto Schenectady, for the short four-block sprint to Sterling Place. As we cross Park Place, Richie and I can see fire lapping out of one window in the rear on the top floor of a building at the corner of Schenectady Avenue and Sterling Place. The fire building is a four-story, 75-foot × 100-foot brick and wood joist "H-type" apartment house. "Rescue 2 to Brooklyn–K," I radio, as I bang my left hand on the pass-through opening to the crew cab where the firefighters ride, the signal for "a job." "Go ahead Rescue" calls the dispatcher. "10-75 for box 1012, top floor fire in a four-story occupied MD" (multiple dwelling), I reply. This last tidbit, top floor fire, will alert the roof firefighters and OV's of Ladders 123 and 111 of the possible need for power saws on the roof. Richie stops the big rig out on Schenectady, even though the building entrance is around the corner on Sterling, because he sees Engine 234 coming down Sterling toward us on the narrow, one-way street. The engine has to have priority access in order to get a hoseline stretched and connect to a hydrant. Everything we need, we can carry with us.

I hustle toward the building entrance with my inside team. Billy Lake has the irons tonight, while Tommy Richardson has the can. At the building entrance we are met by an absolutely frantic woman of indeterminate age. She is screaming at us, "The babies are in there! The babies are in there! Please, God, save the babies!" We take the stairs two at a time now, heading for the top floor. As we round the newel post onto the last flight of stairs, we run right into a wall of inky darkness. The smoke is banked down into the staircase. That will lift as soon as Timmy Stackpole, Rescue's roof man, forces open the bulkhead door and smashes the glass out of the skylight atop the bulkhead over the stairs that lead to the roof, but for now, it complicates our search. Fortunately, since we passed the rear of the building on our approach, I saw where the fire is located, so I know enough to move directly along the left wall toward the left rear. The amount of smoke in the hall signifies that the apartment door must be open, which now becomes a good thing, because we won't have to waste time forcing a heavily fortified apartment door. I find the open door with my left hand and start in, calling for Billy and Tommy to follow my voice. I tell Billy to take the right wall, while I take the left as we make our way into the apartment, so that we don't miss the doors to any rooms.

We crawl a long way down a hallway, without finding any doors, then after about 30 feet, the hallway makes a left turn. We're very close now. The temperature has increased dramatically, and the crackling roar of the fire is very loud, but we still can't see shit, even with both of my lights on. This is really blackout conditions; it means there must be a lot of plastic and foam rubber burning. We continue for about eight feet down this new hall, before Billy calls out, "I got a door, Lieu, I'm going in." "OK, Bill," is all I say, as Tommy and I continue forward, hunting the fire and searching for the trapped "babies." I know Billy will be all right—the fire is still ahead of us, and if we have to retreat, both Tommy and I will know we have to get Billy from the room as we exit—but the small size of the bedrooms in these apartments allows a single person to toss them fairly quickly during the primary search. I didn't see Ladder 123 arriving as we entered the building, but I know either they or 111 will be behind us shortly with reinforcements, as will Engine 234 with the vitally important hoseline. The problem, though, is the city administration under the direction of Mayor David Dinkins has seen fit to cut the staffing on our engine companies from five firefighters to four—leaving only three people to stretch hose, since the driver must stay with the rig, connect to a hydrant, and operate the pumps. That means that at a fire like this, the hoseline will take longer to get into operation.

Tests have shown that this decrease in personnel from four to three has a tremendous impact on speed and stress on members. Instead of a 25% reduction in efficiency as you might assume from a cut of one quarter of the personnel, the measurable impact is more like a 45% increase in time, since the remaining members not only have to stretch additional hose, but free the butts (couplings) from around the newel posts and other obstructions, flake out the hose once it is in position, then return to the street, break the hoseline at the proper coupling, and connect it to the apparatus before water flow can begin. For an eight-length hand stretch around the stairway, such as this one, that additional firefighter goes a long way!

As Tommy and I advance, we begin to see flashes of light over our heads: rollover, tongues of flame traveling ahead of the main body of fire, seeking more fuel and oxygen. We're close now. We come to two doors on the right, illuminated by the flames leaping out of the farther one. I tell Tommy I'm going into the nearer one to search, as he moves into position with the can to begin the battle with the devil. A full can only lasts about a minute, so I know I have to be quick. I exit the bedroom, empty-handed, just as the last of the can is expended. The two and a half gallons has not made the slightest dent in the fire. This building is built largely of wood—wood floors, wooden doors, wooden wall studs, wooden window frames, and in many cases, wooden wall paneling. The can just doesn't do very much, and the fire is not appeased. Just as I exit the second room, I hear a low moan, barely audible really, but definitely a moan, from farther down the hall, past the fire. I ask Tommy if he heard that, but he must have had his back to the sound and missed it. I know we have to go down that hall now, past the fire that is once again licking out into the hallway, raising the temperature and preparing for flashover. I tell Tommy I am going down the hall. "Get that bedroom door closed somehow," I plead with him. If he can close the door, it will buy us some time. The way we usually do this is by reaching into the room with the 6' pike pole and hooking the edge of the door, but Tommy set the hook down in the hall as he used the can, and now the ceiling in the hall has collapsed on top of the hook and him, there is fire in the cockloft, and Tommy is unable to find the hook in the debris. Without pausing, Tommy knows he will have to reach in with his hand and pull the door closed as the inferno surges around him. I crawl forward as Tommy reaches for the knob. I call to him: "Call out if you have any trouble. Let me know if things start to go bad!"—as much to let him know by the sound of my voice how far away I am getting, as to provide instruction. Tommy is a sharp firefighter, soon to be promoted to lieutenant (and eventually chief of operations for the whole FDNY!), who knows how critical his position is to my well-being. He'll let me know if things go sour, hopefully in time for me to retreat. But first he *has* to get that door closed.

When Tom reaches for the knob, a full two and a half feet into the fire, the intensity of the heat through his protective clothing causes his body to recoil in reflexive defensive nature. But he refuses to give up. His "boss" is somewhere in the darkness ahead, depending on him to buy some time. He steels himself for what's coming, changes hands for his next attempt, gauges the distance to the edge of the door, and springs forward with his out-stretched fingers clawing for the edge of the door. The knob makes too small a target for his blind effort. In an effort to shield his face from the flames, Tom has tucked it down, close to his shoulder. He puts his back toward the bulk of the flames, as his fingers close around the edge of the flaming door. Instantly he falls backward now, desperately yanking the wooden shield back with him toward the relative safety of the hall, but debris from the collapsed ceiling prevents him from closing it tightly. It doesn't matter much anyway. As the door strikes the ceiling debris, the impact shakes loose a pile of burn-ing embers that used to be the top of the door. It has burned through!

"Hey, Lieu, you better get back here. The door is burnt away at the top, and the fire's still coming!" I hear Tommy call, as I brush up against what appears to be a stove with my right hand. I crawl forward, sweep-ing blindly with both hands. I brush past a hard wooden chair as I call out, "Hey, is anyone here?" As I move further forward, I run into a wall. I squat back to turn around, and my helmet and mask bump into something hard overhead. As I am trying to turn around Tommy is calling, "Hey, Lieu, get back! It's starting to roll down the hall real good!" "Shit, where's that line?" I think, and then, "Where the hell am I?" There are chairs and other objects all around me as I try to turn, making my progress difficult. I realize I am under a heavy-legged kitchen table. I push a chair out of my way, and head back the way I came, but as I do, I make one last sweep out with my right hand, the one with my light and hand tool in it. The tool strikes something soft, and my heart leaps. Everything else I've touched in this room has been hard—table, chairs, stove, and what I presume to be cabinets. This could be something! I stretch out toward the object, trying to keep my foot against the wall so that I don't get turned around and lose my bearings. "There it is, something soft," I murmur. I grab hold of some fabric and pull it close. With my two lights I make out a diaper and a red and white striped shirt. They are wrapping what looks like a one-year-old child. Great! "Rescue to the 38, 10-45," I radio. The chief acknowledges the 10-45 and instructs his aide to put it out over the department radio in the chief's Suburban right away.

I drop the light and my hand tool, and head for the door with the infant. Then I remember the hysterical woman's cry in the lobby, "Save the *babies!*" I pause for another sweep with my free hand, but feel nothing. There might be more kids in here, but how long can I keep this first child in this poisonous mess? Just then Billy Lake scrambles headfirst into me, nearly knocking me over, since I am leaning over, nearly prone, probing for the missing kid or kids. I tell him we are in the kitchen and that's where I found this one kid, as all hell breaks loose around us. Engine 234 has begun to operate their line, driving the fire back into the room of origin, and quickly hammering it into submission, but the powerful stream from the $^{15}/_{16}"$ solid tip brings down more of the fire-weakened ceiling, and hot water cascades onto us. I do my best to shield the child from this blistering mess and am grateful that most of the busiest urban fire companies use the solid tip, instead of the fog tip, which would produce a lot of steam. With my mask and protective clothing, steam is more of a nuisance than a danger, producing burns at the least protected areas, like at the wristlets of the gloves, around the edge of the mask facepiece, through the Nomex hood, or where the mask compresses the insulating air out of the liner of the turnout coat. To the diaper and T-shirt clad infant, however, steam would be deadly. I scramble back down the hallway, on one hand and both knees, crawling over the steaming debris that was the ceiling, and the door to the fire room. At that juncture, I again find myself going the wrong way against traffic: I have to force my way past a bottleneck of people, the nozzle team of 234, plus the forcible entry team of Ladder 123. "Let me out, I've got a victim!" I shout, and a narrow path opens. I tell 123 that there might be more kids down at the end of the hall, and that the Rescue's irons man is in there, then I continue down that long entrance hall.

Out in the stair landing, the smoke has lifted. The roof team has done its primary job very well. As I rip off my mask facepiece, my helmet crashes to the marble floor. I know that this baby needs oxygen in its lungs right now. The thought of the pocket CPR mask flashes through my mind, as do the statistics on the number of AIDs babies born to crack cocaine addicted mothers at the nearby St. John's Hospital. This kid doesn't have time for that, I decide, and offer a silent prayer, "God protect me," as I begin mouth-to-mouth on my latest tiny victim. I take the stairs at the same pace as on the way up, turning right out of the building, heading for the rescue truck. As I round the corner, a sight too good to be true meets me—an EMS ambulance, alerted by the Brooklyn dispatchers by a call reporting children trapped. I climb up into the back of the ambulance, not missing a beat in the CPR. Between puffs of air into these tiny lungs, I radio Richie Evers, letting him know I'll be going with EMS to St. Mary's Hospital, only three blocks away. The EMT in the back of the "bus" has hooked an infant ambu-bag up to the oxygen tank, and she takes over breathing for the child. Being an EMT, she can't intubate the baby, so we must struggle to maintain the correct neutral head position for the airway as we careen toward St. Mary's. Thank God it's only three blocks! As the bus pulls up outside the emergency room, I jump out the back doors, still performing the chest compressions on the tiny heart, located just beneath the heat-blistered brown skin under my fingertips. I gently place the child on the open stretcher that the ER nurse points towards. Instantly, a team of physicians and nurses surrounds the child. One nurse snips away the striped T-shirt and diaper, while a doctor attempts to place an incredibly small intubation tube. I back away to make room for the other nurses, technicians, and therapists that have been summoned by the "code." Trying to be as unobtrusive as possible, I watch from the edge of the curtained off treatment area as the caregivers perform their part of this team effort to save this tiny baby's short life. The things I can see on all the monitors, as well as the faces of the ER staff, do not bode well. The cardiac monitor does not show anything like a normal sinus rhythm, only a crazy racing of lines that looks like a seismograph during an earthquake. The faces are just as confounded, a sea of taut brows and hard-pressed lips, as everyone appears to be attempting to *will* this child to live.

Tommy Richardson and Billy come into the ER. Billy has found another kid, but he was so far gone that the medics pronounced him dead at the scene. Richie comes in and says he's got my light and helmet as well as my hand tool. Tommy asks to take my mask, which I am still wearing on my back. He'll take it out to the rig and insert a fresh cylinder so we can go back in service. You never know how far away the next run is. I tell Richie to put us 10-8 (available for duty), but I want to hang around here for a while to see what happens with the kid. I guess the Doc overheard me. He says, "No need to hang around, guys. I'm calling it. This one didn't make it." A collective sigh of, "Ah, shit," escapes the lips of a bunch of tough guys. We turn and start back for the rig but then Tommy grabs me and asks, "Lieu, can I get a Doc to look at my hands and head while we're here?" "Yeah, sure, Tom. What's the matter with your head?" Tommy silently leans forward, presenting me with the top of his head. There right in the middle of an incipient bald patch is a blister the size of a half dollar. "What the hell is that?" I ask. He mumbles, "I must've gotten burned when I was trying to pull the door closed." Then he shows me his hands. The blisters on them have broken. Here's a guy with second degree burns to his head and hands, burns he inflicted on himself in order to protect the kids in the apartment and me. God, how I love these guys! They'll do almost anything for each other and me. I want to scream out at the world, "Look at these guys, they're the best this whole world's got!" But I hold it in check. Emotions are not something "tough guys" let surface. Back in quarters, that thought doesn't leave me though. In fact, it is what helps when the inevitable "what if's" start. "What if we had turned out a split second faster? What if we didn't have to slow down for the light on Schenectady Avenue? What if I had just scooped up the first kid, and beat feet for the door, instead of pausing to search for the other kids? What if we'd had three gallons of water in the can instead of two and a half? What if Engine 234 had had five men instead of four?" I don't know any of the answers. Perhaps it's just inevitable that no matter how hard we try, people are going to die. But I do know that those kids on Sterling Place weren't let down by the firefighters on Bergen Street. They were let down by their mother who was down in the street drinking and smoking with friends while her "babies" were playing with matches upstairs and by their father, if he is even in the home. We did our jobs. Now the cops and the courts need to do theirs (fig. 16–1).

Figure 16–1. Battalion Chief Tom Richardson, in a command post tent on the afternoon of September 12, 2001 at the World Trade Center. Tom became Chief of Fire Operations for the FDNY in 2019, but he is one tough firefighter at heart.

CHAPTER 17

Kingston Is Not Just in the Caribbean

A newcomer to Crown Heights might think that Kingston Avenue is named after the city of Kingston on the island of Jamaica. By the 1980s, this is particularly true, given the area's large Caribbean population. The area is the home of New York City's Caribbean Day Parade, held Labor Day weekend. Due to the large number of violent incidents that have occurred on the periphery of the parade over many years, by the 1980s it had gained the nickname of "The Homicide Day Parade." In fact, however, Kingston Avenue is named after a historic old New York State city. Kingston was the capital of New York State in the late 1700s, prior to its being moved to Albany. The major north-south avenues in Central Brooklyn are all named after New York State cities, tracing the route along the old Erie Canal and Hudson River: Buffalo, Rochester, Utica, Schenectady, Troy, Albany, Kingston, Brooklyn, and New York. On October 30, 1990, this fact was certainly not at the top of my "most important things to remember" list, but knowing the proximity to Kingston Avenue from our quarters near Troy Avenue certainly was beneficial.

This particular Thursday night tour is business-as-usual for Rescue 2: several working fires that are "nothing to write home about," just a few more times in and out of the rig, and a couple of more times walking around in smoke. Then things started to pick up. A job comes in for a meat market down in Brownsville around 2:00 a.m. We scramble for the rig and head down Bergen Street the wrong way. A hard right onto Schenectady Avenue, then south six short blocks, brings us to St. John's Place. Sirens screaming and air horn blasting, we swing by the quarters of Engine 234, Ladder 123, and the 38th Battalion. The chief is just turning out now, following us east on St. John's Place. He's just been assigned as the "All Hands Chief." Shortly, we turn right onto Rockaway Avenue, and then pull up just short of the fire building.

We report in to Battalion Chief Richie Fanning of the 44th Battalion, one of the finest gentlemen I've ever met, who directs me to see if we can give Ladder 120 a hand with venting the roof. Normally, roof ventilation is relatively easily handled by the assigned roof firefighters of the first two ladder companies, plus the rescue and squad, but the chief has received reports from the roof that they are having difficulty. The reason is soon apparent. The original wooden roof has been covered over with ¼" steel plates, which are welded in place to form an impenetrable barrier to our standard roof cutting methods: the wood cutting circular saw, or if that fails to start, the axe. Such extraordinary procedures are necessitated by the severity of the crime situation in the neighborhood. The burglars, frustrated by brick walls and bars over the windows, have resorted to cutting through the wooden roof to gain entry.

Normally, aggressively fighting fire in a steel-plated building is a losing proposition. The lack of ventilation results in severe heat and smoke conditions, and the resulting heavy fire, combined with the weight added to the roof, often results in early collapse of the structure. That's why I'm caught off guard by Chief Fanning's insistence that the handlines must make a good push to get in, and that we have to make sure we get a good roof opening. Chief Fanning is a cautious, knowledgeable fire officer, who knows the dangers involved in conducting an interior attack in a building that may be subject to sudden, severe collapse. I know he wouldn't be putting us into this position without good reason. There must be reports of people trapped!

I call my chauffeur, Billy Hewitson, and tell him to bring our metal-cutting saw to the roof. Ladders 120 and 176 have their saws on the roof, but they are still changing over from the wood-cutting blade they normally carry to the special steel-cutting blade that is required here. In the early 1990s, the FDNY is so short on money that we can't afford two saws for each of the 140 ladder companies. Only the rescues have saws equipped with both wood and metal cutting blades at all times. Billy arrives with the saw, and I direct him and Ladder 120's roof firefighter in starting the cuts that will eventually produce a 4' × 8' hole over the fire. The steel just about gobbles up the abrasive cutting discs, and soon both saws have worn their blades down to such a small diameter as to be ineffective. The first two men step away from the work area and proceed to change old blades for new ones. Ladder 176's roof firefighter goes to work, along with the roof firefighter from Ladder 103, which has been special-called by Chief Fanning to assist.

The roof cutting is difficult and dangerous work, conducted in an eerie setting. It is pitch black out, with heavy smoke dancing around us as we work. The only light up on the roof comes from the hand lights that each member has slung under an arm, the flashing and revolving apparatus lights below the roof level at the front of the building, and a dazzling display of sparks being thrown off from the grinding carborundum blades of the saws as they slash and tear at the steel barrier beneath the firefighters' feet. One of the members changing saw blades finds a gallon container partially filled with gasoline near one of the only openings that normally pierces the steel sheath that is now the roof, a 6" × 6" exhaust duct, which is lying on its side on the roof. It seems someone has decided to burn this place down by pouring gasoline into the building from the top and igniting it. I notify Chief Fanning of our finding and our progress with the ventilation. We finally strip off the steel plating and cut through the wooden roof boards, which are being peeled back now, providing some relief for the engine companies trying to push in beneath us. The chief directs me to bring part of my company down to street level to assist in searching the interior.

Bobby Galione, who is a new guy in Rescue 2 at the time, Richie Evers, one of our most senior men, and I make our way down to the first floor, and begin searching for a meat market employee who is reported missing. By the mid-1980s and early 1990s the crime situation in East New York, Brownsville, Bed-Stuy, and the other poorest neighborhoods in the city is so bad that store owners, even those with steel-plated roofs and walls, have taken to locking themselves inside at night to further deter the vultures that will steal anything that is not bolted down—and a few things that are. More than one homeowner in this crime-ravaged region has awoken in the morning and come out of his house to find that all of the aluminum siding within reach of the ground has been stripped off overnight to be sold for scrap to support somebody's crack cocaine habit. The more prosperous shopkeepers don't spend the night in the store themselves. They hire someone. Then, to make sure that the person they hire remains a loyal and trusted employee, they promptly lock that person inside the store with no way out, so the employee can't walk away with the entire contents before the owner returns in the morning. Of course, in the event of fire in the store, the owner is in a lot better shape than the locked-in employee. More than once, Rescue 2 has pulled up to a good store fire on one of the heavily fortified commercial strips—Pitkin Avenue, Fulton Street, or Belmont Avenue—and found a hysterical citizen trapped behind the locked gates, banging on the steel slats and begging to be freed from his or her prison before the flames reached them. Most of the time, we are successful in getting to them first. A few times we weren't. That's what we're facing right now; looking for someone we weren't in time to save. Ladder 120 has been through this same scenario. They know that if there is a civilian in there, he is probably right near the front gates. They try to get as far away from the choking smoke and searing flames, and as close to the salvation of the metal-cutting saw, as they can. 120 has done a thorough "primary" of the store and has not found the missing employee. They direct us toward the back, where the destruction is heavier, and there is more debris to be sorted through.

The Engine companies 231, 283, and 290 have done a good job hitting this fire fast and hard, knocking it down before the destruction of the roof is too severe. It is still hazardous to dig through the debris beneath this charred roof with its extra load of steel plates. Each 4' × 8' sheet weighs over 300 pounds and would act just like a guillotine should it fall on you on edge. But since the engines did their job, now we can do ours. If they had not knocked the fire down so quickly, we would have definitely had a collapse and would need a crane to lift the debris up so we could find the body of the missing employee. Now though, the roof is still hanging, so we can clamber around underneath it, always with a watchful eye for any signs of impending collapse, while we dig through piles of *really* roast beef, sausages, and other meat products that look like a *really* roasted person. Suddenly, the deputy chief of the 15th Division, who's now in command, calls off the search and orders everyone out of the building. The missing man has been found safe and sound across the street, watching the fire. Apparently, he had somehow obtained a key to one of the locks, and secretly kept it for the time he might *really* need it, like tonight, or when some other appropriate opportunity presented itself.

There is still a lot of overhauling of the smoldering debris to be done, but now that the occupant is accounted for, it can be done from a safer distance using the reach of the hose stream. The deputy tells us to "take up." Back at the rig the guys are changing the "smoke tanks": replacing the expended cylinders of breathing air in our masks for new, fully charged ones. You can't afford to wait till you're back in quarters to get a new mask cylinder. More than once we've gone right to another job while on the way back to quarters.

On the way back, Billy comments on "how lucky that guy in the meat market was to have hidden that key." There is no way that 120 would have gotten to him in time; the store was burned right down to the floor all the way to the front gates. Later we find out that the guy had been let out of the store, even before the fire was set, possibly by the owner. The fire apparently was an attempt by the owner to "sell" the business to the insurance company.

Back in quarters now, I jot down a few notes for the CD-15, the operations report or "Second Alarm Report," fill in the fire record journal, wash up, and finally hit the rack. It is now 4:25 a.m. as my head hits the pillow. Before my eyes close, I hear the radio come alive, "Brooklyn to the 38 Battalion, we're transmitting Box 1033. We have a report of a fire on the ground floor of a dry cleaner at Kingston Avenue and President Street, reported people trapped on the second floor." The whole firehouse hears the message as the radio speakers in Rescue 2's quarters are always turned up so that everyone in the building can monitor fire activity in Brooklyn. For a detail from another company, this is quite disruptive to their sleep, but it pays off in an extra 20 seconds head start at times. Tonight is one of those nights where the practice will pay off—big time!

I'm already pounding down the stairs, right behind "Espo" and Kevin Dowdell, when the "tooo-dooo" sounds and the teleprinter starts printing. Billy Hewitson, who was on "house watch" in the kitchen, is already in the cab as the apparatus door goes up. We know this one is close by, and it sounds like a job. As we're leaving quarters, Engine 234 gives the 10-75, "fire in the ground floor of a 2-story brick, store and apartments above." The dispatcher informs the 38 that Ladder 111 will be the second-due truck—Ladder 123 is on "R&R" from the previous second alarm and Ladder 113, normally second due, will be the first-due truck. I know we've got a good two minute head start on 111. Their quarters are about 12 blocks north of ours and we're out fast, so we prepare to act as the second-due truck until they arrive. That means our primary responsibility will be the floor above the fire.

As we turn off of Eastern Parkway onto Kingston Avenue, going the wrong way down the one-way street, the block is shrouded in a thick, inky-tasting veil of black smoke. Ladder 113 has only gotten in about 30 seconds before us, and they are working on forcing entry to the fire store, through the metal scissors gates, as well as forcing the door to the apartment entrance, which is located down a small side yard toward the rear. I report to Chief Farnsworth, covering in the 38th Battalion, at the entrance to the stairs to the second floor. The door is heavily secured, and 113's irons team has just popped it as we arrive. There are neighbors all around, screaming about the people trapped, and we all know this is the real deal. The chief says, "Get upstairs and get 'em!" As the door swings open though, we can see the stairs are blocked by fire. Engine 234 is operating the first hoseline in the front door of the heavily involved dry cleaning store, and the second line is not yet being stretched. I know we can't wait. Mikey Espo is working in with 234, and Dowdell is on 113's aerial enroute to the roof. I turn to Richie and Bobby Galione and order a portable ladder to the front windows as we sprint to the street side. Richie and Bobby grab a 20' straight ladder off of 113's rig, and throw it into a set of double windows on the right side of the building. While they're putting the ladder up, I don my mask facepiece and pull my Nomex hood up: something I rarely did in those days. Usually, since not all the firefighters have hoods, I leave mine down around my neck so that I can better judge the heat conditions they are experiencing. Heat is a precursor to flashover, but now I know I'm going to need the hood for the added protection it will give my head. There are definitely people up there, and we're going to have to go "dancing with the devil," right up to death's door, right till the last second before flashover occurs and the rooms ignite in a ball of flame—before we turn away and retreat. The hood just might save my face from severe scarring burns, if my estimate of how severe conditions are turns out to be wrong by four or five seconds. After that time frame, it probably won't make much difference; it will be a closed coffin funeral.

As soon as the ladder is in position, I start my ascent, moving quickly to the top. I pause momentarily to break the window, top pane first, bottom, and then clear out the horizontal sash. This does two things: makes it easier to get in and out, and also slows us down for five or six seconds, which in this case, is just what I want. When I vent the top pane, the stuff that comes out is really nasty: hot, black, and looking like it's ready to ignite. By taking the time to take the sash, I've given the area a chance to blow. If it were going to backdraft (a smoke explosion caused by allowing oxygen into a super-heated atmosphere), it should have done so before I stuck my head into the opening. Now that the sash is out, I dive in head first through the window. Not exactly the textbook way of entering a window. You know, probe with a tool or your leg to make sure there's a floor there that will hold your weight, before committing. But I make a decision based on conditions that this is the proper thing to do. For one thing, the heaviest fire seems to be in the rear of the first floor, so the floor in the front should be in halfway decent shape. Secondly, the heat coming out of the top of the window I'm about to enter is quite severe, and it's worse the higher up in the window you get. There can be no straddling this windowsill while I poke around for a floor. Finally, there just ain't time for all the niceties of entry—people may be dying right now!

I land on a mattress and find myself wrapped up in bedding and heavy window drapes. It takes a second to free myself and let Richie know what to expect as he enters behind me. I know Bobby and Billy are right behind Richie, so I move out of this room, into the next room toward the rear. I hear human voices, and call to Richie, "I think I hear them; follow me." I'm groping blindly forward, trying to stay in touch with the right-hand wall so that I can find my way back to the window if the shit hits the fan. The smoke is so thick up here that even the two powerful lights that I'm carrying (one clipped to the top buckle of my turnout coat, and the other, larger one slung across my shoulder and under my arm on a seatbelt scavenged from an ADV Chevy Camaro) do little more than waste their batteries. If I put my facepiece right on an object and the light right next to it, I can probably make it out, but that doesn't matter anyway. Right now, the things I'm looking for, I'll know right away when I find them. Fire will show me where it is without my light. Heat I can feel. And people—well, people I can feel, too. I just hope I don't have to tonight. Pressing forward toward the sound of voices, I find the television on, which tells me I'm in the living room, when coupled with the feeling of stuffed chairs and couches. It's a big room, as wide as the entire building I figure, about 25 feet wide and about 12 feet deep. I know the first bedroom we entered wasn't this wide, only about 12 feet, which means there's another room to the left of the first bedroom. I know Richie will get it, since when he came in I said, "I'm going right," and he replied, "I'll take left." I figure with Bobby and Billy coming, they'll back us up, so I head back for the right wall. I find a doorway that leads into about a three-foot-wide hall. The first room down this hall is the kitchen, which I recognize immediately by the refrigerator followed by the stove. I'm sweeping the floor and walls with my gloved hands and the large appliances are easy to pick out. Near the stove I see a glow. It is fire burning up under the sink, probably through a pipe chase for the water and waste pipes. I get on the handie-talkie and call the 38, "Rescue, to the 38. I need a line on the second floor. I've got fire in the kitchen." Chief Farnsworth replies with a laconic "10-4, as soon as the next engine gets in." A fast sweep of the kitchen is aided by the glow from the sink. I know there is little likelihood of finding victims in the kitchen, so I quickly retreat to the hall to continue my search.

I turn left out of the kitchen, pressing deeper into the apartment, heading for the rear. I know this is getting me into someplace I might not be able to get myself out of, and that violates one of my cardinal rules of survival, but I feel in my bones that this is the real deal, and thus, the right move. All those people out in that side yard can't be wrong. The hallway continues back toward the rear, over the heaviest fire, for what seems like a football field, but in reality is only about 20 feet before it makes a left. As I crawl along, my mind is racing, "How far should I go?" I know I'm getting in pretty deep. I've gone about 40 or 50 feet from my entry at the front window, which is now my only escape route. I know there is fire between that window and me. If an aerosol can of roach spray or furniture polish stored under the kitchen sink decides to BLEVE (boiling liquid expanding vapor explosion), it will cut off my escape, filling the kitchen with fire and blocking the hall. I know there is fire in the walls, and from the sound of the power saw beginning its screaming wail overhead, the fire is probably in the cockloft already as well. In a lower crime area, I might be able to escape from any window I find. I've already "bailed out" of a second-story window before, but not in central Brooklyn, where every window that can possibly be reached, and some that can't possibly be, are heavily barred. To rely on one of these windows for escape is to literally bet your life on being right.

I try to judge the heat condition around me, the Nomex hood disguising my normal senses. I extend my right arm overhead and bend my hand back sharply, while pulling down on the sleeve of the turnout coat. This exposes a little of the skin on my wrist, so that the area is exposed to the air. I gauge the temperature there, like a new mother tests the temperature of the milk in a heated baby bottle. I find this method far better than removing the glove, since it only exposes a small amount of non-critical skin, and I don't have to stop to do it or to put the gloves back on. My wrist tells me it's hot, hotter than I should be in if there's no life hazard, but at least the skin didn't crinkle up immediately. I'm debating whether to push on or go back in my mind, when I hear a noise that sounds like a cough up ahead. My mind is made up; I again press forward, my every sense at its state of peak alert. My ears are acutely turned, listening for a repeat of the cough, hoping to use it to home in on the source. Instead all I hear is the snap, crackle, and pop of the fire below me, in the walls around me, and in the cockloft above me. I hope to pick up the sound and vibrations of the 2½" hose stream Engine 234 is operating below me, but I passed that 20 feet back. The line is being slowed by display counters and heavy stock. I hear a low hissing sound straight ahead and head toward it. I see another glow looming ahead, near the source of the sound. I'm in the bathroom, again recognized by the fixtures. I don't hear any faucet running, but my hand encounters scalding hot water, so I sweep the tub in case somebody tried to avoid burning to death by seeking refuge in a water-filled tub. Thank God it's empty! This is not good; I'm running out of building and time. I again call the 38 asking where the line is for the second time. Chief Farnsworth assures me it's on the way. The hissing noise comes from a copper water pipe whose solder joint has melted and given way, because of the fire roaring up into the vanity under the sink. Shit! I think. I hope there are no aerosol cans of hairspray under there.

Backing out of the bathroom, I continue following the right wall about four feet. Suddenly, on my right the heat and crackling increase, indicating a doorway. Now I can make out a faint glow along the floor. Making the turn into the room I see flames now, licking up the wall and along the ceiling in the corner to my right, back toward the bathroom. This room is about to light up—flashover is only seconds away. I make a desperate lunge into the room. By the firelight I have made out the shape of a bed. On the bed I sweep my two arms quickly across the surface, until my right hand encounters a soft mass. Even as I say to myself, "I hope this is not a person," I know that it is. I fairly shout into the radio, "Rescue 2, to Battalion, 10-45, 10-45, second floor, right rear, I need help!" I frantically sweep the rest of the large bed for a child that might be with its parent. The fire overhead is really rolling now, lighting up the entire room, and making the bedding smoke. I see that the rest of the bed is clear as I start dragging 23-year-old Milde Gausintu, a Caribbean immigrant, off the bed, but what I see next horrifies me. There in the corner, on the far side of the bed, lie the figures of a man and another woman, both burning. I drag the woman I'm carrying to the relative "safety" of the doorway to the room, and I'm about to drop her to go back for the other two when Billy Hewitson arrives, followed by Bobby Galione, Richie, and John "Zuke" Zazulka of Squad 1, who have heard the 10-45 report. I tell them about the two victims remaining in the room. Billy makes an almost suicidal entry into the room, just far enough to grab the woman's ankles, and yanks desperately back, hauling a person with him that weighs damn near as much as he does. Then the room flashes over. Flames knock Richie and "Zuke" back before they can reach the man. Now the flames are rolling out into the hall over our heads, preheating this area, before spreading out like an octopus' tentacles, seeking to pull its victims into its grasp. Bobby grabs the first woman's ankles and I grab under her arms and across her chest, and we scramble along the hallway back toward the front windows.

As we crawl back down the hall, trying to stay below the fire overhead, yet get the hell out of there with our victims as quickly as possible, we encounter the engine trying to crawl forward down the same narrow hallway, trying to get water on the fire over our heads. For several moments it is pure pandemonium, as two bodies try to occupy the same space. Then suddenly we are clear, into the large living room. Bobby and I continue through to the front bedroom where we know there is a ladder. As we drag Ms. Gausintu over the bed and prepare to exit the room, another firefighter is trying to enter. We tell him to get back down a rung; we've got a victim for him to carry. We pass Ms. Gausintu out into clean, fresh air, then plunge back into the inky poison. The firefighter on the ladder has his hands full carrying the unconscious woman down, and his balancing act is caught on film, ending up in the next day's newspapers. But our job is still not done.

Billy and Richie have brought out the second victim, and she too is passed down the ladder. Now there is one more person we must get to. With Engine 280 pushing their hoseline down the hall, we jump back into the room where the three victims had been overcome. John Zazulka grabs the victim that the flames had separated him from before, and he and several others drag him out, and start back down the hall. The man's weight, a good 275 pounds, and the fact that his severely burned skin was sliding off in their hands, making it extremely difficult to get a grip on him, slow them down. The two women were each at least partially clothed. This poor son-of-a-bitch was stuck in the room when it flashed over, and all his clothing, as well as the first few layers of skin, are burnt off. They finally reach the open space of the living room. One of the members in the street has already brought up a resuscitator, and we begin to apply it as soon as we reach an open area. By now, every window on the second floor has been vented, as well as the roof itself, and since the hoselines have knocked down the fire, conditions begin to improve. I shine my light on the victim as the members work on him. I want to see if there is any way to make the task of removing him any simpler and faster. What the light and the improved visibility reveal is not a pretty sight. This man is a charred remnant of his former self. I say to no one's surprise, "He's a Code 1." Everyone knows what it means—he's dead. There's no chance of reviving him.

The two women are another story. We got to them before flashover, and though I know one of their lower legs was also burning, and both breathed in tremendous amounts of highly toxic products of combustion—smoke, carbon monoxide, hydrogen cyanide, and oxides of nitrogen—they were, as the saying goes, "only smoked, and not roasted." Later, we will find out that the first woman removed, Milde Gausintu, will live. The other woman, removed only seconds later, does not make it. That is the difference seconds can make in a fire: life and death. If it weren't for the radio blasting in Rescue 2's quarters, another woman would have died as well. A working smoke detector in the apartment would have perhaps blasted all three occupants out of bed in time to escape. A working sprinkler system in the dry cleaners would have saved everyone a lot of grief, including the store owner. The fire would never have killed those people and even the store would have suffered only minor damage instead of total destruction. As firefighting professionals, we know that's the answer, but we don't have the political clout to make it happen.

CHAPTER 18

The Troy Avenue Boys

It's a beautiful, young spring evening in 1991, when the days grow longer and warmer, and all seems right with the world—until 7:08 p.m. Rescue 2 is out buying the evening meal. The rig is parked on St. John's Place at Kingston Avenue in Crown Heights, and I'm monitoring the department radio while the rest of the troops do the shopping. Radio traffic is quite lively, as it often is on a warm evening, with a lot of car fires, brush fires, false alarms, and other routine incidents. I hear a box go out in the Canarsie section that sounds like it has some possibilities of being worthwhile, and I start to pay additional attention to the chatter that makes up everyday radio traffic. When I hear the Brooklyn dispatcher start to add companies to the initial assignment, due to the dispatcher receiving additional phone calls reporting the fire, I get on the handie-talkie and call the troops. "It sounds like a job out in Canarsie. They're filling out the box." In the store, the guys drop their groceries on the checkout counter and tell the clerk, "We'll be back later," as they head out the door. They're just climbing on the rig when the dispatcher calls, "Brooklyn to the Rescue, start out for Canarsie, East 95th Street, Avenues J to K. Fire in a private dwelling. We are now getting a report of a child trapped." I acknowledge as the rig is already pulling out into traffic. It is about a 10-minute ride from Crown Heights out to Canarsie, and we pull into the block just as Ladder 170's outside vent firefighter comes out the second floor front window with a horribly burned baby.

The main body of fire was on the second floor, with some extension to the attic, but Engine Company 257 has beaten it down now, and the trucks have completed the primary search. The only victim is this 18-month-old child, who is now being given mouth-to-mouth resuscitation, "the breath of life," by her rescuer. Still climbing down from the cab of our rig, I call the chief on the handie-talkie, "Rescue to the 58, can we do anything for you?" Chief Frank Montagna of the 58th Battalion replies, "We've got the fire, Rescue, but see if you can help 170. They've got a 10-45 burn victim." Lee Ielpi is already at the back of the rig, about to get dressed, and instantly he springs inside the crew compartment and grabs the burn kit and resuscitator, then sprints down to the front of the building where the member from L170 has just stepped out of the Tower Ladder basket at ground level. He lays the child on the lawn to continue his work and make room for others to assist. The 58 Battalion aide has requested an ambulance from the dispatcher, and an ETA. Quickly, we begin treating this poor child's injuries, pouring 1000cc bags of sterile water over the baby's charred skin to cool the injury and flush off contaminants. We've got an infant Bag-Valve mask hooked up to the resuscitator now, and we use it in place of mouth-to-mouth. Not only does it do a superior job, supplying nearly 100% oxygen to fire-seared lungs, it reduces the risk to the firefighters of catching contagious diseases. The members work feverishly over this little baby, still cooling the broiling hot skin, performing CPR, and checking vital signs. It appears that this kid might have a chance to survive, if only an ambulance would show up to get her to the hospital.

The 58th Battalion aide calls over the handie-talkie, "58 Alpha to the 58, Chief. EMS says we'll get the first bus available but they have no ETA." That means that there are no ambulances available right now, and as soon as one gets free, they'll send it. But it could be quite a while, and it could end up coming from quite a distance. (Note: This is actually an improvement from the way things were with EMS. I know many times on a busy night in the late 1970s and early 1980s it was not uncommon to have a three-hour backlog for ambulances for even the most life-threatening cases!)

I make a quick estimate of the situation. This kid is showing signs of life. A pulse has been detected and there are occasional coughs between ventilations. I ask Chief Montagna for permission to transport the child to the hospital in the back of the Rescue truck. He knows the severity of the situation, and with no estimate of when a real ambulance will arrive, says to go for it. The rig is on the corner, 150 feet from our position. The entire company joins in a concerted effort to keep this kid alive until a higher hand can intervene. Lee races back to the rig, starts it up, and with the aid of a police officer, backs it out onto Avenue J, pointing towards the main street, Remsen Avenue. At the same time, a short backboard is slid under her, and one man carries the baby strapped to the board while another administers ventilation. The rest carry the resuscitator and burn kit, and continue dousing her burns. At the rig, I make an unusual decision. Normally the officer always sits in the cab, controlling the siren, monitoring the radio and giving directions to the chauffeur. In this case though, I feel I might be needed more in the back with the victim. I used to be an advanced emergency medical technician, which on this particular tour made me glad for all the time I spent in training. I knew Lee didn't need me up front, but this baby might need me in back. So I told a firefighter to take my place up front, to sound the siren and air horn, handle the radio, and act as a second pair of eyes, watching for kamikazes trying to ram us, as we began our possible life and death run for the hospital.

The ride in the back of the rig is a lot different than it is up front. To begin with, we're all forced to stand, clinging to overhead grab rails, as the rig bobs and weaves its way around traffic, in order to continue to work on the baby who is laid atop the deck above an inside tool compartment. The bouncing around we take makes it difficult to perform delicate tasks, like trying to monitor this tiny one's pulse. Also, the limited view out the side windows makes it difficult to gauge our progress, so I am unable to tell exactly how far we've gone, or judge how much further we have to go. This leads to frustration. Finally, after what seems like an eternity, we arrive at Kings County Hospital. Brookdale Hospital was actually closer to the fire than Kings County, or KCH as it's known, but KCH has a very good burn unit, which Brookdale lacks. Also, while Brookdale is very good with many injuries, especially knife and gunshot wounds, KCH is *the* place in Brooklyn where seriously injured cops and firefighters are brought. I figure if it's good enough for us, it's what this kid needs now.

As we pull into KCH, the entire loading operation goes in reverse, and respirations are continuously administered as we clamber down out of the rig and up the walk into the emergency room. Lee has called ahead via radio, and the dispatcher has alerted the ER staff, who wait outside the door for us. Quickly they relieve us of our charge and begin the process of advanced life support. An airway is placed to ensure the trachea does not swell closed and suffocate the child, while intravenous tubes begin the process of replacing the body fluids that are literally pouring out of all the burned flesh. The firefighters continue to do all they can, pouring sterile water, restraining the child who is now regaining consciousness, squeezing the bag-valve mask that is now hooked up on the express track to the lungs via the airway, until finally the ER is overcrowded. The staff of KCH finally has to shoo us away, saying that we must clear out in order to let them do their job. "Hey, we know how it feels. We know you want to help; you brought her this far, but now you have to step outside. Thanks."

The ride back to quarters is a solemn event, almost as dreary as a funeral itself. If we were just allowed to stay and watch, we'd all feel better, but of course that was not possible. Not only would it have impeded the ER staff, who are busy enough without a bunch of firefighters underfoot, but there are over two million more people in the rest of Brooklyn that might need your services in another second's notice. It's time to go back in service.

The resuscitator and all its parts are just laid up on the shelf where the baby was laid out before. The back of the rig will require a major cleaning, as everything is covered with blood and body fluids. I tell Lee to stop by a bodega on the way back to quarters, so we can pick up some bleach to disinfect the resuscitator as well as the back of the rig.

We pull over to the curb on Troy Avenue, at the south side of Eastern Parkway, where there is a fairly large, well-stocked bodega. The four men in the back decide to finish the shopping for the meal here since the place is open, and chances are the place we were shopping at earlier is already closed. They go inside together, leaving me and Lee in the cab discussing the baby's chances of survival and possibility of living something like a normal life if she does make it.

On the corner is a large group of young men horsing around, bragging quite loudly about their latest exploits: who had hit who the hardest, who had bedded who, and passing around a large bottle of either gin or vodka. We had been sitting in the cab for four or five minutes when one of the group decided that the firefighters look like they need a hit from his bottle of magic elixir. He proceeds to come over and offer it to me, "Hey 'fireman,' how you doin'? Want a sip of our gin? It's goooood!" "Nah, no thanks," I say, "We're on duty, we can't drink." "Oh, go ahead," he says, "Ain't no one gonna know. Come on, hang with the Troy Avenue Boys." He is now joined by several more of the group, each saying something like, "Yeah, man, have a drink with us." Again I refuse as tactfully as possible. I don't want to insult this large group, for to do so could mean that we and the other companies will be kept up all night chasing false alarms and minor fires set by people who felt slighted that the firefighters think that "'they're' better than us."

Suddenly a woman in a van pulls up next to Lee's window and yells to us. "They're stealing your tools out of the back of your truck!" Instantly it hits me. This is a set up! These guys are keeping our attention occupied so someone else can rip us off. Lee has jumped out of the cab and is heading for the back. Before I get out, I call the rest of the guys on the handie-talkie and tell them to get out here, we've got a problem! For a moment I look at the department radio, and I am about to call the dispatcher to get the cops here, but I look at the odds. There are at least 15 of these young men, 17 to 25 years old, with an unknown number and type of weapons, against just six of us, and we're split up and unarmed. The cops would probably arrive in time to keep our bones from being picked clean by the rats. No, my only hope is to try to talk our way out of this mess. I jump down from the cab and start toward the back of the rig. I see a guy walking swiftly away from the rig, carrying two sets of irons, Halligan Tools and axes, which are mounted in brackets just inside the rear of the rig. Lee is trotting up behind him, determined that he not get away. I see Lee's arm go up. He's about to cold cock this guy, so I yell, "NOOOOO, Lee!" All I can picture is how badly we are about to get our asses kicked if Lee hits this guy. The gang of youths has swarmed past me now, towards the two. I see the leader, with the gin bottle raised high over his head. As he rushes at Lee and the thief, he swings it in a vicious arc that connects squarely with its target; the temple of the tool thief's head! It drops him like a sack of potatoes, without breaking the bottle, and only spilling a few drops. Now the mob has set on the man, each one kicking and punching him mercilessly. "Hey, fool," they taunt him. "Why you stealing from the firemans? They's the only ones in this city that's here to help us. Don't you never let us catch you stealing from them firemans again! You hear?" The leader hands over the Halligans, just as the rest of the Rescue 2 men run up. They don't know what just happened and are apprehensive, as well they might be. It's only when they see the look of relief on my face, and the smiles and pats on the back for Lee and me from the group surrounding us, that they realize everything is under control.

My heart is pounding as I haul myself back up into the front of the rig, from which only minutes ago I had alighted, fully expecting to get the beating of a lifetime. I see the group of youths return to their corner turf, boasting about who hit the thief the hardest before they sent him on his way with a lesson he won't soon forget. They are still passing what's left of the gin around, and suddenly my mouth feels as dry as the Sahara. I would like to join them in a toast of thanks, but that's out of the question. I keep thinking about how that situation went totally 180 degrees from the way I was envisioning the outcome as it swirled around me. It shows you can't judge a book by its cover. This has been one hell of an unusual tour so far but it's only 9:00 p.m. There are 12 hours left before the change of tours tomorrow morning. An awful lot can happen in 12 hours in a borough the size of Brooklyn with its incredible cast of characters.

CHAPTER 19

The Soundest Sleeper You Ever Met!

Every fire company in the City of New York has its own particular personality, its own mode of operating, its own reputation, and its own soul. The soul of Rescue Co. 2 is depicted in its mascot: the most tenacious bulldog you have ever seen. This tenacity is the reason the company has the excellent reputation it does. They have earned it the hard way, with blood, sweat, and tears.

Unfortunately, this kind of tenacity, this devotion to completion of the mission, earned the company the enmity and animosity of some of the other companies in the department, even ones that had never worked with the company. As a result, there were many firefighters and even company officers who had "attitudes" toward the company and resented it when the company showed up at a fire or rescue operation. These other units generally saw the rescue firefighters as "competition"—someone that was trying to embarrass them, or show them up—instead of seeing what the rescue was really trying to do: to embarrass or show up the fire. We always thought that the fire was the enemy, not fellow firefighters.

At times, the reputation of Rescue 2 caused the officers of the unit some headaches, which I'm sure all of us would have preferred to do without. We had some companies that would actually try to keep us out of their operating area at fires, telling us "we have enough help in here, we don't need you," even though the primary search was nowhere near complete yet. Shortly before I was assigned, one of Rescue 2's firefighters had actually gotten into a shoving match with an officer from another company and was transferred out of the unit as punishment. The officers must react quickly to this attempt to keep us from an area: usually one officer to another. Most of the time, I could make them see the light: "Look, guy, we're only doing what the chief told us to do. Either let us in to do our job or go tell the chief why you don't want us in there." That usually sufficed, although even then I'd get resistance, and it would be decision time. I would get an idea of what conditions were like in the area, whether the searches were going to be done soon or not, how much fire remained, and what the occupancy looked like: easily searched or heavily loaded. If it appeared that the area was lightly loaded, the fire was pretty much under control, and the company did in fact have enough help on hand to complete the task in a reasonable time, then I'd back the troops off. On the other hand, there were times when conditions didn't look as good to me as the officer trying to keep us out perceived them to be. Then I would just ignore him, do what had to be done, leave, and thank you very much.

Of course, there being five firefighters in a rescue company and only one officer means that there are times when the officer will not have all of their firefighters working right alongside of him or her. That is one of the best features of a rescue, though, from a chief's point of view. The rescue arrives at a critical point in many fires, just as all of the first alarm units are committed. In large fires, the chief may need information from a large number of remote locations to decide where and how much help to deploy. The rescue companies' officer and five firefighters are all very experienced, radio-equipped firefighters. This is not always the case with the ladder companies, where in the 1980s and early 90s two of the members were not radio-equipped, and the crew could very well have two or three members with only a year or two of experience. So it is that when the rescue company arrives, very often the chief in charge will order us to split up and give them reports from different sections of the building, or the adjoining buildings (called exposures since they are exposed to flames from the original building).

One evening in the winter of 1991–92, we decided to pick up our meal at a new location, instead of the usual Crown Heights meat markets or bodegas. A new supermarket had opened up in Bushwick, located near the Brooklyn-Queens border. It was a very large store, operated by one of the national chains, and resembled the suburban stores that many of us have come to get used to, with a large selection of meats and even a seafood section: and most importantly, it was open late on Sunday nights. Many of the local stores in Crown Heights and Bed-Stuy had taken to closing soon after dusk due to the rampant crime in their neighborhoods. A person walking home after dark with two armloads of groceries is an easy "mark" for a robber. This new store had a large suburban style parking lot in front that was patrolled by armed guards, so shoppers could feel safe even after dark and drive away in the relative safety of their cars. It also meant that if we had a fire at the start of the night tour, that we would still be able to buy a decent meal by the time we finally got around to shopping at 9:00 p.m. or so.

On this particular Sunday evening, we had just picked up the meal and were headed back down Myrtle Avenue when the dispatcher transmitted Box 743 at Evergreen Avenue and Stanhope Street. This was a manual pull box, designated a DRB (discretionary response box), because of the tremendous number of false alarms that have been pulled from it. A DRB box gives the assigned battalion chief the discretion as to whether they will respond, or remain in quarters or other activity, while monitoring the radio to find out what the units discover at the box. Rescue 2 was not assigned to this box, but since we were located only about a quarter of a mile away, and it was on our way home, I made a comment to the chauffeur to watch out for the responding units. I also said, "head over in that direction."

No sooner had I said that than the 28th Battalion comes on the air and tells the dispatcher "We're taking in Box 743." I recognize the voice of Battalion Chief Craig Shelley. Craig is a newly promoted BC, covering in the 28 for a few weeks of an assigned chief's vacation. He had been a firefighter in Rescue 2 and had come back to Rescue 2 as a Lieutenant, although it was on a UFO (until further orders) basis. In fact, it was Shelley's spot that I had taken in Rescue 2 when he was promoted to captain. I am wondering to myself why he's taking in a DRB when the dispatcher's voice crackles over the radio "Brooklyn to the 28, you're getting 3 and 2, Engine 218's your third engine, we're receiving reports of a vacant building fire on Stanhope between Evergreen and Bushwick." As we come in line with a vacant lot on Myrtle Avenue, we get a view to the west out from under the elevated train tracks. A cloud of black smoke is clearly visible silhouetted against the gray winter sky. Well, now I know why Shelley was taking in the DRB—he must have seen the smoke!

As the first due engine, Engine 217, transmits the 10-75 signal, we are already rolling along Myrtle Avenue toward Evergreen Avenue. Turning into Evergreen, we stop just short of Stanhope Street. These streets are pretty tight, and we don't want to block out any needed ladder apparatus. The fire building is a three-story wood frame, vacant multiple dwelling, about 40' wide and probably 60' deep, and about three buildings in off the corner. There is heavy fire visible on all three floors, out the front door, and behind the sheets of galvanized steel meant to seal the building from the elements and keep out derelicts and vagrants. Engine 217 has a hydrant right across the street and is already operating its New Yorker Multiversal Nozzle through the front door while its crew is stretching a handline into exposure 4. The fire building is already fully involved, and the fire is spreading down the row of 10 other attached wood frame buildings by way of the common cock-loft beneath the roof. Ladder 111 is already beginning to work on peeling the razor-sharp sheets of "tin" off the windows in preparation for using their master stream on the upper floors.

I report in to Chief Shelley for orders, and I know exactly where we will be going even before he says it. The top floor of exposure 4 is a critical location that has to be searched. The interior team of Ladder 112 is already en route into this building, but fire is coming through the walls and ceilings at the first- and second-floor levels already, in addition to the fire in the cockloft. They are going to need assistance. As we scramble up the wooden stairs in exposure 4, the smoke is pulsing out of every crack on the stairs and in the walls, as though it were being pumped by a high-pressure compressor: which, in fact, it is. A fire's heat causes gases to rapidly expand, increasing tremendously in volume and pressure.

We reach the top floor and spread out, searching for overcome occupants and checking for any fire that has extended to the rooms. Our first acts are to locate and clear out as many windows at the front and rear as we can. (There are no side windows because of the attached buildings on each side.) This is

done not just to relieve the smoke condition, but more importantly to provide escape routes in the event fire blows down from the cockloft or burns
through the wooden wall separating the staircase from the roaring inferno
on the other side. An engine arrives with a hoseline, gets water, and we start
pulling ceilings in all the top floor rooms along the wall that abuts the fire
building. At first, the flames in the cockloft begin to blow down as the hooks
punch up through the ceiling, but the water stream pushes it back. The
problem is there is only one line up here yet, and we have fire in the entire
cockloft across six rooms. As the line "stuffs" the fire back in one opening, it
blows out of the remaining. We need more lines up here.

Chief Harry Rogers of the 57 Battalion, a very senior battalion chief and
a great fire officer, arrives to take command of operations in the exposure.
He quickly determines the severity of conditions and radios the command
post for help. Deputy Chief Driscoll of the 11th Division is in command
now and assures Chief Rogers that as soon as the second alarm units arrive,
he'll send another line to the top floor. As more of the ceiling is pulled, the
extent of the fire in the cockloft is evident; this fire is way past this building,
moving rapidly down the tinder box row of houses. Chief Rogers radios the
deputy to send the third alarm and forget that second line to our building,
send it to exposure 4A (the next building to the right of the fire building).
He then tells me, "John, take some of your guys and get next door. Let me
know if we can stop it there, or if we have to skip that building too."

I take Mike Pena and Bobby LaRocco with me and hustle down the
stairs. As we go, it is plain to see this building is lost. There is fire burning in
the stair treads and ceilings on each floor as we go down. I yell back up the
stairs to "watch your asses up there! There's fire breaking out into the stairs
underneath you." I would like to remain behind to be sure everyone heeds
the warning and gets out safely, but there's no time. Besides they're all big
boys who know how to take care of themselves—I hope.

The occupants of exposure 4A are already fleeing, trying to carry all
their worldly possessions on their backs. Anyone who has lived in Bushwick
for any length of time knows how fast and furious the fire spread in these
attached buildings can be. This area was virtually decimated in a wave of fires
in 1976 and 1977 that can only be described as a slow-motion firestorm.

Whole blocks of these wood frame firetraps were burnt out at a time, their occupants escaping with only the clothes on their backs if they escaped at all. Now, at the first sign of a serious fire anywhere in the row, occupants up and down the row can be seen carrying suitcases, TVs, and other small and not-so-small appliances (mostly their own, but not always, to safety).

As I reach the pitch-black top floor of exposure 4A, there is already a firefighter up there. I send Mike and Rocky toward the front to get those rooms searched and windows vented, but I tell them to hold off on pulling the ceilings yet. We don't have a charged line. Instead, I say, "Just make small examination holes, to see if there's fire there yet." As I move toward the back to get those windows, I bump into the lone firefighter, who is busy going from room to room with his hook checking that area. I ask him who he is, so that in case something goes wrong, I know who I have to account for, but through the mask facepiece all I get is a loud mumble. I repeat the question twice and still can't decipher the reply. I'm expecting to hear the number of one of the area ladder companies: one-twelve, one-twenty-four, one-eleven, or one-o-eight. Finally, as I shine my flashlight right on the front of his helmet, I can understand the shouted reply, "Chief Shelly!" Well, hell! No wonder I couldn't figure it out. Who expects to find a chief up in a nasty area like that, pulling ceilings no less? It seems that when the deputy sent him to go check out exposure 4A, there were no "truckies" available to go with him, so the chief decided to grab a hook off one of the parked aerials as he went by! From what we can tell, there is a good fire rolling along in the cockloft, but nothing like in the building I had just left. The chief has called for a line and I call my roofman, Billy Hewitson, and tell him to drop back to exposure 4A with the saw and get us some vent. If we get a line up here right away, and a couple of big holes in the roof, we just might stop this fire before we lose the whole block!

As Engine 222 brings its hoseline up to the top floor, I tell the troops to start getting the ceilings down, concentrating on the rear rooms where most of the fire is. Just then I get a call from the command post. "Rescue 2, get all of your men together and report to the command post." My mind races trying to figure out what the hell is so important at the command post that they're pulling us out of a critical location without anybody there to replace us. The only thing I can think of is that we are being relieved, sent home. I wonder to myself whether any of my men got into a "pissing contest" with an officer somewhere, and now the chief wants us out of the way to avoid further problems. For a moment I feel just like Capt. Morton of the USS *Reluctant*: the character Jimmy Cagney plays in one of my favorite movies, "Mister Roberts." Usually I identify with Lt. (J.G.) Douglas Roberts, Henry Fonda's character in that movie, but now I feel what Capt. Morton felt after returning from the Port Director's office: "Kicked out!" Kicked out! I feel a chill go down my spine as I exit the fire building.

By now, the third alarm units are arriving and are being put to work in the exposures, trying to hold the fire to the area that is already involved. Tower Ladder 111 and Ladder 112 have peeled much of the tin off the windows of the original fire building, and 111's and 124's towers and the deck pipe from Engine 217 have knocked down most of the heavy fire, at least in the front of the building. I report in with the members to the command post, not know-ing what to expect. Deputy Assistant Chief Donald Ruland is now running the show. I don't know much about the man, so I don't know what to expect. The chief comes out from behind the command board and tells me, "Lieu, I want you to take your guys and get into the fire building and give me a good search." Instantly, my opinion of this man is formed—he's fucking nuts! The wooden building in question has been burning fiercely for at least 15 minutes now, and it is showing the effects. We all know that a frame building exposed to heavy fire like this for 20 minutes is *supposed* to fall down. That's one reason why there are no hoselines inside it. Using the heavy deck pipe and tower ladder streams keeps your people a safe distance away from the collapse zone. They also throw about 6,000 pounds of water per minute into an already weakened structure, and that's 6,000 pounds per minute apiece!

The chief must have seen the look in my eyes questioning his sanity. I'm thinking, "I'd rather be kicked out at this point, than be sent into a vacant building that I *expect* to land on top of me." He quickly points out one of the sorriest looking homeless people I've ever seen, really bedraggled and soak-ing wet. The chief continues, "That guy just came walking out of the front door there, and he says there are between 20 and 30 people still sleeping in the cellar in there!" Now my expression changes again to one of just plain thunderstruck. "How the hell can there be 20 people in there?" I ask myself. For that matter, how the hell did that guy get out of there? The main body of fire in the front of the building might be knocked down, but there is still plenty of fire in the back of the building, and a heavy smoke condition pours from every opening that is not venting fire.

Turning to the rescue guys, I make a quick decision. I really don't like the idea of entering this free-standing pile of rubbish. I really think that there is a good chance of it falling down on top of us, burying us in the debris. I think back to my first year on the job, as a probie in 290 Engine, to two terribly similar situations: fires in vacant three-story, wood frames just like this. One fell down on top of Battalion Chief Frank Tuttlemondo and the members of Engine 227, killing the chief and severely injuring several members of 227. Just months earlier another three-story vacant frame fell on top of Engine 332, killing Lt. Robert Dolney and severely injuring firefighters Patrick Quinn and Steve Fillipelli. Now we have to go into this one to get 20 sleeping homeless people. I *really* don't like it!

I decide not to take everybody inside. If this thing falls down, I want somebody outside coming to get us right away. I turn to Pete Bondy, Mike Pena, and Billy Hewitson and explain my fears. I tell them I want them and the rest of the company to be *our* rescue team. I will take Rocky with me and do the search. If we find somebody, I'll call for them as reinforcements if needed, but I want them to start working on creating a second way out of that cellar in case we need it. Just to the left of the entrance stoop is a debris-filled areaway that leads down under the stoop. I figure this leads down to an exterior cellar entrance, and that's where I would like to enter from, but the amount of debris there will take 20 to 30 minutes to clear away. I tell the guys to start on it anyway, while Rocky and I head inside (fig. 19-1).

We feel our way along the entrance hallway for the stair to the upper floors. The flooring itself is very bad here; burned away in some spots and just rotted away in others. We try to move our feet along 16 inches at a time, duck-walking along on top of the somewhat more substantial floor joists. We find the stair on the right and proceed past it to the staircase downto the cellar. The door frames are still burning and there is fire up inside the ceiling joists, but we don't have a handline with us or the time to do

Figure 19-1. Fr. Bob LaRocco and the author head for a vacant frame on Stanhope Street in order to search the cellar for reported squatters.

anything about it. The good news is that there is not a lot of heat coming up from the cellar, only smoke, so the fire is probably only confined to the cellar ceiling areas. We should be able to get down and move along underneath it without too much difficulty, as long as it doesn't collapse on our heads!

Rocky and I make our way down through the still smoking stairway into the cellar. Here we find the water to be about three feet deep and rising, courtesy of the three master streams that have dumped 2,400 to 2,600 gallons per minute into the upper floors for the past 15 to 20 minutes. I remind Rocky to be extra careful. I don't want him or me to step into any holes beneath the water that could twist an ankle or even drown us if it were deep enough. Visibility above the water level is actually surprisingly good. The fire is, as I surmised, burning overhead, but the smoke is able to escape up through the porous floor. There seems to be a pretty good supply of fresh air as well, coming from the rear. Rocky and I come across numerous makeshift sleeping areas, from a king-size mattress and box spring set floating around, to a pair of hammocks slung from the ceiling. We can see how there might be 20 or 30 people sleeping here but find absolutely no sign of them still being here. At the back wall we find a three-foot by three-foot hole that had been cut through to the rear yard, probably the route that the vagrants used to get in and out with the tin covering the front openings. We do another sweep of the cellar as we return toward the front of the building. We hear Mike and the guys pounding on something that sounds like some very solid steel. Our primary search of the cellar is negative, meaning there are no more live human beings in this hellhole. There may be people beneath the surface of the charcoal black water, but we haven't bumped into them as we waded through it. Besides, if they're under water, they're dead! We'll find them during the secondary search, after the water is pumped out. Right now, I'm more concerned with preventing any further casualties, namely Rocky and myself. I give Rocky the word, "That's it for now. Let's get the hell out of here." We have been hearing creaking and groaning noises the whole time we are in the building. We have pushed our luck far enough. We scramble back up the interior stairs. I call over the handie-talkie to Mike to forget about the front door, we're coming out. Then I call to the command post, "Primary in the cellar is negative, we're on our way out."

At the command post, I report back to Chief Ruland and Deputy Chief Driscoll. I explain that it looks like the original report of 20 to 30 people in the cellar is inaccurate. There may well have been 20 to 30 people there when our informant passed out in his drug and/or alcohol induced haze, but they had all fled before he had awoken. It seems he was so unconscious that the noise of everyone waking and fleeing, the subsequent pounding of the "truckies" as they beat and pulled on the sheet metal covering the window and doors (which sounds a lot like thunder), and even the smoke and noise of the master streams pounding into the structure were not enough to awaken this dozing denizen of this dungeon. It seems that when the water level finally rose up to the two-foot deep level and reached his mattress, he finally sensed something was wrong. He then arose in the pitch darkness and smoke, stumbled to the stairs and made his way outside, emerging in full view of the command post staff.

I inquire of the chief as to the whereabouts of this character. "What do you want with him?" Chief Driscoll asks, thinking I might mean him harm. "I just want to meet somebody that can sleep through all that commotion," I say. "Maybe he has some tips on how to fall asleep when I'm wide awake. Besides, if he can take the kind of a feed he did in that cellar, maybe we ought to sign him up for a job." Chief Ruland gives me a questioning look, like. "What are you talking about?" Chief Driscoll gives half a grin, more of a smirk, but the rest of the guys around me get a kick out of it.

By now, the fire extension down the row of homes has been stopped and the situation is looking much better. There are three tower ladders in operation, "hydraulically overhauling" the fire building and opening up the fascia into the cockloft of exposures 4 and 4A. Handlines are stretched into exposures 4B, 4C, and 4D, but from the looks of things, the fire has not gotten past 4C. Fourth alarm companies are staged behind the command post, so there is plenty of help available. I ask the deputy if he has anything else for us because if not I'm going to take the guys back to the apparatus to change air bottles and take a blow. The deputy says to hold on for a second while he checks with Chief Ruland, the IC, but we should be able to take up. A minute or so later, he comes over and says, "The chief wants you to hang around awhile, just in case." I tell him we'll be back at the rig. The water in the cellar came over the top of my ¾-length hip boots a couple of times and my legs and feet are soaked. I say a small silent prayer of thanks that it's not the middle of February and five degrees out, as Rocky takes my mask to replace the cylinder. We spend the next hour hanging around, waiting for something else to go wrong and eating the melted ice cream we had bought before the box came in. Then another 10-75 comes in farther west on DeKalb Avenue, and we're back on the road again!

CHAPTER 20

The Thirsty Corrections Officer

Rescue 2 responded to an unusual incident that had some very special circumstances and left several of the members commenting on the length to which some people will go to save a buck. At about 5:30 p.m. in June 1992, a warm summer day, we received an alarm for what we call a "man in machine" run. In the past, we have had people with arms caught in printing presses, legs caught in rock-crushing machines, and many other severe entrapments. As these things go, this particular run was not particularly difficult or deadly, but it was rather ironic. You see, the run was to the Brooklyn House of Detention (the jail), where a person was reported entrapped. En route we were picturing all sorts of scenarios involving prisoners enmeshed in bars, while trying to escape, until the dispatcher advised us via radio that the victim was a corrections officer who had her hand stuck inside a soda machine! That sounded rather mundane compared to what we were expecting, but it sure turned out to be interesting! Engine 226, Ladder 110, and the 31st Battalion had arrived several minutes ahead of us and, having sized up the situation, began the normal removal attempts which quickly proved futile.

It seems that when the corrections officer (CO) had attempted to purchase a soda from a can vending machine and the machine balked at delivering. Rather than allowing herself to be ripped off by the machine, she decided to take things into her own hands (and arm). She heard the can begin to drop, and then stop. Squatting down in front of the machine, she inserted her hand into the slot where the cans fall. Reaching farther and farther up inside, she arrived at the spot where the reluctant can was stuck. As she attempted to maneuver the can out of its position, the operating mechanism suddenly slipped, pinning her wrist between two rollers. The rollers were crushing the bones of the wrist as the machine attempted to completely close, as it was programmed to do. By the time that we arrived, the first due companies had attempted to remove the victim by lubricating the wrist area with soap, a task made very difficult by the position of the victim inside a very tight space with her arm extended so far up inside the machine that her shoulder was wedged against the door of the machine. This attempt failed and caused the victim great discomfort. One member of Engine 226 was able to get his arm up inside the machine, parallel to the victim's, while holding a small screwdriver. He used this to try to pry the rollers apart but this too was unsuccessful. At this point, BC Craig Shelley, who was now covering in the 31st Battalion, called a halt to further attempts at removal pending the rescue's arrival. Chief Shelley's background in Rescue 2 prompted him not to risk further injury to the victim with less certain approaches.

This tour, I had some of the senior firefighters in the company working, all with good analytical minds: Pete Bondy, John Barbagallo, Dave Van Vorst, Tim Higgins, and Bruce Howard. Bondy is a registered nurse (RN). I took Pete and Bruce up with me to size up the situation, leaving the rest of the company to bring up tools after we got an idea of what we faced. One of the unusual aspects of the situation was that the area we would be working in was inside a prisoner area on the 10th floor of the building, which had to be kept secure. We had to be identified, pass through sets of double interlocking doors, and be escorted throughout the area. There would simply be no running back and forth to the rig for extra tools. Just about a month prior, we had built a new cart to carry our air-powered tools: air chisel, air drill, "whizzer" saw, bits, hoses, and two 90-cubic-foot air bottles, manifolded together to a regulator. After sizing up the scene, I called for the remaining members to bring up the air cart, as well as a Sawzall, extension cord, and tool kit. We also asked for small (4-inch × 4-inch and 4-inch × 8-inch) air bags. All the tools would have to be kept secure and accounted for before leaving.

One of our first thoughts was to open a similar machine nearby with a key and see exactly how that mechanism would behave if we were to disconnect power. Would the machine loosen or tighten its grip? But no keys were available on site and the vendor was located at least one and a half hours away. The correction officer was complaining of loss of feeling in the hand and wrist now, both due to the lack of circulation and the 35-degree temperature inside the refrigerated area where her hand was caught. Sensing that this situation had a potential for serious injury, including loss of use of the hand, I decided this was not the time for finessing. We would go with brute force.

We began by cutting through the lighted plastic front that constituted three-quarters of the front of the machine with the Sawzall. We did this with the electricity on since I didn't know what the machine would do if we disconnected power. If you don't know the answer, don't guess with someone's life or limb at stake: at least, if there is any other choice. Timmy Higgins then used the air chisel to cut through the inner steel door to the refrigerated compartment where the cans are stored. Since we were going to destroy the door no matter what we did, Tim made a nice, big hole about two feet wide and three feet high. This would give us plenty of room to maneuver tools inside. Even with this large an opening, however, we could only see the officer's fingertips from above. The sheet metal collecting chute that guides the cans from the various rows to the center discharge point was in the way. Tim started on it with the air chisel, but soon the size of the tool prevented him from getting an effective angle on several parts. Bruce saw the problem and stepped right in with the correct tool, again the Sawzall, fitted this time with a long, 10-inch metal cutting blade that allowed him to complete cutting the chute into several easily removed pieces. Once this was out of the way, we could see most of the entrapped hand. I asked Pete Bondy, one of two nurses working that day, to assess the hand for pulse, movement, and nerve sensation. Pete confirmed our suspicion. The hand was without sensation, although he could detect a pulse just above where the rollers gripped the wrist. I handed Pete my "officers tool," a modified miniature Halligan tool, to see if he could pry the offending rollers apart. No dice. But the fork end, turned sideways, did fit very snugly, acting as a chock to ensure the rollers couldn't close any farther. This stabilized the situation and allowed us to attempt other approaches.

Now that we had the rollers secured, we could experiment a little by killing the electric power. Hopefully, that might loosen the pressure on the rollers. At the very least, it turned off the refrigeration unit. No such luck on releasing the pressure, because this was an electromechanical connection, which works on 220V electric current to open and close mechanically. "How about trying to put more coins in and operate the machine again, causing the rollers to reopen?" you might ask. Good idea! "Plug it back in and let's try that," I said to John Barbagallo, who was acting as an overall scene observer and idea consultant. Still no dice. The machine couldn't complete the old cycle with the hand (and now the officer's tool) in the way, so it couldn't begin a new cycle. It looked like further disassembly was needed.

Again I turned to Pete, John, and Dave for ideas. Our new air tool cart was sitting right there. I was thinking of using the small air bag that can displace about 1,000 lbs. Dave wanted to try the impact wrench first, though. He thought he saw a clear enough access to the nuts that hold the end of the rollers in place to be able to get a wrench on them and back them off. Then, hopefully, the rollers would budge. Disassembly is usually better than brute force, where some of the forces you apply may have other, unplanned effects. Control is more important than sheer power in many circumstances. Now, a corrections department captain appeared with a Polaroid camera, snapping pictures all around us. I figured this might be good for future drill use in the firehouse and mentioned that we'd like a copy. "Oh yeah, no problem," was the reply. I also figured that the pictures might end up on an employee bulletin board, since the officer with her hand stuck was not at all happy that he was taking them.

Dave backed off two of the nuts on the end of the rollers that we could get to, but the rods barely budged, although there was definitely some more play. The officer's tool was looser, and the corrections officer said the pain was less sharp. "OK, let's try the bags." "Which one?" asked Tim. "Can you get them both in there?" I asked. "Sure, Lieu, I might be able to stack them," (place one on top of the other, which increases the distance that an object can be moved, but limits the force applied to the limit that the smaller bag can exert, which is under 1,000 lbs). "Try it," I said. No good. Still not enough force, although the rollers did seem to deflect. "Try them alongside each other," Pete suggested. Dave VanVorst was operating the inflator controls, and he looked for my decision. I nodded to him and he deflated both bags. That was a no-brainer since the first plan didn't work; we had to try something different. But it's very important in an operation like this that all commands be given by one person to avoid any confusion from conflicting orders that might, for example, have someone deflate the bags just as someone else pulled the chock of the officer's tool out of the way to reposition it. The sudden movement could cause further injury. Dave was just following protocol; we all knew trying them side-by-side was the next logical step. That's why Chief Shelley had been careful to address all of his comments and suggestions to me, instead of directly to these guys he has worked so closely with over the years. It's called the chain of command (fig. 20–1).

With Tim and Bruce carefully positioning each bag as near to the injured hand as they dare, I gave Dave the "open red" command, meaning to inflate the larger bag attached to the red hose. Movement! Still not enough

Figure 20–1. Tim Higgins—who was on the scene to assist a woman caught in a vending machine—prepares to enter a confined space with a full body harness.

though. Timmy said "Give us some more on the red." I nodded, but then the relief valve on the controller popped off with a loud hiss of compressed air, announcing the red bag was inflated to its maximum working limit. The air bags are like steel belted radial tires: on a car, as little as 32 psi of air pressure can support a car that weighs several thousand pounds. This is a factor of how many square inches of surface area are in contact with an object (in the car's case, the road; in our case, the rollers) multiplied by how many pounds per square inch of air pressure is inside the container (bag or tire.) The result of this calculation determines how much weight can be supported. Bruce called out, "Ready for black," meaning he had the smaller bag in position and was ready for it to be inflated. I called to Dave, "Up on black," and he slowly opened the valve for the black hose. "That's it," Bruce called and immediately Dave closed the valve, locking in the pressure where it was. Timmy quickly slipped a wooden wedge in where the officer's tool had been. The tool was no longer large enough to hold the rollers apart, should something unforeseen happen. Bruce and Pete aided the officer in pulling the arm out of the machine and then turned her over to the waiting EMS paramedics.

As she was taken away on a stretcher, several of her fellow correction officers who had been keeping up a steady commentary all along were heard inquiring, "What flavor did you want? I'll bring it to you in the hospital," and "Do you want me to get your dollar back?" "Hey, are the cops going to meet her at the hospital to get her for theft of services? I don't know, isn't that vandalism or something?" Firefighters, being the caring, considerate souls that they are, refrained from joining in on this verbal abuse—for about three seconds.

By the time we got packed up, escorted out to the free world, and back to quarters, it was 7:00 p.m. I reflected on how expensive that can of soda was going to be. Not only did it cost the City of New York an hour of overtime for this group of firefighters, plus the cost of the destroyed machine, the correction officer was likely to be out of work on medical leave and was on her way to Bellevue Hospital for examination by a team of the nation's best neurologists and microsurgeons. That was some expensive can of Coke! I hope it was worth it to this embarrassed officer.

A few weeks later, we were in downtown Brooklyn again, for a drill at the New York City Transit Museum on Jay Street, where we practice lifting subway cars off of people (usually attempted suicides). We used the air cart and the air bags again, and this prompted me to think of the soda machine incident. I said to John Barbagallo, "Stop by the jail, John. I want to see if we can get those pictures." I was led inside the fortress-like entrance to the deputy warden's office, who greeted me warmly and again thanked us for our assistance. I asked about getting copies of the pictures to use for our training, but the warden was suddenly guarded. "Oh, I can't do that!" he exclaimed. "Why not?" "Oh, they've been confiscated. They're evidence." "Of what?" I asked. "Is she in trouble over this?" "Oh, no," he replied. "But she's suing the state for the injury to her hand, calling it a line of duty injury." Don't that beat all!

CHAPTER 21

Sometimes, I'd Rather Be Lucky Than Good

Heavy timber truss construction has been one of the more deadly creations in the firefighters' world for many years. Timber truss roof failures have killed many firefighters, including six in Brooklyn in 1978 at the Waldbaum's Supermarket fire. The six who died were part of a larger group working to vent the roof of the burning supermarket when a single truss failed, dropping 12 firefighters into a blazing cockloft. Six of them were lucky enough to fall quickly through the flames and out the bottom of the ceiling, into the aisles below. They survived that nightmare. The other six were not so lucky and got hung up in the roaring inferno. Waldbaum's opened a lot of eyes in the fire service to the hazards of trusses and resulted in the FDNY's first written policy on how to deal with fires in this type of construction: stay well away if fire involves the trusses.

What makes the heavy timber truss in general, and the bowstring type in particular, so dangerous is that it is prone to sudden collapse, and the collapse produces a very large opening very quickly. Almost instantly, the failure of one single truss can open up an area 40 feet wide and up to 100 feet long. Anyone working on top of a collapsing truss is thrown into the flames below. Anyone working underneath it is crushed by tons of flaming debris. We also learned of a third way trusses kill firefighters in 1992, when a collapsing bowstring truss roof buried two more Brooklyn firefighters.

I am working in Rescue 2 that night, writing my roll call for the tour, when the voice of Battalion Chief Artie Lakiotes of the 40th Battalion comes over the department radio, "40 Battalion to Brooklyn, K." "Go ahead, 40," answered the dispatcher. "Transmit the box for 60th Street and Fort Hamilton Parkway." ("Transmit the box" means the officer arriving on the scene wants two engine companies, two ladder companies, and a battalion chief to respond.) "We have a fire in a vacant garage," continues the chief, which brings to my mind a mental image of a one-car, wood frame garage— a fire that can look pretty serious, but which is well within the capability of two engines and two ladders. ("Transmit the box" is one step below a 10-75 signal in FDNY policy.) If the fire escalates at all, Rescue 2 goes, so I pay attention to the radio traffic, even though I say to myself, "Well, the chief is already on the scene. He knows what he needs." I figure that if a lieutenant had asked for the "transmit the box," the chief might arrive and upgrade it to a 10-75, but he must know what he's got.

A minute and a half later, Engine Company 247 comes on the radio and the officer transmits a *screaming* 10-75 for the box. You can tell from his voice that this guy is excited. I mention to Mike Esposito as we head out of the kitchen for the apparatus, "It must be that guy's first night as a lieutenant," thinking he doesn't realize that the chief is already on the scene, and he is reacting to a well-involved one-car garage. As the rig turns left onto Bergen Street, I mention to Billy Hewitson, the chauffeur, "It sounds like a one-car garage fire, Billy," knowing that he'll know that there's not enough fire in a one-car garage for there to still be anything left burning by the time Rescue 2 rolls halfway across Brooklyn, and Billy will be very careful as we respond. As we cross the first intersection, Troy Avenue, Battalion 40 radios the dispatcher, "Transmit a second alarm for box 2688. Have an additional tower ladder respond, and come in off 60th Street onto 10th Ave." I look at Billy as he glances at me. So much for the one-car garage theory!

Fort Hamilton Parkway and 60th Street is a long run for Rescue 2, down Empire Boulevard, right past the Brooklyn dispatchers' office, left onto Ocean Avenue, a right onto Parkside Avenue, around the traffic circle, across Ocean Parkway, and onto Fort Hamilton. As we cross McDonald Avenue, Ladder 149 comes on the radio, "149 to Brooklyn, urgent!" "Go ahead, 149," comes the reply from Warren Fuchs, one of the sharpest, most senior dispatchers on the job. "We've just had a major collapse and we've got firefighters buried! Transmit a third alarm, give us two more rescue companies and a bunch of ambulances!" The message turns my blood into ice water. "Holy shit," I mutter as I turn to call back through the pass-through into the crew compartment. "You guys heard that, right?" I verify. Pete Martin is closest to the pass-through. "Pete, you come with me when we pull up," I tell him. "Have Espo come too. Tell Terry (Coyle) and John (Kiernan) to stay with Billy until we see what tools we'll need." "Right, Lieu," is all the reply needed. This is a good, solid crew tonight. I feel confident we'll be able to handle anything we encounter.

As we approach the scene, the 40 aide is calling us on the handie-talkie, and directs us into a spot on 61st St., just south of a pile of smoking and flaming debris. Several large, heavy timber trusses are visible in the pile. Chief Lakiotes is standing just behind a crowd of firefighters that resembles a pile of ants swarming over some crumbs at a picnic. He sees us pull in and waves for me to come meet him at his command post. There are two tower ladders, 149 and 172, blasting away at the shell of a building that looks to have been about 60 feet deep and 100 feet wide: up until about four minutes ago, anyway.

"Rescue, we had a wall and roof collapse here that buried a couple of guys. We got two guys out so far, and they're on their way to Maimonides [Hospital]. I've got a roll call started, and so far, it looks like they might have been the only two we're missing. Just to be sure though, I want you to do a real thorough debris examination. Make sure we don't have anyone else under that mess." "Yes sir, chief. Where were these guys when it collapsed and what unit were they with?" "Engine 247. They were right over here, setting up a Stang [a large portable master stream nozzle that can apply up to 1,000 gallons per minute] in the street when the wall just blew out. It buried both of them." "10-4, chief. Did anybody see anyone else nearby?" "No, just those two so far, but look anyway." "You got it, chief." I turn to Pete and Espo, who heard most of the conversation and say, "OK, that's the deal. We're doing a secondary search here, really thorough." Espo sees some obvious void spaces in under the remains of the roof and suggests we start there. That will require shutting down the two tower ladders that are working away at the smoldering ruins. I turn back to the chief and tell him our plans, but he quickly cancels that idea. "There was no one in the building or on the roof when this thing came down. We were doing strictly an outside operation. I knew it was coming down and thought we had everybody back far enough. That's why they were setting up the Stang, so we could stay way back, but that wall caught them. All I want is to make sure there's no one else under this wall right here," he says, pointing to a 100-foot long × 25-foot wide pile of smoking cinder blocks, brick, mortar, and splintered timber. "OK, chief. We got it."

Turning back to Espo and Pete, I call Billy and the rest of the company over. I brief them all on what has happened, what our orders are, and how I plan to approach it. The masonry rubble is pretty well broken up. There are some sections of wood roof mixed in with the brick and block. We shouldn't need any heavy lifting tools or even the jackhammers for this job. It's strictly bull work, moving handfuls or shovels full of debris at a time: long, arduous, physical labor. Our normal complement of firefighting tools will suffice for the most part: Halligan tools, axes, hooks, and the wood cutting saws. With the third alarm companies in, plus two more rescue companies, we'll have plenty of manpower. We just need to organize things. Rescue 4 and Rescue 5 arrive shortly, and I confer with their officers to let them know the game plan, then go back over to the command post to get the bosses' OK. Since we have no specific information pointing us to a particular spot, there's no point in doing selected debris removal. The first and second alarm companies have dug through all the debris around where the first two firefighters were removed. We're going to have to start at one edge and simply pick every brick, stick and block out of the way, until we move it all away up to the remains of the building wall (fig. 21-1).

We start clearing rubble from the near side of the street—a line of about 50 guys, passing debris by hand back to other guys, who deposit it in a pile on the far side of the street. We have cleared about 10 feet in depth along the length of the wall when the Deputy calls off the operation. Roll calls of all the units on the scene have been conducted twice, and both have verified that the original two victims were the only firefighters involved. We're

Figure 21-1. Rescue 2 pauses after the collapse on 60th St. From left are Terry Coyle, John Kiernan, Mike Esposito, Lt. Norman, Pete Martin, and Billy Hewitson.

ordered to "take up." The tower ladders will continue soaking down the ruins for a while longer, but everyone else is sent back in service. I call the dispatcher to go 10-8, but we are directed to take in a 10-75 on Livonia Avenue in East New York: the far opposite side of Brooklyn. Back up Fort Hamilton Parkway we go, heading for Linden Boulevard. About five minutes into our trip, Battalion 44 gives his preliminary report. "We're using all hands. We have a fire on the second floor of a two-story brick 25-foot by 60-foot, semi-attached private dwelling. One line is stretched and operating. The second line is standing fast. Trucks are opening up. Primary search is complete and negative. Probably will hold. You can 10-2 the Rescue." "10-4, 44, all hands operating, probably will hold. Brooklyn to the Rescue, you can remain in service," answers Warren Fuchs, dispatcher 120, who is an expert on all things Brooklyn and fire. The rest of the evening continues at a rapid pace: a 10-75 in Williamsburg, an automatic alarm at St. John's Hospital, a car accident on Utica Avenue and Kings Highway, and another 10-75 in East New York, but nothing of any consequence.

Around midnight, we go back over to the 40th Battalion for an all hands in a Queen Anne style private house off New Utrecht Avenue. The fire began in the basement and extended up the balloon frame walls and interior stairs to the attic, three floors above. We split up. Espo and Billy go down into the basement while the rest of us go to the top floor to assist with the searches and opening up. The engines have done a good job pushing into the basement and getting two more lines in operation on the upper floors. Ladder 109 has taken a pretty good beating during the primary search, so we do the secondary. There was not really much fire on the upper floors. The trucks opened the walls at the top floor real early, and the engines got water going into the hidden voids before the fire got a good head start, so we finish the search and head downstairs. In the street I report that, "The secondary (search) is negative on the first and second floor and attic." Chief Lakiotes thanks us and says, "You can take up, Rescue." I ask him how the guys are from the collapse. "They're pretty broken up, but they'll live. One guy's got a broken arm, broken ribs [a punctured lung was discovered later], and some pretty bad face burns. The other guy's got a broken leg, arm, and collarbone." "No shit? They're really banged up." "Yeah, but they're lucky. You should have seen that wall come down. It was like an explosion." "What happened?" "John, I'll tell you, I've never seen anything like it. We were driving by, delivering the bag [inter-departmental mail], when I saw smoke coming out of the side door. I told the aide to pull around the block and asked to transmit the box. This was a vacant auto dealership that we'd drilled on, so I knew the building. I went in the overhead door and saw five separate rubbish piles burning. If I'd had even just a booster line with me, this would've been an 18 [a one-engine and one-ladder operation]. As I was waiting for the rigs to arrive, one of these rubbish piles really took off. It had a couple of tires in it. The fire climbed up and hit the underside of the truss roof, then ran from one end of the building to the other in under a minute. Engine 247 was just coming around the corner and fire was blowing out the door on 61st Street. They gave the 10-75. I tell you, this was impressive. In under a minute, we had fire out every opening. That's when I sent the second. The building went from a couple of rubbish fires, to on the ground, in under 10 minutes! Amazing!" I wholeheartedly agree with the chief; that is pretty amazing. What's even more amazing was our next truss fire on Richardson Street in Red Hook, about three months later.

In late summer of 1992, an evening thunderstorm rolled over the city, right at the evening change of tours at 6:00 p.m. The sudden downpour and large number of lightning strikes, coupled with the high winds, generated several back-to-back runs for Rescue 2 and other Brooklyn units; power outages caused automatic fire alarms to malfunction, while power lines came down and trees were blown over onto houses.

Lee Ielpi and the rest of the boys in Rescue 2 take it all in stride. The hectic pace of going from one run to the next comes and goes, almost as quickly as the storm itself, and by 9:30 p.m. we're back in the firehouse eating dinner. The responses from 6:00 to 9:00 p.m. have prevented us from shopping; all the supermarkets are closed in Crown Heights by 8:00, so we have decided to get take-out food. Very often when this happens, we end up over on Seventh Avenue in Park Slope, where there is a pretty good selection of restaurants in the "yuppiefied" neighborhood that usually stay open until 10 or 11 p.m. on weekends. Instead of ordering one meal for everyone, usually we "roll our own," with each person choosing whatever they like, from Italian dinners to Chinese, American, or even Thai restaurants. I usually go with shrimp from a joint called the Grand Canyon. We all get our meals to go but Lee talks me into getting a slice of pizza while we're waiting. Tonight, for a change, we get our food without having to tell the restaurateurs, "We have to go! Hold our food 'til we get back," as so often happens. We even get to eat it in peace. That's really something to comment on. After the early evening spurt of activity, we expect something more to happen. It does at 1:45 a.m.

The teleprinter in the kitchen goes off with a surprising message: "Verbal 10-75, Box 1376, Richards Street and Seabring Street." Surprising, because we hadn't heard any radio traffic prior to it, and because the 32nd Battalion is one of the slower battalions. A "verbal 10-75" means a fire right near the firehouse. Lee, as usual, is sitting in the cab even as I make my way down the stairs. Out we go, down Albany Avenue, onto Eastern Parkway, straight across the Grand Army Plaza, to Prospect Park West, right onto Ninth Street, and across the Gowanus canal into South Brooklyn. As we are turning onto Ninth Street, the 32nd Battalion aide gives his first progress report, "We're using all hands for a fire in a one-story, 150-foot × 100-foot warehouse with a timber truss roof. The fire appears to be in the roof. We're having a hard time getting water on the fire. Request two extra trucks." We continue along Ninth Street onto Hamilton Avenue, around the Battery Tunnel, onto Richards Street. As we come down the block, the entire street is blanketed in heavy smoke, the source of which appears to be a large warehouse diagonally across the street from the quarters of Engine 202, Ladder 101, and Battalion 32.

I am surprised when we pull up to the building to see two hoselines stretched through the front door of this building. I was anticipating seeing Ladder 131's tower ladder in operation in a defensive outside attack: given the fact that the building is known to have a timber truss roof and knowing the history of such roofs. As I report in to the chief at the front door, my surprise deepens further. I can look in across a large open area at street level that is pretty much free of smoke. That means all the fire is definitely in the cockloft, right up with the trusses, which is not a good thing. There are two or three teams of firefighters inside working with 20-foot-long hooks, which are heavy, cumbersome tools, trying to punch holes through a metal ceiling 20 feet over their heads. The hoselines are standing by, unable to put a drop of water on the seat of the blaze. This really makes me nervous, since the response time from Rescue 2's quarters to here is at least 10 minutes. That means the fire has been free-burning for at least that long—which is not a good sign in a truss roof building! The chief tells us that the fire seems to be a relatively localized fire in the truss area, but they can't get water on it. He wants us to look for another approach, since the hooks are only able to punch tiny one-inch diameter holes in the sheet metal. I suggest positioning the tower ladder in front of the open garage door, and extending the basket inside the building, to work on the ceiling out of the basket. The chief says he's already working on that, but he has to move several other rigs, including the first pumper that is supplying the handlines, to do so. "Try to find something else," he orders. I brief the troops and split us up into three teams, two inside and one outside, to look for access into the trussloft. Lee and I head for the back right side of the place, where there are a couple of small offices nestled in against the back wall. I call for a 16-foot extension ladder to be brought in off one of the ladder trucks.

We can see that the roof of the offices is used for storage: a bunch of old car parts, and tires. If the roof can hold that, maybe it can support a ladder that we can place up against the wall near the metal ceiling. Sure enough, the office roof is decked with plywood. We put the ladder up, and Lee starts to wail away at a seam of the sheet metal ceiling with his Halligan. Soon, he has torn and bent away enough of the steel to create a sufficiently large hole to allow a person to squeeze through. Standing on the top rung, he hoists himself up through the opening into the trussloft. I ask if that's the right location, if we can reach the fire from there. Lee replies, "I don't know, Lieu. I can hear it, but I can't see shit. There's no floor up here. I'll have to try to walk the trusses to find it." I call for a handline to be brought to our location anyway, with about five extra lengths of hose so that we can cover the whole area. This thing's been burning for too long as it is. I don't want to have any delay, if we can avoid it. Little did I know how long it had been burning!

I clamber up to join Lee, to see for myself what this looks like. As I enter the hole, I have to ask myself if I was really doing this. Was I really climbing up into a trussloft of a heavy timber truss roof that has been burning for at least 15 minutes, knowing that the thing could collapse at any minute? Here I am a Lieutenant in Rescue 2 of the FDNY, a con-tributing editor for *Firehouse Magazine*, someone who has written articles about the dangers of truss roof building collapse, someone who obviously knows better—and yet here I am, not only *not* taking a defensive exterior position, but actually climbing up into the truss area! I think for a second of my own reaction when I first heard of the deaths of five firefighters from Hackensack, New Jersey two years earlier, as a result of the collapse of the truss roof of the Hackensack Ford dealer. I remember thinking then, "How did those guys not know that was going to fall down, and what were they doing there?" Now I find myself doing the same thing. As I sit on the edge of this massive timber truss, my legs dangling out of the ceiling below, waiting for the nozzle to be passed up to me and Lee, it dawns on me that this is different.

The smoke here is very thick, with a real creosote sting to it, but it is not very hot. I call for someone to pass up a six-foot hook. Taking the hook, I stand upright on the truss and thrust the hook skyward. I find the roof sloping upward to my left. I move in that direction along the truss, up to about 40 feet from the wall, where the hook no longer touches the roof. Now I can tell which direction the roof pitches, and I know that there is no fire that we can make out down low along the ceiling line. I call for Lee to follow me with the line. It's time to go hunt this thing down and kill it. Espo follows me up the ladder into the trussloft. The rest of the company, and the engine who had stretched the hose line, start to follow but I tell them all to stay down below on the office roof. There's no fire where we are, so they should be safe there in the event of collapse. I want to use the minimum amount of personnel with the maximum amount of supervision: me and Lee! We move out to the middle of the truss, which I recognize by a huge steel rod about 2½ inches in diameter that descends vertically through the smoke and penetrates the middle of the bottom beam that Lee and I are crawling along. I recognize this as the center support of a king post truss; in effect it pulls the top and bottom of the assembly together—a giant bolt that keeps the ends from pushing out. I know now that the highest point should be right directly overhead: the spot where the most heat should be. We still can't see any fire yet, but we've definitely gotten closer. The crackling gets louder, or the fire gets much bigger. I tell Lee to open up the line for just a second or two, directed at the ceiling just overhead and off to our right, to see what kind of reaction that produces. We must be careful not to create too much steam (which could scald us in this unventilated space), but the solid tip nozzle helps to minimize the chances of that happening. No reaction occurs when Lee opens, then shuts, the nozzle. We're not under it yet. I estimate by the sound that the fire is still off to our right, but the question is, how do we get there? The trusses in a building like this are usually about 20 feet apart.

Spanning the trusses at the ceiling level are joists, two by eights, that support the sheet metal ceiling. I can't see this, but I know there has to be some support for the ceiling, and I remember squeezing between some beams as we climbed into the trussloft. I probe out in front of us with the hook. Sure enough, there are the joists below; now the trick will be to crawl out along them to reach the next truss. This is not as easy as it sounds. We still can't see shit, even with our lights on. The joists run parallel to the direction we want to go, and they're spaced about 16 inches apart. We have to crawl along them on our hands and knees, while dragging the hoseline and praying that our weight is not the straw that breaks the camel's back—snapping one of these old joists and plunging us 25 feet to the concrete floor below. I tell Lee to stay on the first truss, while I go across, so we don't overload any joists, but he tells me no. He's got the nozzle and he should go first, while I stay where I am to feed him hose. He's right again, as usual. I wait until he calls that he's across before I start over. Lee calls back that we're definitely heading in the right direction. It's getting hotter and louder. I call for Espo to come up to the position that I had just left to feed hose. Lee tries the hoseline on the ceiling again, and this time we can hear a slight reaction, a little hiss and maybe a little less crackling. We're headed in the right direction, so we move up another set of trusses.

We hear the screeching of power saws really close now. Somebody is cutting the roof over our heads. That's good news for us as it should improve visibility, although I'm not sure it's a good tactical move, since I don't know how anybody could be standing on this roof with its steep pitch, and I don't think anyone *should* be standing on the roof with its potential for truss collapse. But then I remind myself that Lee and I are actually up here in the middle of this thing! Who's crazier? We crawl forward to our fourth set of trusses when suddenly the fire appears directly overhead. The ventilation has let the smoke lift and given the fire fresh oxygen, so it really starts to take off. It's a good thing we're right there with the line. Lee opens the nozzle and starts to work on the fire. The scalding hot water cascades back down on us. With mobility being limited to staying on the joist, we are prevented from moving out of the way, so there's no choice but to stay there and take it and keep the nozzle going continually to cool the overhead. After about two minutes of this, the water is much cooler as it runs down our necks, so Lee shuts down the nozzle and we listen.

There is silence except for the sounds of our breathing. That's good. No more crackling means no more open flame. Just then Lee's vibralert (low air alarm) goes off and mine follows three seconds later. We're running low on air. I call Espo and Terry Hatton to relieve us so we can get down before we run out of air. The hole in the roof is letting the smoke out, but it's slow. This building is very big. I radio to the deputy chief of the 10th Division, who has assumed command, that all visible fire is knocked down and we're continuing to wet down pockets. Lee shows Espo where to continue to operate the nozzle and we head back across the open joists. Lee goes and gets two fresh air cylinders and we prepare to return to the hole in the ceiling. But, as we are about to re-enter, Terry radios down that everything looks real good up there. The smoke has lifted and they've soaked the burning underside of the roof thoroughly. The deputy tells us we can bring our guys down, but he wants me to go back up and give it another look. That's good, because I really want to see what it looks like, now that I can actually see. Lee and I go back up, and this time, with visibility, the trip takes us only about two minutes. What I see when I arrive at the nozzle sends a sudden chill down my spine. Terry is shining the beam of his flashlight on the remains of the charred roof. I add my own lights to the effort, as we sweep the beams along the roof purlins, which span the trusses and support the roof deck. The purlins on both sides of the truss are completely burned away, disintegrated for at least six feet in all directions. Worse yet, the heavy timber truss itself is also severely damaged. The massive 12 × 12 timbers have been burned so severely near the tops that they appear to be little larger than four-by-fours. I don't know what is holding this thing up. I give the area one last look for evidence of any remaining hidden fire. No wisps of smoke or even steam appear. I tell everyone to back off to the next truss and turn out their lights. I wait a few minutes for my eyes to acclimate to the darkness and look for the faint red glow of any embers. Seeing none, I tell everyone to head back to the hole. We drag the nozzle back to the relative safety of the next truss that was far less seriously damaged and leave it there. I want to tell the deputy about this in person, and we'll have to keep an eye on this roof for a while to prevent a rekindle, but I don't want anyone going past that truss. The nozzle marks the end of the safe zone.

In the street I comment to Lee, "We got lucky in there." Lee's answer doesn't really reassure me, "Yeah, Lieu, we were pretty close to a collapse." "Should we have been up there?" I ask. "Well, we couldn't tell how bad it was from down below. Somebody had to go and find out." Now I think of the stories I've heard about Lee's service in Vietnam, including a Bronze Star: how he'd been one of the guys to enter the tunnels chasing Viet Cong. I'm sure that's the kind of attitude it took to get yourself to crawl into a tunnel with all kinds of booby traps and people looking to kill you. Then Lee says, "Sometimes I'd rather be lucky than good." This was one of those times.

CHAPTER 22

Only the Lucky Ones Get to Retire!

One of the problems with having a job like a firefighter is that someday it has to end. While many people count the days until they can retire, firefighters are rarely among them. They enjoy going to work, being with their second family, sharing the kitchen table and the humor that surrounds it lest anyone let the job and the tragedies sink in. The place is full of gallows humor that few outsiders can understand. Most of us live for the job. A case in point: John Barbagallo.

John was another alumnus of Sheffield Avenue, working in Engine 290 until he decided to try something different. In the early 1960s when John first was assigned there, 290 was still relatively middle class and had not yet developed into the busiest engine in the city. At the time, that title was held by Engine 58 in Harlem. That company was doing so much work that it was known as "The Fire Factory." John wanted to be part of that, and transferred "uptown" where he met Ray Downey, forming a bond that would last the rest of their lives. They became part of what was to be known as "The Steam Team," for when they showed up at a fire, it soon turned to steam, drowned under their relentless attack. When Ray became the company commander of Rescue 2, John followed, becoming Ray's chauffeur. Now, after more than 33 years, it was drawing to a close. John was retiring.

The last tour in the firehouse for a senior member like John is a time for friends to pay their respects and to celebrate the experiences they have shared—the good times and the bad. Extra food is brought in as nearby companies stop by to pay their respects. Off duty and retired people that the honoree worked with going back so many years make an appearance to wish him well, and let him know that "there is life after the fire department." Tonight is John's last tour. He tells me in the cab that "only the lucky ones get to retire." He drove the apparatus around Brooklyn on his farewell tour. He wouldn't have missed it for the world.

The rest of the world doesn't know that an event like this is happening in the firehouses around the neighborhood; the beat of the city continues, the car wrecks, the false alarms, and the everyday emergencies continue unabated. So it is on this poignant night. In between runs, John is roasted by those who know and love him, but every time the *tooo-dooo* sounds, John still sprints for the cab. John is different from most chauffeurs in the FDNY. He likes to drive fully geared up—turnout coat on and buckled, boots on and pulled up—every time he gets in the rig. He doesn't want to be delayed for even a second by having to stop and get dressed at the scene. He does this every time he gets in the cab, dead of winter to the hottest days of the year, going to a 5th alarm or going shopping for the meal. John is always ready.

The night progresses slowly until about midnight, when the teleprinter sends us to Box 658. John is at the housewatch and sings out "Nostrand and Quincy, guys, Nostrand and Quincy for a fire." It seems that the only time we go to that box is when John is working, and he calls it out with relish. It is usually a job. Now John navigates the American La France through the streets of Bed–Stuy that he has come to know so well: one more time to Nostrand and Quincy.

We turn off Marcy Avenue, into the block behind Ladder 111. The fire is on the third floor of a brownstone, blowing out the front windows. Ladder 111 is setting up their basket now. People are hanging out the top floor windows and the staircase is blocked by fire. There are no fire escapes on the front of a brownstone, so the basket is their only potential salvation unless Engine 214 can get the fire knocked down in time to allow the inside team of Rescue 2 and Ladder 102 to get to them. We are pushing hard, but the fire in the hallway and stairs makes it impossible to get up the stairs. "Where's the water?" Engine 214's officer is practically screaming at his chauffeur via the handie-talkie, demanding that he charge the line now! But the nozzle remains empty, not a hint of the air being forced through the line ahead of the water that is so desperately needed. Engine 214's chauffeur swears to the officer that the line is completely charged, but that means nothing. We still have no water.

Engine 235 arrives with a second line, and when they call for water, it soon arrives, the line is bled of air, and now 214 has to back out of the way and let 235 take over the fire floor. They are pissed. Now we can get up the stair, but when we do, we find the fire has gotten a good foothold on the top floor. We need at least two lines up there. The chief transmits a second alarm. Engine 214 is now tracing their line back to the pumper, looking for the source of the problem that is keeping their line as dead as a doornail. They find it out in the street. Ladder 111 has set their outrigger jacks, large hydraulic pads that support the weight of the boom as it operates off the side of the truck, right down on top of 214's hoseline. When the chauffeur charged the line, the water flowed right up to the jack and stopped. Now they have to shut down the line, break the couplings on each side of the outrigger, and replace that length with another before they can get water to their nozzle. By the time that is accomplished, Engine 217 has stretched a third line to the top floor.

Engine 214 is left with only one room of fire in the rear apartment. We avoid giving them a hard time, knowing that to do so will only inflame the situation and invite bad karma to visit us. They are a good company, and it was just one of those myriad things that can go wrong when stress is so high and time is of the essence. The fact that the city has just taken one firefighter away from the engine companies, in a budget-saving move, is certainly a factor, and it goes to show us how we all have to be little more careful in the days of four-firefighter engines. Finding problems with the hose stretch, such as a line in the path of an outrigger, is one of the prime duties of the fifth firefighter that we previously had on all engines. We head back to the firehouse, satisfied that we had at least gotten a decent job for Barbagallo's last tour. If only we knew.

The *tooo-dooo* went off shortly before 4 a.m., and again John calls out "Nostrand and—." I say to myself, "Damn!" thinking he's going to say Quincy Street again, when he says, "Fulton Street, fire in a store." Oh, well, that has a lot of potential. John pushes the rig hard crossing Atlantic Avenue, then west on Fulton Street. We pass by the fire building, pulling down the block to leave room for the ladder trucks in front of the building. The fire is in a store on the ground floor of another four-story brownstone, which has apparently been converted from its original configuration, and now each of the upper floors appears to be its own apartment. Heavy fire is present behind the roll-up storefront gates. We report in to Acting Battalion Chief Tom Narbutt, working in the 57th Battalion for the night. Ladder 111 is setting up in front of the fire building and getting ready to cut the gates. Narbutt directs us and Ladder 132, the second due truck, to get upstairs and search the apartments. I survey the building and don't like what I see. Every window on the upper floors is pushing heavy smoke under pressure. We are going to have a lot of fire up there.

As Pete Bondy and Dave Van Vorst force the door to the entrance hallway for the apartments, I look at the mailboxes alongside the entrance door for a clue to how many occupants we may have to search for. All have names written in the little slots; it looks like they are all occupied. While Dave and Pete finish the forcible entry, I pull my hood up over my head and my earflaps down on my helmet, thinking that they are both going to be needed. As I do so, I see Lt. Ed Geraghty of 132 do the same. Eddie had been a firefighter in Rescue 1, and has his own hood, even though the job has not yet issued them to everyone. Once inside the staircase, my premonition is proven correct. There are several holes in the wall between the stair and the store on our right, and fire is blowing out of the holes into the staircase. I hear several minor explosions on the store side of the wall and recall that the sign over the storefront said Beauty Parlor Supply. It is probably aerosol cans of hair spray BLEVE'ing: the rocketing cans punching holes through the plaster wall. Wonderful.

When I reach the top of the stair, it is nasty, hot, and the lights are out. I crawl headfirst into Pete Bondy, who tells me that Ladder 132 went left, toward the rear apartment, while Dave went right, toward the front apartment and the stair to the top two floors. You can't see your hand in front of your face, but this is where we do our best work. "Okay, let's get up to the third floor and see if we can get a primary there before this turns to shit," I say. As we get to the door to the front apartment, there is heavy fire in there already. Fire is also burning up through the floorboards around us. This is *really* not good. I call into the apartment for Dave. No answer. Maybe he's already made the turn and started up the stair. Just then there is another explosion in the apartment next to us, this one must have been a whole case of aerosol cans BLEVE'ing, for suddenly a massive ball of flame blows out of the apartment door, filling the hall and stair with fire. "SHIT!! Where's Dave?" I try to get to the base of the stair to see up, but the fire is too intense and drives me back, hood, earflaps, and all. "Dave, where are you?" I scream through the facepiece, and again in the radio microphone. No answer. "Rescue to Irons, where are you?" No answer. Never have I been so afraid in my life; I have lost one of my guys! I scramble back to the head of the stair to get a hoseline; we have to go get Dave. The staircase itself is worse than when we came up. The hallway at the stair landing is a madhouse; Ladder 132 is scrambling back from their location in the rear, reporting heavy fire there and calling for a hose line to the rear.

As I reach the landing, a nozzle appears before me. I grab the shoulder of whoever is holding it, and shout "Let's go, you're coming with me, we have a firefighter trapped!" as I pull him in the direction of the front apartment and stair. "Who's trapped?" the firefighter asks. I tell him, "Dave VanVorst of Rescue 2," continuing to pull him forward. Suddenly, the figure stops, pulling back out of my grip. I am about to lunge at him, to pull the hoseline where we so urgently need water, ready to advance into the gates of hell, when he says "No, I'm not!"

It seems that when Dave first made the turn at the head of the stair, he made it as far as the front apartment doorway and realized we would need a hoseline to be able to get above, so he went to get one. Somehow, he passed me and Bondy in the inky black chaos that the floor above such a heavy fire creates, and went down the stair, encountering Engine 214 in the staircase. He described the layout and conditions to the boss, and started to guide them up the stair. That was when the BLEVE's reached their peak. Realizing that Dave was alive changed my whole perspective. Now this is a really shitty place to be: it is obvious that these are not apartments but storage areas for the beauty supply place downstairs. I take a head count. I have Bondy and now Dave and 132 is already making their way down behind 214. "Let's get the fuck out of here!" does not have to be said twice.

When we get down to the street, conditions look like we had been inside for an hour: there is fire not only on all four floors of our building, but in the two buildings on each side, traveling down the block in two rows of stores. That is not supposed to happen. A brownstone should have solid 12-inch brick walls between it and its neighbor. That is what has kept Bedford-Stuyvesant such a vibrant neighborhood compared to Bushwick, for example, where the rows of wood frame buildings allowed a single fire to destroy huge swathes of buildings at a time. It takes a 4th alarm assignment to finally bring this one under control. When the fire is finally extinguished, part of the reason for this dramatic escalation becomes evident: the rear first floor wall of the brownstone was removed, opening the beauty supply store into the back of one large building, which wrapped right around it on both sides and the rear. In effect, it was all one big open space. We were beaten before we even got the call. Rescue 2 hung around for another hour in case any other surprises developed. It was 9:00 in the morning by the time we got back to quarters at the end of the shift. We sent John Barbagallo off in the best fashion we know how. His statement about only the lucky ones get to retire would prove to be so profound (fig. 22–1).

Figure 22–1. John Barbagallo goes out with a bang. From left to right are Lt. Norman, Mickey McDonald, John Barbagallo, Dave VanVorst, Mike Esposito, and Pete Bondy. (*Courtesy of BC Don Forsyth, Orange County Fire Authority*)

CHAPTER 23

An Unusual Smile

Rescue Co. 2 is having a slow, late fall evening at 8:00 p.m. in 1992. The 15-hour-long night tour began two hours ago, the evening meal has been completed, and the company is now out on the road driving around Crown Heights, Brooklyn, looking for an abandoned, derelict vehicle (ADV) dumped on the streets either by its owner or the thieves who stole it, usually after stripping it of everything of value, to cut up. We'll use the car to practice vehicle extrication techniques. Just as the tools are unloaded, the radio crackles to life:

> "Brooklyn to the Rescue. Respond to Box 885, at 262 Reid Avenue near Halsey Street. We're receiving calls for a fire on the second floor of a multiple dwelling with reports of people trapped. Rescue 2 acknowledge."

"Rescue 2 on the way!" I hurriedly acknowledge as the Hurst tool is stowed and the rig roars to life. We make a left from Park Place onto Ralph Avenue heading north, then swing hard left onto Fulton Street. As we do, Engine Company 214 radios an urgent, "214 to Brooklyn. 10-75 on Box 885. Fire in a four-story brick MD [multiple dwelling]."

We pull up to the corner of Reid Avenue and Macon Street just as Ladder Company 111 pulls into the fire block on Reid Avenue from Halsey

Street. We stop our rig short of the fire block in order to leave room for the second due ladder truck, Ladder 176, which has not yet arrived. Fire is blowing fiercely out of the two right-side windows on the second floor of a four-story old law tenement. It is an old, railroad flat design with two apartments per floor, which means that the rooms in each apartment are arranged one behind the other like the cars of a train. We can tell this by the absence of a fire escape on the front of the building. It means that the building's occupants will be forced out onto a single metal fire escape hanging off the rear wall of the building, or down the smoky, heat-filled interior wooden stairs. It also means that we will have to use these same limited means to get past the occupants and the fire to search for other people overcome inside the building.

We follow Lt. Patty Concannon and Ladder 111's inside team through the front door and head for the open interior stairway. As we ascend the first flight of stairs, we pass several fleeing occupants, who tell us about people trapped above. We know that 111 will handle the search of the fire apartment. Jack Theobald, normally assigned to Rescue 1 but detailed to Rescue 2 for the night, is our roofman, while the other four firefighters with me are my regular crew from Rescue 2. I take most of my team up above the fire, leaving our can man, Bob Galione, to assist on the fire floor. As we reach the third floor, I poke my head into the apartment directly over the fire. (The line of apartments vertically aligned with the fire are called the fire line apartments—that is where fire extension is most likely, via the pipe chases that run up the building supplying all the sinks and toilets above.) Normally, John Kiernan and Bruce Howard, the floor-above team, would handle the search here, while I assisted on the fire floor. But with the second due ladder company not in yet, and reports of people trapped above, I know we'll have to spread out until reinforcements arrive.

The door to the right-hand apartment at the head of the stairs, which is directly over the seat of the main blaze, is open. It only takes me five to 10 seconds to see that while there is fire on this floor, it is still relatively small and located toward the front of the apartment, three rooms away from the doorway. I report the fire on the third floor to the chief of the 37th Battalion and request an additional hoseline be stretched to handle that fire. The open door is a sign that the apartment's occupants may have already escaped, but that must be verified by physically searching all the accessible areas of the apartment, just to be sure. I pull the door closed, being careful not to lock it, and tell my crew what I plan to do. I want John Kiernan to make a primary search here while I take Bruce with me to the top floor. Since this door is open already, John won't have any need for forcible entry, which is often a two-person task, and we don't know whether the top floor apartments are locked, so I want Bruce with me.

The top floor is very often worse than the middle floors because of the tendency for heat to rise through any vertical openings like stairways and pipe chases and then spread out as the heat hits the top floor and roof; therefore, this area must be searched very early in a fire. We say the heat is "mushrooming," due to the shape of the heat pattern. Bruce and I quickly make our way up the stairs to the top floor landing. The heat has really banked down on the top floor, so I assume that the roof team has not yet gotten to the roof and smashed open the bulkhead door over the stair, what we call "venting." I know that Bruce and I are in a dangerous spot, two floors above an uncontrolled fire, but I'm not worried. Engine 214 has their line on the fire floor and should be getting water very soon.

The door to the top-floor apartment that is on the same side of the hall as the fire apartment is open; but rather than take both of us in to search it immediately, I tell Bruce to force the door to the other apartment across the hall first. That way, if conditions in the hall deteriorate, we'll hopefully have an area of refuge, someplace to hide from the flames. Then I plunge into the inky darkness of the fire-line apartment. I don't go five feet before I realize that there is heavy fire in this top floor already, and from the crackling coming from the cockloft overhead, in that area as well! "How the hell can that be?" I wonder to myself, but I get an answer over the handie-talkie almost as if someone was reading my mind. "111 roof to Ladder 111. There's heavy fire in a shaft on the exposure 4 side, K." Ah, so that's it—a shaft fire! These old buildings were often built with six-foot square light and air shafts between the adjoining buildings, with windows from several rooms in each apartment facing onto the wooden walled shafts. The shafts run from the first floor up through the roof. They were intended to allow outside air and sunlight into the inner rooms of each apartment, but their presence is an absolute nightmare when a serious fire reaches any of the windows. The closely spaced wooden walls are the perfect arrangement for extremely rapid vertical flame spread, which then rapidly extends through the windows into the upper floors above the fire, as well as into the adjoining building that shares the shaft with the fire building. Okay, we'll really have to hustle now. There is already fire on three floors and in the cockloft and we haven't even begun the primary search! I urgently call the battalion to report the conditions on the top floor. I also call for another hoseline to the top floor. I return to the landing to check on Bruce's progress in forcing the "off apartment" door. (As in the situation here, fire often spreads very rapidly up the vertical line of apartments above the fire apartment. The apartments across the hall from the fire in a vertical line are referred to as the "off apartments." They are considered a "safer space" or an "area of refuge" since theoretically there is a fire-rated partition separating it from the fire.) It's a well-secured door, and he has made no real progress by himself using only conventional forcible entry with the Halligan tool and flathead axe. I tell him to go down to the third floor and get the Rabbit Tool from John and come back to the fourth floor.

As Bruce descends to the third floor, Engine 233 is stretching a line to the top floor with Capt. Serge Giovina in command. They are followed by Lt. John Fox and the inside team of Squad Co. 1. I begin to notice a change in conditions as the line reaches the top floor. It's getting very hot now, uncomfortably hot. I ask John to have his men work on the off-apartment door as I describe conditions in the fire line apartment to Captain Giovina. I know that we need water right away, as well as several members with hooks to get the ceiling down, so that we can hit the fire in the cockloft. We also need roof ventilation badly, especially over the stairs. I radio to the roof to "please, get the roof," and I'm assured that they're doing their best, but the bulkhead door is very heavily secured so burglars can't enter the building from the roof.

Suddenly a blast of heat knocks us all to the floor. Fire is now venting up the stair from the third-floor landing. Bruce is down there, screaming for us to get down. Fire is roaring out of an unnoticed front door of a third-floor apartment. The fire's intensity is too severe to permit him to approach the door to close it. I consider our predicament for a moment, estimating how much damage we will sustain if we attempt to descend the stair through the fireball. I quickly rule that option out as too painful. Meanwhile Captain Giovina has been urgently calling for water in his still as yet uncharged line. If we get water immediately, we ought to be able to fight our way back down the stairs behind the stream. That option is destroyed when the hoseline is charged. The fire has already burned through the line on the stair, allowing a useless flood of hot water on the third-floor landing, but not a drop at the nozzle on the fourth floor. I turn back to the two apartment doors, trying to prod everyone into the apartments for some refuge, but there is very little refuge to be had. The door to the off apartment is still locked, and the severe conditions in the hallway prevent anyone from rising up off the floor to work on forcing it. Our only retreat is into the fire line apartment that is already roaring with fire as well.

Everyone hurriedly crowds into the apartment and Serge closes the door. Believe it or not, it's better inside here with the fire than out in the hall. John Fox is on the handie-talkie now telling the Battalion of our plight. John is a really tough guy: a Vietnam-era combat Marine. He has been in some tough positions before but doesn't like this one. There are eight of us cut off by fire, and we need help. I crawl quickly to the back of the building, looking for a window. In my pocket I carry 40 feet of ⅜" life rope, which I consider my "last resort." But we're not that desperate yet. I know that there should be a fire escape on the back of the building, accessible from at least one window in this apartment. Probing blindly in the pitch darkness, I finally locate a window, and as I am about to smash the top pane, I am startled by what I see. The entire fire escape outside is bathed in flames, as the voracious blaze blasts out of the second and third floor windows below. Climbing out on the fire escape would be like throwing a fresh steak on the barbecue. Now we're really in trouble, with fire above us, below us, and all around us. I don't dare break the window now, as doing so will likely make the situation much worse, letting fire and fresh air in to intensify the blaze that's already in our apartment.

I give a "mayday" message. "Mayday, Mayday, Mayday, Rescue 2 to Battalion. We're trapped on the top floor, right rear." Receiving no acknowledgment, I repeat it, this time addressing it to my own members. Bobby Galione quickly replies, "Can to Rescue, 10-4, we're on our way." Battalion 37 also replies that he has help on the way. Unfortunately, I have no way of knowing how long that help will take to get there, or how much time we have left before we are overrun by fire or the blazing ceiling above collapses on top of us. I consider using the rope for a moment. I know that I could single-slide down it through the fire venting below, if I don't deploy the rope until just as I go over the edge out the window. I would probably get burned, but if I go fast enough, I figure that shouldn't be too bad. The rope might even burn through as I get to the second or third floor, but it's better than trying to imitate a pigeon from the fourth floor. The only problem is that even if I were successful, there are seven other guys up here, and I know that at least the four engine firefighters don't have ropes and harnesses. No, the rope is not an option. We're in this together. If you go, we go.

Down on the second floor, Engine 214 is moving through the original fire apartment and has just darkened down the last room. Bobby Galione is in the process of searching the rooms that have been knocked down when he hears the "mayday" over the handie-talkie. Now he tells the Engine that they have to withdraw their line and go to our aid. With the noise of the stream operating, 214 had missed our radio messages and were unaware of our predicament. They initially ignore Bob's directions to "Get your line out of here and get upstairs." They know that the fire floor is their responsibility, and the floor above is the second-due engine's job. They assume, for a moment at least, that the second and third due engines will be up above and shortly have things under control. There is still mopping up to do on the fire floor. They have no way of knowing that Engine 233 is trapped by a wall of flame on all sides. But Bob knows. He and John Kiernan, who was on the third floor, tell them to get their asses in gear because firefighters are trapped! 214 immediately springs into action. They rapidly withdraw their line from the original fire apartment and ascend to the third floor. A fourth line is waiting for water on the stairway, but there is no time to wait. Engine 214 pushes past the engine with the dry line and begins to drive the fire back into the third-floor apartment. (The third line was stretched into exposure 4 to stop fire that was spreading into that separate but attached building on the right side via windows in the shaft.) After driving the fire off the landing, 214 directs their stream up the stair to the top floor landing, knocking down most of the fire in the public hall there, before pushing on into their second fully involved apartment in less than 10 minutes. The members are nearing the limits of endurance but know that since they are the only line putting water on the fire, they must press on.

The sound of 214's stream striking the top floor door sounds better than any world class band or orchestra! After a few seconds, the door to the public hall is opened, just a crack at first, to check conditions. The hallway is still as hot as a blast furnace, but at least there is no fire out there, which is not the case inside the apartment that we now occupy. Somebody says, "Let's go for it!" and without another word all eight of us are out of that kitchen like a shot, scrambling down the stairs. I meet John Kiernan and Bruce and Bobby coming up the stairs. They are charging hard, looking for anyone who has not gotten out on their own. They are followed by the fourth hoseline from the stairway, which now has water. I assure them that all of the firefighters have preceded me down the stair, but then tell them that we still haven't gotten a primary search of the off apartment. Now Bruce has the Rabbit Tool, a small, hand-operated hydraulic jack, which makes fast work of the otherwise recalcitrant door. I order Bobby and John to get into the fire line apartment and open up the ceilings, to give the engine a chance to hit the fire in the cockloft as they move through that apartment. They will search that area as the engine knocks down the fire there. Bruce and I have to get into the off apartment. This is the last place the missing occupants can be that has not either been searched or overrun by fire.

Mindful of the immediate past, and of the fire in the cockloft over our heads, I call to the Battalion to get a tower ladder or an aerial to the top floor front left windows in case we need it as an escape route. I also direct one man to stay at the door to the hall and keep us apprised of conditions in the hall. Moving towards the front of the building, I follow the left wall. I know that with this apartment layout, this should lead me to the front as directly as possible. Visibility in this apartment is zero. I've only gone through two rooms, and that has been a challenge. The place is clogged with furniture, and the walls seem to be tapering in, the rooms getting narrower. Just then the vibralert (low air alarm) on my mask starts to sound, telling me that under "normal" fireground circumstances I would have about four minutes of air left. But this is not your "normal" fireground. There's something wrong here. We can't find the way through the apartment to the front windows. I consider suspending the search. I know we are getting in over our heads again. I'm low on air and there's a lot of fire in the cockloft overhead. I can't discern a clear path through all the clutter of the apartment, and it seems as if my mind is playing tricks. I can't for the life of me figure out why the walls would be tapering in to such a narrow hallway, barely two feet wide. But we still haven't "made" the front three rooms. Just then my decision is made for me. The deputy chief has ordered everyone out of the building. The right-hand top floor apartment is still blowing fire out the front windows. The cockloft fire is now blowing out the entire front of the cornice, and fire has spread within the ceiling joist bays to the left side of the building on the second and third floors. It's time to give it up. We've been outflanked by the fire right from the start and now it's time to call in the "Water Monster"—the tower ladder's heavy caliber stream.

It is almost with relief that I back out of the apartment. I fear that there still may be someone in those last few unsearched rooms, but I also know that there is nothing I can do to reach them now. As I descend to the street, I pass that damned third floor. Why hadn't I checked for that other door? I know that's not an uncommon feature of an old law tenement. This layout was unusual though. The doors weren't at the end of the hall, opposite the base of the stair, as is most common. Instead there was a small alcove at the front, with the doors off on each side. That minor deviation from typical architecture of the era was enough to nearly toast eight firefighters.

The tower of Ladder 111 is at work now, throwing nearly 800 gallons of water per minute into the building. It is soon joined by Ladder 124's bucket as well. The two towers do their job and the fire is knocked down rather quickly, but not before the roof collapses into the top floor and the front wall of the top floor crashes into the street. The deputy has made this an outside operation for now, and we are ordered to "take up," and go back in service for another alarm.

The next day, in daylight, we go back to the scene to complete the search of the three rooms we did not reach in time. The building is now only three stories high. There was no one else in the building. We tear off the remains of the roof, board by board, in our search, right down to the charred remains of the fourth floor. Eventually, the owner decides that he likes having a three-story building more than he likes spending money to build back up to the original height. The building sat there like that for nearly 20 years, a three-story building in a row of attached four-story buildings, looking like a chipped tooth in an otherwise neat smile. An unusual smile to be sure, but one that comes from eight lucky firefighters. I, for one, smile every time I pass by and think of our close call and how the men of Engine 214 bailed us out.

CHAPTER 24

"Precious"

In the middle of a recreation complex of ball fields, an indoor pool, and a school playground sits Rescue 2's quarters at 1472 Bergen St. This complex is intended to provide recreational opportunities for the more than 5,000 residents of the Albany Houses—a complex of nine, 14-story public housing projects that has an almost unmatched record in the New York Fire Department. In 1990, the amount of gunfire heard from these projects is incredible: at times, especially after dark, rivaling Beirut. It is not at all uncommon to hear the "mutts" rip off a magazine on full automatic, usually fired from the rooftop of one of the buildings and often directed toward the lighted baseball field adjoining the firehouse. While we didn't feel they were specifically aiming at us, several shots struck the rear windows, prompting Captain Downey to requisition bulletproof glass for those windows. In the four years that I spent on Bergen Street, the glass never came, so people just tended to stay away from the windows at night. (There was nothing to see anyway because we had lined the rear fence with plywood, to make it harder for a sniper to get a bead on anyone silhouetted in the light at night.)

As I said, this one complex of nine buildings has accounted for an incredible record in the fire department. There has been at least one department medal awarded for a rescue of at least one of its occupants every single year for at least four years. For a building that is built of poured concrete, where the fire rarely threatens occupants who are not inside the fire apartment, that's saying a lot about the number of fires in these buildings. It is one reason why the fire companies are so active in the Crown Heights/Weeksville area: Engine 234, Ladder 123, and Battalion 38; Engine 214 and Ladder 111; as well as Rescue 2. Each building is approximately 100' × 100' and star-shaped with five points, and has eight to 10 apartments per floor, mostly two bedrooms, but with some having three bedrooms. The stairs, elevators, and trash chutes are located together in a center core configuration, with a circular hallway leading around the core to provide access to the apartment doors. This circular hall is uniformly lined with glazed terra cotta block and can be very confusing, especially when it's "lights out"—smoke banked down to the floor, or even the floor below. This confusion is compounded by the scissor stair arrangement, which has the stair landing alternate to opposite sides of the hall on every other floor. That makes it very difficult to accurately gauge the distance from the stair door to the door of the fire apartment if you are not absolutely certain of the fire location. The apartment layout itself serves to complicate the difficulties of escape for the occupants. Since the buildings are star-shaped, and the entrance is from the circular interior hall, each apartment has a wedge shape, like a slice of pie. The kitchen and living room are located right inside the entrance (and sole exit) door, with the bedrooms further back, toward the wider part of the wedge. Since the kitchen and living room account for a very large percentage of sources of ignition (stoves, televisions, and other appliances), it's no surprise that many people find themselves trapped in a back bedroom, with their only exit blocked by flame in the kitchen or living room. Such was the case for "Precious" on Nov. 12, 1992.

The alarm for Box 920 at 1400 Bergen Street came in at 0300 hours—3:00 a.m. in the morning on the dot—reporting a fire in apartment 13 E. I hate the short runs to 1400, 1430, and 1440 Bergen, since they are only about 500 feet from the firehouse, which doesn't give me a lot of time to finish buckling up the turnout gear, don the mask, strap on the flashlight that I carry on a sling, double-check the alarm information on the teleprinter ticket, report our arrival (10-84) via radio, and size up the building from the exterior. At times, if there is other traffic already on the radio, I have to ask the chauffeur to give the 10-84 before he gets out of the rig so I can get out as soon as we stop. That, of course, doesn't make the chauffeur very happy—he is already going to be somewhat delayed since he not only has to park the rig, but then go to the back and lock it up after everyone gets out (or else the "mutts" will have a field day looting all the tools while we're inside); then he has to don his mask, grab his tools, and follow the rest of us into the fire building.

Tonight, however, the radio is free of other traffic as we approach, so I personally give the message, "Rescue to Brooklyn, 10-84 at 920." The chauffeur, Richie Evers, has one less thing to deal with. As we sprint the 150 feet or so to the building entrance, we all look for some hint of what lies ahead: smoke or flames visible in the windows above, falling glass, trapped people calling for help, an agitated crowd in the elevator lobby, or smoke coming from the ground floor trash compactor chute. There is nothing at all to indicate anything but a minor fire or maybe even a false alarm. One thing that is unusual is that both elevator cars are down in the lobby, and both are working. City housing projects in general are notorious for the number of out of service elevators, but the Albany Houses are even more infamous in this regard. We have hiked the 14 stories many times for alarms when both elevators are out of service, so this is a pleasant surprise. We all pile into the nearest elevator. Billy Lake has the "can," Timmy Higgins has the floor above irons, Bobby LaRocco has the floor above hook, and Bobby Galione has the roof—a seasoned, aggressive crew. Being first due at a project requires us to alter our usual positions. Timmy will assist with forcing the apartment door on the fire floor instead of proceeding directly to the floor above as he would if a ladder company were in ahead of us. We'll also need the extra help in searching.

As the elevator doors close, I push the call button for 11 and then reach around behind me to turn on my mask cylinder as everyone else does the same. The elevator car starts to move as I make sure my facepiece is handy, in case the elevator door opens into a hallway full of smoke or flame, as I have seen when a compactor chute fire encounters a chute door that has been ripped off by vandals. Against all odds, the elevator proceeds uninterrupted to the 11th floor where we all pile out. Billy, Timmy, and I head left for one staircase, Rocky and Bobby go right for the other. I pass apartment 11A as I enter the stair and note that 11E is on the other side of the core. Still, there is no indication of any problem. At the 13th floor, Timmy pulls open the stairwell door and there, directly across the hall, is the fire.

The door to 13E is wide open, and we can see fire rolling out of the kitchen into the hall and living room as we don our facepieces. I scan the hallway around us for signs of the apartments' occupants, but there is no indication of anyone being in the hall. "That's strange," I think. Usually we're met on the floor below by throngs of shouting and screaming people whenever there's a "real fire," telling us what to do and how to do it. If there are people trapped, their actions can border on the hysterical. At this fire, I haven't seen a soul since we got off the rig. I call out to Billy and Timmy, "Be careful, it might just be a vacant apartment," but as I put my face to the floor, I can see furniture in the living room, behind the wall of flames.

Timmy calls through his facepiece, even as I make this same judgment, "I don't think so, Lieu." Billy Lake goes to work with the can as I radio, "Rescue to the 38 [Battalion], 10-75, chief, we got a kitchen and living room going." The 2½ gallons of water in the extinguisher can do a lot in the hands of an experienced guy like Billy, but it can't work miracles. We'll need Engine 234 and their 2½-inch hoseline. Billy initially hits the ceiling straight in front of the door, knocking the fire back into the apartment, then quickly scoots in farther, hitting the flaming materials immediately on each side of the door, creating a sort of "tunnel" through the flames. Since the ceiling above is poured concrete, we don't have to be concerned about fire travelling in voids over us as we would in a building with wood floors. Timmy and I see our opening and lunge through the gap in the flames that Billy has created. I figure that once we're through the flames, we can get back into the bedrooms to do our search while Engine 234 is putting the fire out with the hoseline that they will stretch from the standpipe on the 12th floor. Timmy and I make our way through the living room which is extremely cluttered, with two pull-out sofa beds and a fold-up bed filling nearly every inch of floor space. We hurriedly search these sleeping areas, "tossing" them for victims.

Visibility here is pretty good, with the glow from the flames overhead actually aiding our efforts. We complete the "primary" here in about

15 seconds and move down the right-hand wall that carries us farther along the hallway to the bedrooms. Timmy enters the first door on the right and I tell him I'm going on to the next room. I find an incredibly small bedroom, no bigger than some walk-in closets, with two beds jammed into it. I toss them both, but both are empty. Billy yells over the radio, "Hey, Boss, you better get out now. The can's empty and 234 and 123 ain't up here yet!" As I exit the room, I see the "tunnel" diminishing. Two and a half gallons has bought us our entry but not our exit. The fire is back out in the hall and living room.

I call to Timmy, "It looks like we're going to have to sit this one out and wait for the engine. Find a room with a door and a window. You got your rope?" Tim calls back, "Yeah, Lieu, I got a good spot here!"

With the room door closed between us and the fire, chances are pretty good we'll be able to wait out the fire, breathing air from our masks while the closed door protects us from flame. As a last resort though, I want to make sure that there is a window to get out, and that Timmy has his 50-foot personal rope and harness, which if things come to it, would let us rappel down to safety on the 11th or 12th floor. Everyone in Rescue 2 wears some type of personal harness, even men that the department has not gotten around to issuing one to yet. Not all of them carry their own rope, though, since it is bulky and doesn't fit well in the coat pocket. They usually assume that the lifesaving rope carried by the roof man will be able to reach them. But I prefer to carry my own rope. It's my last-ditch insurance policy: sort of a personal fire escape.

Assured that Timmy is safe, I weigh my options. I can't stay where I am because there is no door on this "closet" nor is there a window. I can join Timmy and sit it out there, or I can take a shot at making it to the last bedroom at the end of the hall. I look back toward the living room. The first of the sofa beds has lit up, and fire is now moving back toward us, but I estimate it is 20 feet away. I should have plenty of time to crawl the 10 feet down the hall to the next room. I start for the doorway. Billy calls over the radio, "Are you guys all right in there? There's still no line up here yet!" Engine 234 radios that they're on the 10th floor, walking up. The elevator stalled on them at the fourth floor. It figures! I crawl my way blindly into the end bedroom and immediately close the door behind me. Just as quickly as I close the door, I realize that there is something wrong with it: I can still hear the crackling of the fire almost as clearly as I could in the hall. Instinctively I know that's not right. It should have gotten quieter! I stand up to feel what's going on, and I discover much to my dismay that there is a hole through the top of the hollow-core wooden door big enough to shoot basketballs through. The heat is very intense when I stand up, and I quickly drop down to my hands and knees again. I radio out to no one in particular, but everyone in radio contact, "Would you hurry up with that line, guys? I'm in a back bedroom without a door."

I start searching for a window, just in case. I don't want to open it yet, for fear of drawing the fire toward the oxygen supply, but I want to be sure it opens. As I crawl along the floor, I bump into a bed. I give it a fast toss across the surface and then underneath, but, like the other eight that Timmy and I have searched so far, it too is empty. I make a left down the wall, still feeling for the window, and bump into the headboard of yet another bed. I move to the side and toss it also. Again empty! But this mattress is different from the others, lower, closer to the floor. "Bunk bed," my training shouts at me. I feel the headboard again, and my hand follows it up, to a set of top rails. I jump up now, standing on the bottom mattress, trying to sweep all the way back across the top mattress to the far side. My feet slip off the bottom piece of foam (wrapped in bedding), and I have to make another shot at it from the other end, since I can only feel an arm's length at a time. Just as I am almost convinced that this bed too is empty, my gloved hand brushes the foot of a person. Of course, I can't tell this for sure yet, since I can't see diddly squat, but I know it wasn't just a blanket or pillow. I "guesstimate" where the object was that my hand brushed, and I pull myself back up by the top rail of the bunk bed, leaping upward while stretching forward, trying to propel myself up far enough to get my waist over the rail, six feet off the floor, no easy task dressed in turnout gear and a mask. (This is one reason why I so vociferously decry Federal Judge Charles Sifton's decision to throw out the five-foot and eight-foot wall climbs from the Fire Department's entrance exam, claiming it unfairly discriminated against women candidates. Hell, we actually have to do these things, and we have to do them dressed in our "work clothes," not in gym shorts and sneakers!)

On my second attempt, I get myself up far enough to be able to sweep the rest of the bed. Sure enough, there's the foot. I follow the leg up to the torso and then head. Laying half on and half off the bed on my belly, I clumsily wrap one arm under the child's armpit and pull her up toward me as I push back off the bed with my feet and free hand. Carrying the little girl, I tumble backwards onto the other bed. With my left hand, I key the microphone on the handie-talkie and radio my discovery to the 38th Battalion, "Rescue 2 to the 38, 10-45, 10-45!!!" Then I call, "Billy, how are we doing with that line? I gotta get out of here!"

I flashback two and a half years to a fire just six blocks away where I had found another lifeless body of a child and wasted precious seconds, looking for a second child that was also reported trapped. That child died, and the question of whether I did the right thing looking for the second child instead of immediately taking off with the first victim has haunted my dreams.

Now, that memory propels me forward toward the wall of flame. I decide to head for the door with the child. Thankfully, since the response was so short, I didn't buckle the bottom buckle of my turnout coat. I unbuckle my mask waist strap and pull the storm flap snaps of my coat apart and find I can just fit the child's head and upper body up under my coat. As I crawl back toward the door, I keep looking for the fire, or the closed door that Tim has found. As I approach the entrance to the living room, as if on cue, the "tunnel" reappears in the wall of flame. Richie Evers and the can man from Ladder 123 have brought up two more extinguishers, another five gallons of precious water and are beating back a path into the apartment. I crawl through it with the child tucked up against my chest, shielded by my coat.

Once at the public hall, I scramble over Engine 234, who are poised at the doorway, bleeding the air out of their line. In another 30 seconds their 2½" line will have made short work of this two-room fire, but this child could not have waited that long. Lying unconscious on that top bunk, she has been sucking in all the poisons the fire produced. I nearly leap down the flight of stairs, seeking an open space on the 12th floor in which to begin

CPR. I rip off my helmet and facepiece and reach for the child's head to begin mouth-to-mouth. I suddenly remember the pocket CPR mask in my coat pocket. I've made this same decision before and not used the mask, but now I remember the look in my wife's eyes when she gave me the mask as a Christmas present, before the job issued them. I decide that it's worth the few seconds it will take to get it out this time. I check the child's carotid pulse but feel none. I check again at the sternum, still nothing. Using the pocket mask, I give two quick breaths and then begin one-handed CPR. Rocky joins me after checking the floor above for fire or someone overcome, but the six-inch thick concrete floor has confined the fire to the fire apartment. He radios the battalion to get EMS and have a resuscitator meet us in the lobby. I give a few more cycles of CPR and check the pulse again. This time I think I feel one! Rocky calls and tells me the elevator is ready. I shuck off the SCBA, leaving it and my helmet and gloves in the hallway, then scoop up the child and sprint the few steps to the elevator, continuing CPR as I go.

In the elevator, Rocky asks if I need a "blow," relief in CPR, but my adrenaline is pumping now and I say, "No, I've got her." As the doors open, we charge for the street heading for the nearest pumper, which happens to be Engine 227. The chauffeur there has heard the 10-45 report and the request for a resuscitator, and as we run toward him, he pulls the ambu-bag and oxygen out of the compartment. I gently lay the lifeless child on the cold, black asphalt alongside the throbbing pumper and again check for a pulse. "Yes!!" This time I'm sure of it. There is a pulse: weak, but steady. "This kid might make it!" Rocky takes over the ambu-bag and starts squeezing life-giving oxygen into the tiny, smoke-wracked lungs. I monitor her pulse and check her pupils for reaction to light with the Sunlance flashlight that I keep clipped to the top buckle of my turnout coat, just in case I lose my main light. The pupils contract well when I shine the light in them, a sign the brain is still functioning. Great! (See fig. 24–1.)

When EMS shows up, I carry the still unconscious child to the ambulance and place her onto the stretcher. The child is now breathing on her own, but very weakly. A paramedic examines the child's airway and sees signs of inhalation burns. That top bunk was really hot! It is a terrible place for a four-year-old. I back out of the bus and take a few moments to calm

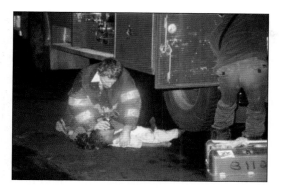

Figure 24–1. The author prepares to give mouth-to-mouth to four-year-old Precious Lucas, as Engine 227's chauffeur gets out the resuscitator. She would make a full recovery. (*Courtesy of Bob Welch*)

down before rejoining the company. The medic intubates the child before the ambulance pulls away, heading for Kings County Hospital. The child is later transferred to the New York Hospital Burn Center in Manhattan. She will spend two weeks in the burn center, but eventually makes a full recovery. This was a good save. I doubt if she would have survived the extra three or four minutes it would have taken before Engine 234 knocked the fire down for us to begin our search. I know some OSHA inspector somewhere will read this story and nearly have a shit fit because of the way things were done. Intentionally passing an uncontrolled fire. Not remaining locked in a death grip with my partner, Timmy Higgins, as we conducted our search. Leaving Billy Lake alone at the door without a partner. Heaven forbid! Well, that's just tough shit, pal. This is what we do. This is what it takes to save a life sometimes. We take calculated risks, not suicidal actions, and most of the time we're right.

The final twist to this story occurs about eight years later. I meet Billy Lake and some of the boys from Rescue 2 at the Department's annual memorial service on Riverside Drive. I ask Billy how things are going and how some of the older guys that I had worked with are doing. After giving me an update on Richie Evers and Pete Bondy and some of my other favorite people, he casually mentioned, "Oh, yeah, hey Chief, remember that job we had at 1400 Bergen, where you made the grab of that little girl?" "Yeah, sure," I reply, expecting to hear something about Precious, like maybe she's a star in junior high school or something similar. Billy continues, "We had another job in almost the exact same apartment last month—the same thing we had the first time, fire at the front door, people trapped, the whole works. How did it go?" I inquire. "Yeah, it went good," Billy says, somewhat subdued. "Ask him how he's getting fucked," Bobby Galione calls from the edge of the group. I look at Billy with a "What gives?" look. He then tells me he made almost exactly the same moves I did—went past the fire, got a kid and brought her back past the fire before any hoseline was in place—but the officer didn't want to write it up for official recognition. "What kind of bullshit is that?" I ask. "Who was the officer?" "Aaaahh, just some covering asshole who couldn't be bothered with the paperwork," he replies. "Was he there? Did he see what you did?" I ask. "Well, I can say that he was in the elevator with us going up, but that is all I can say for sure. After that I don't know. I never saw him again until we got down to the street." "What a royal screwing!" I think to myself. Here's a guy that does a great job, saves a life, and gets nothing for it because some dipshit doesn't want to do some paperwork, or wasn't in the right location himself! I want to give Billy a medal, but I know he'd want no part of it if it didn't come the right way. Everybody's supposed to "do the right thing."

CHAPTER 25

Planes... Something That Big Isn't Supposed to Move, Let Alone Fly!

Every time I take a flight, it reminds me of how amazing it is that something that weighs 150 tons can move around, let alone get off the ground. It also reminds me of how short and truly fragile life is! Growing up in the shadow (literally) of John F. Kennedy International Airport—or Idlewild, as it was called in my youth—airplanes were just something that were there, like the air they fly in. Planes passing very low overhead were a constant, if annoying, presence with their ear-shattering roar. You learned to time your speech into sound bites so that you could pick up where you left off when the plane had passed. During peak times, this goes on every 45 to 50 seconds! More than a few close calls have occurred in the crowded skies along the approaches to runways 13R–31L, leaving many Inwood residents wondering when, not if, one of these things would land in their tomato garden.

Growing up in the fire service in this environment just about guarantees that aircraft crash rescue will be part of your study. My hometown fire department, the Inwood Fire Department, was well aware of this potential and had prepared as best they could. Since the town was also home to six bulk oil storage plants supplying gasoline, jet fuel, and home heating oil to much of Long Island and its 2 million residents, the availability of firefighting foam was a major step forward in such preparedness. Over the years, the department had been involved in several actual crashes, including a U.S. Navy patrol bomber that crashed during WWII about 300 feet from what would later be my home, and the crash of an Alitalia Airliner at Idlewild in 1954, which killed 26 of the 32 people aboard. My father, Butch, used to tell us about the Alitalia crash. He had just completed his term as chief in Inwood at the time, and it reminded him of his service aboard two aircraft carriers during the war.

My first exposure to the havoc that is an airline disaster would come 20 years after my father's last, on June 24, 1975, when Eastern Airlines Flight 66 attempted to land during a thunderstorm on JFK's runway 4L-22R. The plane was struck by a severe wind shear while on final approach and was shoved down into the ground at over 125 miles per hour, about 400 yards short of the runway. The impact tore the plane apart, throwing debris, bodies, and burning fuel over a wide swath of swampland as well as the heavily traveled Rockaway Boulevard, the main access route from Queens into Inwood and the Rockaway peninsula. At the time, I was working in Tri-County Fire Equipment's shop. When the radio went off announcing a "mutual aid to the Port Authority" for a "plane crash on Rockaway Boulevard," everyone in the shop scrambled for the trucks for the ride to the firehouse. Going up Rockaway Boulevard that day was an adventure. The thunderstorm was still going full force. The sky was pitch black and the wind whipped rain and hail into our faces as we clung to the backstep rails. As we went north up Rockaway, a New York City pumper passed us going like a bat out of hell in the other direction.

When we got to the scene, the Port Authority's crash trucks had nearly all the fire extinguished, but they were about out of water and needed help in resupplying. They asked us to pitch in and give a city engine company a hand in stretching a 3½-in. supply line by hand through the debris. As I walked along Rockaway Boulevard humping hose, I thought of all the times I had been stuck in rush hour traffic at that exact spot because of the holiday shoppers driving between Mays, E.J. Korvettes, and T.S.S. department stores about a mile ahead. If this plane had come down there, it would have wiped 50 or 60 cars off the roadway. "It could have been worse," I say to no one in particular. That was before I found out how bad it was already. Our next assignment was to form a human chain and move out through the swamp to locate any possible survivors. We formed up shoulder-to-shoulder and started eastward from Rockaway Boulevard, working out toward the initial impact point. Right away it got bad. There was a small berm on the swamp side of the roadway that acted almost like a dam, stopping a lot of the debris and bodies.

The first group that I came to was about eight or nine people who were stripped naked and shorn of hair by the impact and fireball. Their skin had a smooth, waxy appearance, so much so that when I first viewed them from 20 feet away, I thought that they were mannequins. For a few moments, I thought that maybe this was a cargo plane and there was nothing but mannequins on board. "Why the hell would anyone fly a plane full of mannequins around?" I asked myself. But as I approached, and the peculiar odor of burnt flesh wafted toward me, I suddenly knew these were no mannequins. Some of the bodies were hardly touched. You had to actually check them for vital signs to be sure they weren't still alive. Others, however, were so horribly mangled, that not only was there no doubt they were dead, it was nearly impossible to recognize them as human beings.

We were now asked to mark the locations of all the bodies and body parts we discovered and bring the bodies up to the boulevard, where the Medical Examiner's Office was establishing a temporary morgue. Picking up these mangled and burned bodies was a messy business. The skin, quite literally, came off in our hands. At that time, most of us volunteer firefighters used the fluorescent orange "fireball" gloves, which were rubber-coated and were horrible for firefighting. They trapped perspiration in them, which turned to steam with a only a bit of heat. If you grabbed something hot, they often melted to it, sticking your hand in contact with the hot object. Either way, the hand burns were not fun. They were ideal for handling bloody body parts, however, since they were very thick rubber on the exterior and washed off easily. New York City firefighters, on the other hand, often wore leather or canvas gloves. They absorbed both the blood and body fluids, and had to be thrown away. As a result, the mutual aid firefighters were doing a lot of the hands-on work removing the bodies, until we met up with one FDNY unit. They wanted us to give them a hand with one particularly messy body in a drainage ditch.

As I and another fellow grabbed the wrists and ankles, one of the FDNY truckies took a pike pole, grabbed the victim we were working on under the stomach, and yanked. The body split in two pieces, spewing bloody entrails all over those closest: me and a fellow from the Lawrence-Cedarhurst FD. We dropped the extremities we were carrying in shock. I must have turned green, because this old-time New York City truckie calmly asked, "Hey, kid, you got kind of a fuzzy, tickling feeling in the back of your throat?" I just nodded and he advised me, "Well then, swallow hard, 'cause that's your asshole trying to get out, and if you let it, it'll turn you inside out." And then he practically fell down with laughter. For my part, I didn't see the humor in the situation, since he had caused it. His officer read him the "riot act" and sent him back to their rig to get a pair of gloves. Another crew from Inwood relieved me after about four hours of this. That was enough for me. I had seen up close and personal what a high impact airplane crash did to human flesh and bone. It cured me of flying for about six years.

Later that night back in Inwood, we were sitting around the firehouse swapping stories about what we had seen and done. By the 1990s, this process was formalized and called critical stress debriefing. We had only been back in quarters about an hour when we received an alarm for a building fire at a local brickyard. The fire was in a 20-foot × 60-foot Quonset hut used as an equipment garage. I was the officer on the first line, and we moved in and did a good job knocking the fire down. About halfway back we encounter an old front-end loader whose tires are being stubborn about going out. I tell my brother, Warren, who is the nozzleman on the backup line behind us, to go on past us to overhaul the far end of the steel structure while we work on the tires.

As we drown the smoldering rubber with our line, conditions start to improve as the smoke lifts and we can just about see. Joe Baal is working with us now, overhauling shelves alongside the tractor with a hook. He pulls on some smoldering material that falls to the floor. Instantly, the entire area around us is engulfed in flame. I nearly scream at Vinnie McCross to open the nozzle on fog. I'd seen this before out at the Nassau County Fire Academy in Bethpage. That's gasoline burning! Vinnie opens the nozzle and starts to back away from the flames. I start pushing him forward toward them, and he resists. He didn't see the other line go past us, but I know they're ahead of us somewhere. I wish that we had foam in this line, but that's not going to happen. The only chance we have is to push the fuel to one side, so the other line can get back to safety. Joe Baal crawls in alongside Vinnie. He's seen the gas tank on the front-end loader, and now he tells Vinnie to get water on it, to prevent a BLEVE. As Vinnie turns the nozzle onto the tank, the fire is directed with it. Now we can see the source of the problem. Gasoline is dripping from a seam that has opened up due to the tank's expanding. We are dangerously close to BLEVE! The fog pattern "captures" the dripping gas, and pushes it against the steel wall of the shed, where it produces an intense flame and a throaty roar, but does little damage. Warren and his line appear out of the smoke, pushing hard back for our end of the hut. They are coming to save us! They had a back door that they could have gone out in the far end wall, but when the thing lit up, they knew the fire was centered on where they last saw us. So it was "charge" time for them, just as it was for us. When the gasoline finally ran out, the fire burned itself out.

That was enough excitement for one day for me! I went home to bed and slept like a log. No second thoughts or nightmares, which I attribute to doing everything that was humanly possible throughout the day. I had nothing to be ashamed of. Oh yeah—that New York City engine that was going south on Rockaway Boulevard as we arrived at the plane crash—that was Engine 302, the first unit on the scene. They had pulled up to that mess and miraculously found seven survivors, some still strapped in their seats. Realizing that these people were in extremely critical condition and knowing the likely delay before ambulances could get through, the officer ordered the chauffer to drive through the burning debris to the tail section. Then they and the members of Ladder 155 loaded the survivors onto the hosebed of Engine 302's rig and the rig took off for Peninsula General Hospital on the Rockaway Peninsula. All of these people survived—the only survivors of this disaster, which claimed 113 lives: the deadliest single U.S. aviation incident to that date. Days later, on July 1, the City of New York laid off some of the members of Engine 302 and Ladder 155 as part of its response to a budget crisis. What a great way to build employee dedication!

My next plane crash would not be for another 14 years: on September 1989, U.S. Air's Flight 5050 (see Chapter 12). After that, they seemed to get more frequent. Having been to two major airplane crashes, I was becoming something of an "expert" on such things. After being assigned to Rescue Co. 2 in 1989, I made it a point to drill the members fairly regularly on plane crashes. This was particularly important, since Rescue 2 is one of the two heavy rescue companies assigned on Queens Box 269 (the "crash box" in the tower at JFK Airport), usually arriving as the first Rescue on scene.

On the morning of March 12, 1991, Rescue 2 received a teleprinter message to respond with the rest of the normal third alarm assignment to Box 269. As we responded east on Atlantic Avenue, I switched the radio over to the Queens frequency expecting to hear an announcement that the box had been transmitted for an in-flight emergency on an inbound jet, which would give us time for the 12-minute ride to the airport; or even more likely, that the box was being sent for a drill. As soon as I switched over the radio though, I knew this was the real thing. JFK tower was relaying to the Queens dispatcher a confirmed crash on take-off of a DC-8. Well, we'd soon see if training had paid off.

Arriving at the staging area off Federal Circle, we were directed into a line of vehicles that was just leaving for the crash site. Everything that moves on the tarmac or aprons must be escorted by a Port Authority vehicle that is in contact with the tower since the airport is still open for landings of planes that can't be diverted. As we turn east along the taxiway for Runway 22, the same one that Eastern Flight 66 crashed on in 1975, a column of gray-black smoke is visible. The color of the smoke is somewhat heartening. It is a good sign in a way, as opposed to the massive clouds of solid black smoke of burning jet fuel that I am expecting to see from a plane that crashed on take-off, which means it would have had full fuel tanks.

Approaching the scene, I can see three of the airport's F-28 crash rigs pouring AFFF foam onto the fuselage of a large airliner. The fuselage appears nearly intact, although we pass engines and whole landing gear assemblies as we make our way in. Richie Evers is the "wheelman" this tour, and he needs little direction. He knows we need to get close enough to make the "toolbox" available, since we might need virtually anything on board the apparatus. Yet we want to stay far enough away to not be an exposure hazard or impede the crash trucks maneuvering. Heavy brown smoke pours from an open door on the right side of the plane at the rear, just ahead of the tail. I yell into the back for everyone to take their masks. This is going to be an interior cabin fire, a rare type of crash, and it looks like a real bitch!

Crash truck personnel have extinguished a fuel fire around the wing root, without cabin burn through—a real testament to their efficiency. The plane is resting on its belly. The landing gear and engines have been knocked off in the crash, and there is definitely heavy fire in the interior. I radio Division 13 for orders. His reply is, "Rescue, give the Port Authority guys a hand trying to move that handline in through the rear door over there." That was what I was hoping to hear. There seems to be a chance to make a difference here.

I tell Richie to meet us at the tail, and take John Kiernan and Mikey Esposito to the far side of the plane. I see a crash crew member trying to get a door open on the near side and tell Espo and Kiernan to help him. Around the far side, there are two crash crew guys directing a 1½" fog nozzle into a cloud of really ugly brown and yellow smoke from the doorway. They don't have SCBA on and the smoke is like a solid wall, so that's as far as they can go. I ask them if they made a search of the interior for survivors. They look at me like I have three heads. Then one says, "Guy, this is a cargo jet. There's only the crew on board and they were all up front and got out already."

"Oh, that's good news," I reply.

I ask them to shut the nozzle down to let us see if we can get in to locate the fire. "Be my guest, but you won't get anywhere," says the nozzleman. Mikey Espo has been waiting for this opportunity. He props the Halligan against the fuselage, fork end down, and steps on the adz end—an improvised ladder to climb the four or so feet up to cabin floor level. Immediately comes the universal expression of bedevilment, "What the fuck?" I call to Mikey, "What's up, Mike?" "Lieu, I don't know what it is, there's a solid wall right inside the door here about a foot inside." Well, that's why the crash guys said we wouldn't get anywhere. Still, this made no sense. Why have a door that leads to nowhere? After 30 or 40 seconds of probing, Mikey calls down, "Hey, Lieu, I think I found a way in." I see him slide forward to his right, so I climb up next calling in, "Mikey, be careful. There's nobody in here, so take it easy." Mike's reply is to call for the nozzle: "I can see fire in here ahead of me." The crash guys pass the nozzle up. I move forward and find that there is a small opening about a foot wide at the forward edge of this interior wall. Now, I'm not a large person (or at least I wasn't in 1992), but I can't fit through this space with my SCBA bottle on my back. I realize Mike must have done the "reduced profile maneuver" to get in here, where the wearer slides the SCBA strap off the right shoulder and twists the bottle around to the left side, in line with the body, in order to pass through a tight space. This is a very useful technique in escaping from a room whose normal exit points such as doors and windows are blocked by fire or a collapse. It is not a good idea to use this to get oneself *into* the fire area, since that means escape from such a space is likely to be very difficult if the red devil is trying to bite you on your ass. I do the reduced profile, and squeeze through the opening—after all, Mikey is in there!

Once through, I find myself in a space about two feet wide that seems to traverse the width of the airplane, about 10 feet. I hear other firefighters pounding on metal on the far side from where we entered. John Kiernan starts to come into the space. I can't see him, but I hear him cursing about the bottle getting hung up. "This is crazy," I think. "If this thing lights up, we'll never get out." I order John to back out and keep the exit clear. Now I ask Mikey where the fire is because I can't see shit, but I know it's there. I can hear it crackling and feel the heat of free-burning fire. Working by touch, Mikey pulls me to the left side of the plane and then tells me to stand up and look along the ceiling. There seems to be a solid metal wall in front of us, but sure enough, at the ceiling level there is about an eight-inch space. "What the hell?" And just like he said, 20 feet in front of us is the fire. With both of our large lights shining along the "wall," it slowly dawns on me what this is all about. This is a *cargo* jet. These are cargo containers. They are inserted into the plane through a large door near the front of the plane and then slid on rollers down the interior to the rear, until the whole fuselage is full. These aluminum shells are igloo-shaped (in fact, that is their nickname), designed to take maximum advantage of the space inside an airplane that is stripped of its seats and overhead luggage racks. They fill the entire width of the plane except for about three inches on each side and the aforementioned eight inches on the top (and make that an Irish eight inches!).

Mikey is working the nozzle back and forth across the top of the igloos now, trying to knock down fire. It's a losing battle. We really can't hit the main body from here. It is burning within the inside of the igloos, and unbeknownst to us, in the cargo hold that runs along the belly of the plane beneath us. I really don't like this. Just then the 13 Division calls, "Rescue 2, what's your location?" I tell him, "We are inside the tail of the plane hitting fire in the middle with a handline, Chief." "Well, get out of there now!" he orders. "I don't want anyone in this thing. All the crew are out and I want to hit this with a tower ladder." Well, you don't have to tell me twice! "C'mon, Mikey, let's get the hell out of here." "But, Lieu, I'm still hitting fire." "You heard the boss, let's go!"

Back outside, we can see the futility of our actions. The whole plane is belching this really nasty looking brown smoke, and the fire has burned a hole through the aluminum skin, right where Mikey saw the fire. I report in to the Division, who is just briefing Assistant Chief Donald Burns. I describe the conditions inside the plane—how the fire has this void space to travel in all along the sides, but which we can't easily get streams into. The department's foam coordinator, Battalion Chief Pete Valenzano, has arrived, since our rigs are now applying flouroprotein foam on top of the fuel, which continues to leak from the plane's tanks in the wings and belly. AFFF is fine for a fast knockdown, but if I have to work in a fuel spill, as we do now, I want a blanket of thick flouroprotein on top of the AFFF, for flashback and burnback resistance. Pete and I have taught the department's Foam Coordinator's Course together since 1985, and we both agree on this fact. Now he asks if I have any thoughts about using high expansion foam to fill the cavities around the igloos to slow the fire spread. "I think that just might work!" is my reply. Pete goes to Chief Burns with the idea, and the next thing you know, a large high expansion foam generator is being mounted on the basket of Ladder 155's tower. Unlike AFFF, which is 97% water and 3% foam concentrate, mixed about six times that volume with air, high expansion foam consists of about 99% water, 1% foam concentrate, and about 1,000 times that volume of air blown into the solution by a fan (generator). It produces a very light set of bubbles that can be used to fill an enclosed space to cool the burning fuel and exclude air from the space.

Meanwhile, Division 13 orders Rescue 2 to get up on one of the wings and try to cut through the plane's skin so that we can get a stream of water going ahead of the advancing fire. The idea is to make a trench cut ahead of the fire to create a fire break. I tell Mikey and Richie to get two metal cutting saws off the rig and meet me on the wing. Now its Mikey's turn to look at me like I've got three heads.

"Are you sure you want the saws, Lieu? You sure you don't want to try the air chisel?"

"Well, bring both, Mike, but let's get moving." I see him head for the rig and I can see he's not happy. When he and Richie return with the tools, I'm standing on the left wing, which is no easy feat. The wings have been covered with AFFF, since the fuel tanks are leaking fuel that drips and puddles under the wings, but that makes the aluminum wings as slick as ice. I tell Richie to pass me up one of the saws; I'll hold it while he climbs up. Now Richie gives me the "What the hell" look.

"What's the matter with you guys?" I ask. Richie blurts right out, "Hey, Lieu, what about all this jet fuel we're standing in, are you sure you want to use a saw here? Remember that training tape you showed us about the plane crash that lit up and burned all those guys when they used a saw?"

Okay, so that's it. I know exactly what they're talking about. One of the training films we use in the Foam Coordinators Course is "The Foam Film" a 1976 film about all types of foam operations. The section I use most in Rescue 2 is a clip that shows an actual cargo plane crash in Los Angeles where the plane has not ignited initially. It does ignite in a ball of flame when the fire department begins to cut into the top of the cockpit with a power saw that is nearly identical to the one I want these guys to cut into this plane with now. The part of the film that sticks in everybody's mind is the aftermath scene, showing severely burned firefighters being treated, their skin literally falling off their hands and arms. Everybody in Rescue 2 has seen it several times—hence, the "what the hell look!"

I explain to them that I really do know what I'm doing. The plane in the film was a DC-3 that used high-octane aviation gasoline (AvGas) for fuel. We were standing on a tank of jet fuel. That made it safe! "Oh yeah, sure!" I have to go on, explaining, "Guys, the flashpoint of AvGas is 45 degrees below zero, the flashpoint of jet fuel is over 100 degrees above zero, and it's only about 35 degrees out here right now. And, we have a thick blanket of flouroprotein foam down that I'll have them apply around us constantly as we work. You couldn't ignite that spill with a cigarette lighter if you tried. Now, let's see if we can't get a hole in here to get ahead of this fire."

Their faces are still skeptical, but they climb up onto the wing. I have to butt myself up against Espo so he can work the saw without sliding off the slippery wing.

"I sure hope you know what you're talking about, Lieu," he says as he cranks up the saw. "Yeah," Richie chimes in, "we just better not end up as some training film, where a bunch of guys sit around at some firehouse saying, 'Boy, look at what those assholes did just before it lit up on them'."

I have to smile at the thought of this. I know where they're coming from. It's not until our hole is made and we climb down off the wing to make room for Engine 302 with a handline that the boys begin to relax. Now they're all happy. They're not going to be on some training tape after all (fig. 25-1).

The operation that day was a long, grueling battle, with an enemy that had all the advantages, lots of nooks and crannies to hide in, with a thick skin to protect it. The impact of the crash was enough to twist all of the aircraft's doors out of alignment, except for the one at the tail that actually popped open. Gaining access to the plane's interior took us the whole afternoon,

Figure 25-1. The smoldering wreckage of a DC-8 frustrated all efforts to get ahead of the fire.

using most of our, and Rescue 4's, tools, a huge forklift, and the assistance of the plane's mechanics. Eventually, the roof of nearly the entire length of the plane burned away, and we got the huge 10-foot × 12-foot cargo loading door open. Only then could we begin moving down the interior of the cargo hold with a handline knocking down spot fires. In the process, John Kiernan broke his nose when the Halligan tool he was using to pry some of the aluminum skin away from the airframe, so that a hoseline could be operated into the lower belly hold, sprung out at him like a spring-loaded ram. All in all, it was what Richie terms another "no impact fire," meaning that despite busting our asses for over eight hours, we had made no impact on the eventual outcome. We might as well have stayed on Bergen Street. The entire contents of the plane, mostly U.S. mail bound for overseas, was a total loss, as was the plane itself. It would take a whole week to finish overhauling the smoldering debris, salvage what mail could be saved, and clear the wreckage. Oh well, you win some and you lose some. On second thought—with plane crashes, you don't win many.

As tough as each of these incidents has been, none of them can truly compare to the total devastation that accompanied my next crash—TWA's Flight 800, which blew up at 13,000 feet, about 10 miles off the coast of Long Island on July 17, 1996.

I was at home that night when I first heard about the accident via a news broadcast. About an hour later, I received a telephone call from the FDNY's Special Operations Command asking about my status as a diver, was I medically able to dive (no head colds or other ailments that might prevent my ears from equalizing), and my availability for what could turn into a long-term operation. When I replied that I was available in all respects, the officer asked who else from Rescue 1 I would prefer to have with me. Right away I knew who I wanted: Paul Hashagen and Neil Fredericksen, both experienced rescue dive instructors, plus Cliff Stabner and Hank Molle, if available. After a few other questions about when I was next scheduled to work, and what size dive gear and how heavy a weight belt I use, the officer advised me to rendezvous with the rest of the team at the commuter carpooling lot at Exit 49 on the Long Island Expressway at 4:00 a.m. I asked the officer to have Hashagen and Molle call me if they would be going, since they live just one town over from me. I then set the alarm for 3:00 a.m. and went to bed. Naturally, I was so wired with the thought of being part of such a large operation that I didn't get any sleep. We were to be prepared to dive at first light.

I was up before the alarm went off. I tried to quietly slip out of bed and turn off the alarm before it woke my wife, Jeanne, but it was no use. She was awake also. For the 10th time, she urged me to "Be careful!" I told her not to worry, that this would be much the same as some of the sport dives I had done. I knew this was untrue, of course. I knew what plane crashes looked like; the miles of electrical wires that run through a plane could ensnare a diver, the pieces of jagged metal could rip a diver's suit, or worse yet, their buoyancy compensator (BC), which allows the diver to inject air into a bladder to help raise them to the surface. This is not all that much different than a lot of what we do in New York Harbor though, where we are diving for cars that have gone off a pier and the whole harbor floor is a tangle of rotting wooden pilings, littered with shopping carts, fishing lines, and the waste products of seven and a half million people. Another potential problem was highlighted on one of the news flashes describing the incident. The plane had gone down about 10 miles off the eastern end of Long Island in an area known to shark fishermen as "Mako Alley." That one had me thinking, even though, as I said, I've dived in similar waters before. At 3:15, Paul Hashagen pulled up in front of my house. I kissed Jeanne goodbye, grabbed my personal gear, and we both hopped into my Chevy Cavalier. Paul and I had a long-standing agreement, since he was my regular chauffeur at work, I would do the driving to and from work, while he did the rest of the driving. That way neither of us could criticize the other's driving.

We pulled into the carpool lot already filling with commuters heading toward the "rat race" at 4:00 a.m. on the dot. The SOC Scuba support van and a FDNY Chief's car were there already, along with the private cars of the rest of the team. I surveyed the group: Paul and I, Neil Fredericksen, Cliff Stabner (Hank couldn't make it), Phil Ruvolo, Terry Hatton (both Lieutenants); Phil, the "Scuba Weasel" Quatrochi, and John Emma, two of the FDNY's lead dive instructors; Richie Euler from Rescue 4; Bobby Haring from Rescue 5; and about six others that I did not know personally.

En route from our staging area in the commuter lot to the Coast Guard Base at Moriches Inlet, I quietly asked Paul Hashagen what he knew of the other divers. Paul had trained many of these guys, and I wanted his evaluation of their capabilities. Paul gave me the rundown on all of the guys he knew, but there were two or three that he hadn't really dived with. All those he knew were good, solid divers.

About three miles from the Coast Guard base, our caravan was met by a Suffolk County Police car, which would escort us into the base. Overnight, the whole media circus had arrived, along with throngs of curious onlookers, hoping for a glimpse of some morbid spectacle. The little two-lane country road leading into the base was packed solid with vehicles converging on what had been a sleepy little backwater (literally) outpost of the U.S. Coast Guard.

Once inside the security checkpoint, we are directed to parking spots in the middle of a large marshy field. I don't want to get stuck in this mess, so I try to ease our way onto some higher ground, but the U.S. Coast Guardsman manning this post was having none of that. "Pull in there where I told you, pal!" he admonishes. "I've got dozens of trucks coming that I have to find spots for."

After parking our non-essential vehicles out there at the perimeter, the group gathers around the Scuba Support van, which has been driven right down to the pier. Battalion Chiefs John Paolillo and Craig Shelley are already there with an inspector from the NYPD's Emergency Services Unit (ESU), getting a briefing from a Coast Guard Officer. Chief Paolillo pauses a moment and tells me to take charge of getting everybody outfitted with gear, then getting the gear checked out and the crew broken up into "dive buddy" pairs. The guys set about preparing for a long day's work as the dawn breaks over the pier (fig. 25-2).

The guys are just about finished squaring away the equipment when Chief Shelley calls us together for his briefing. Shelley has been out to the crash site most of the night, aboard the FDNY fireboat *Kevin C. Kane*. He quickly dispels any hope that we would be doing any real rescue work. "The whole ocean is covered in debris," he says. "The biggest piece I saw was about three feet by four feet. There is no hope of survivors from what I've seen." Well, that's what I expected. Shelley tells us to load our gear onto

Figure 25-2. FDNY divers check over their gear before loading aboard the fireboat Kevin Kane.

the fireboat. The *Kane* is a 52-foot boat, one of the newest in the FDNY flotilla at the time, but it is not designed for this application. We dive off of it back in New York Harbor, but then it's usually just a handful of people: four divers and one or two tenders, in addition to the boat's five-man crew of one officer, a pilot who "drives" the boat, a marine engineer who operates the motor and pumps, and two firefighters who do double duty as deck hands. Now, however, we are looking for places to store gear and find space for 23 extra divers and their supervisors. The boat is pretty crowded.

Shelley and Paolillo explain the game plan. The Suffolk County, Nassau County, and New York City Police Departments, as well as the Coast Guard, all have boats out at the crash site. Actually, it is not one specific site; it is several huge patches of floating debris. They are picking up any floating bodies and large pieces of debris that they find. They don't need us for that. The Coast Guard wants us for diving on the main body of wreckage, if there is such a thing. The problem is there is no good fix on where this plane actually went down. (That's because the 747 blew up in midair at 13,000 feet. Parts of the plane, including the nose, fell near the site of the blast, while the bulk of the fuselage continued flying on for a short period, actually climbing before it too fell into the ocean.) The Coast Guard Commanding Officer wants to hold the divers at the dock until the main body of wreckage is discovered.

Consulting our charts of the area, we discover that the water depth in this area is a relatively flat bottom, varying from 100 to 110 feet deep. Butch Foley from Rescue 4 has joined us. Butch lives nearby and he knows the waters well. He fishes and dives here regularly. He tells us it's probably a flat sandy bottom with visibility that runs from 10–15 feet at the bottom. That complicates things. The 110-foot mark is pretty deep for our gear, which is meant for relatively shallow, shorter duration rescue dives in the harbor, between 30 and 70 feet deep. Our single air cylinders don't hold enough air to allow us very much "bottom time." The depth makes you use air when swimming down and back up, and, if that's not enough, it requires that we do a decompression stop, where we hang 15 feet below the surface, to let our bodies exhale the nitrogen that builds up in your tissues and bloodstream when breathing compressed air at this depth. The decompression stop allows the body to bleed off the nitrogen slowly, to prevent the "bends." It also means we will only be able to do one dive a day per diver, due to the potential nitrogen buildup. I question each of the FDNY divers about the last time they have been down to 110 feet or more. All assure me that they have been that deep in the past year, although nobody has brought their dive logs with them to allow me to verify that. Another problem with this scene is the limited visibility, which means that each diver will only cover a very small area during their 10-minute bottom time. I had hoped to hear we would have 50–60 feet of visibility instead of 10–15 feet; that way we would be more efficient. Well, the Coast Guard Commander is right, we can't afford to burn out our divers just jumping into the water in the middle of a six-square-mile debris field. We need better targeting information.

Unfortunately, there is not a lot of technical help on scene yet. The Navy is sending several ships equipped for this purpose, including the USS *Grasp* and the USS *Grapple*, two specially designed underwater search and recovery vessels, but they are coming from Norfolk, Virginia and are several days away. The U.S. Geological Survey Office has a special ocean-mapping ship, the *Rude* (pronounced "Rudy," like our mayor's name), that is much closer, coming out of a job just off the southern Massachusetts coast, but they too are still several hours away. The only underwater search capability on the scene is the New York City Police Harbor Unit, which has just bought a portable side-scanning sonar unit, a remotely operated video camera unit, and a sonar "pinger" to listen for the "black boxes"—the cockpit voice and data recorders. They are going to put all this aboard one of their 37-foot harbor launches. The black boxes (actually, they are bright orange so they can more easily be seen after a crash) are equipped with an automatic sonar emitter that is supposed to activate if they get submerged. We hope that by locating the black boxes, which are mounted in the rear of the fuselage, it will help us locate the main part of the plane. (I have heard of people who always ask for seats at the rear of the plane when they fly so they hopefully will be found along with the black boxes!)

Chief Shelley leads me over to the NYPD launch. I am to be the FDNY liaison aboard the launch. The PD will go out and search for likely locations, and when something is found, the divers, either PD or FD, will then investigate further. I am there to provide coordination and communication between our units. As I get to the PD launch, I'm a little surprised to see my brother David, who is a NYPD ESU cop assigned to lower Manhattan. "What are you doing here?" we both ask each other. Dave explains that when the first reports of a plane down in the water came over, the PD ordered all of their on-duty divers to report to their Harbor Unit base, and they took them out by boat for the 90-mile ride in the dark. I ask what it looks like, if there's anything to do. He replies that it looks like someone shredded an airplane into tiny pieces and cast them out on the water. (This event was a real Norman affair. Not only have David and I responded, but also later I would find out that my other brothers, Joseph and Warren, had responded out the previous night with Robert Addona and other members of the Inwood Fire Department's dive team. They remained staged at the Center Moriches Fire House most of the night, but then the Coast Guard made the decision that they did not want to use any "volunteer" divers.)

Once the search gear was stowed aboard the launch, we made our way out to the debris fields. Ten miles is not that far on land, but when the land disappears over the horizon and all you can see in any direction is water, and you are in a 37-foot boat, the area we are talking about seems mind-boggling. The sea is incredibly still, eerily calm. I've been out on lakes that were not this flat! I know that the coating of jet fuel on the surface (oil on troubled waters) is responsible for most of this effect, but I can't help but believe that at least part of it is the ocean showing its grief over what has happened, and trying its best to make our job easier. It's spooky. As we pass through these huge areas of charred flotsam and jetsam, I am struck that this is all that remains of a huge jumbo jet full of hundreds of people. Occasionally, we pass a piece of debris that is recognizable and stop to pick it up. A child's sneaker here, a gift box that is still wrapped, on its way to a loved one who will now only associate the gift with sorrow, a briefcase with business contracts that mean nothing now to the executive who placed so much hope in them, a wallet with pictures of a young couple.

Occasionally, we pass bodies. We don't have the room to bring them aboard, or the necessary facilities to properly handle them. We call for other launches, usually Suffolk PD, to take them ashore. The explosion caused the deaths of all 230 people on board, but autopsies reveal that not all of them died instantly: some of them likely survived the terrifying two-mile plunge, until they hit the ocean's surface.

We have been stopping periodically to let Detective Alan Kane dip the pinger in and listen for the black boxes. So far, we have had no luck. Kane seems to think it might be because of the boat's steel hull, which might be masking some of the sonar's reflection. We put a 12-foot fiberglass boat over with Alan and another cop in it, to get away from the bigger boat. The two men move off about 75 feet and begin listening, hoping for that faint noise that will signal we are nearing the right location. They search for 15 or 20 minutes without success. They are about to head back toward us when the biggest damn shark I have ever seen swims right up alongside their boat. I am about to call out to the rest of the crew, whose attention is fixed on some debris on the other side, when the thing submerges. Now, I'm no shark expert or anything, but I know one thing about this one—it's BIG! As it passed by the 12-foot long punt, there was at least two feet of shark past each end of the boat! I can't help but think of the line in the movie *Jaws* where Roy Scheider has just seen the shark up close while chumming and tells Captain Quint, "You're gonna need a bigger boat!"

The pinger has not found any indication of the black box's whereabouts. Inspector Gianelli, the NYPD's officer in command on the launch, decides to deploy the side scanning sonar, that is towed behind the boat and is linked to a computer in the wheelhouse. The *Rude* is on scene now and is using its far more powerful units to search for any anomalies on the ocean bottom.

We begin to do the same, making 10-mile long sweeps at about five miles an hour. Since our unit can only survey about a 200-yard wide swath with each pass, at this rate it'll take days, if not weeks, to fully chart the possible impact area.

We keep making our sonar runs for the rest of the day. At one point, Al Kane thinks he has detected a slight disturbance on the ocean floor. It doesn't seem to be large enough to be a major piece of wreckage, so there is no point in calling out the divers, who have been sitting dockside all day waiting for a hit. Kane decides it might be worth a look with the robot camera. We return to the site and deploy the remotely operated device, which is a large yellow cylinder, about the size of a small trashcan. It is equipped with miniature propellers, to move it side to side and up and down, as well as forward or backward. Kane guides the camera toward where the sonar had indicated a variation in the smooth ocean bottom, but the camera shows nothing but a slight hump in the sand, no sign of anything manmade.

On the way up, the robot has a major malfunction. The video feed is lost, and the remote controls don't respond. The device has to be winched aboard. One less tool in the toolbox to help in the search. It is now 8:00 p.m. and twilight is descending. The pilot of the harbor launch wants to get back into the base before dark, since the inlet is a particularly treacherous body of water. The channel that the vessel must navigate through is very narrow and twisting, with shallow mud flats that creep right up to the channel. The pilot doesn't want to take a chance of missing a buoy in the dark, unfamiliar waterway and running aground. We've spent 14 and a half hours searching a vast area of ocean with nothing to show for it. We head for home. After a debriefing and getting something to eat under some hastily erected tents, I rejoin the rest of the FDNY entourage. They look like hell. They have spent the day on the fireboat and nearby dock waiting for the call to come help, which never came. They are restless, anxious, and sunburned. Deputy Chief Jim Bullock, head of the Special Operations Command, calls everyone together for a meeting. He tells us that the *Rude* and another National Oceanographic and Atmosphere Administration vessel with side-scanning sonar will continue mapping the sea bottom all night and will hopefully have something for us to dive to the next day. The chief wants us all back at the dock ready to go tomorrow at 6:00 a.m. Since it's now almost 10:00 at night, he tells us all to go grab some sleep.

The federal government arrived en masse while I was out "yachting," and the entire compound has filled with command post vehicles and the staff of numerous federal, state, and local agencies: the NTSB, FBI, FAA, and all the rest of the alphabet soup of bureaucracy. There are so many vehicles pouring in that the scene has turned into a major construction site. Dump trucks and bulldozers are busily at work paving over marshland with gravel and sand to create a larger parking area, helicopter landing pads, and logistical support areas. Banks of portable toilets have been trucked in and utility crews are busily installing new water, power, and electrical lines to meet the demands of the swelling population of hundreds of rescue, recovery, and investigative workers. Paul and I decide to head for our own beds for the night, instead of staying in the rooms the authorities have arranged for us at Brookhaven National Laboratory. This will mean an extra 40 minutes on the ride each way, but we both agree it'll be worth it to sleep in our own beds, without the distraction of dozens of others coming and going all night.

On the way out, we have to run a gauntlet of media reporters sticking microphones through our open windows asking for any information about the incident. "Did you see any bodies? What shape is the wreckage in? Do you know what caused the crash?" Rumors fly as we drive through. "Did you hear about the missile?" is one snippet of conversation I overhear. "Missile," I say to myself. "Well, that's a possibility." Some asshole terrorist with a grudge against the United States, TWA, or both, gets an anti-aircraft missile onto a boat just beyond the horizon and waits for a fat lumbering 747 to pass by and blow it and 200 or so people into Kingdom Come. Shit, it makes more sense to me than a perfectly good 747 just deciding to blow itself up a half-hour into a flight, especially in the post-Gulf War era with Saddam Hussein still running U.N. weapons inspectors out of his country every time he feels like it. That's something to sleep on.

The next day dawns bright and clear, promising a repeat of the previous day. Overnight, additional resources have arrived, including a group of U.S. Navy divers from an Explosives Ordinance Detachment (EOD). Chief Bullock and Chief Nagel from the NYC Mayor's Office of Emergency

Management have arranged to have us teamed up with the EOD guys for the operation. The sonar scans overnight have located several potential wreckage fields. The EOD guys have all the high tech "toys" that can speed up an operation like this, including hand-held underwater sonar units that a diver can use to home in on a target in low visibility conditions like we face.

The problem is there are only eight of them, and they face similar bottom-time restrictions as we do. Once they have worked on the bottom and surfaced, medically they cannot do another bottom dive for 24 hours. The idea is to have them use their guys to locate and examine possible targets, and then if conditions warrant, we will use our 24 divers as "mules," doing the less technical tasks of bring up the bodies. Our main mission is body recovery. We are not to disturb any wreckage more than necessary to remove bodies. The NTSB and other investigators will want as much left intact as possible to aid their task of determining the cause of the explosion.

We leave the inlet just after dawn, we on our 52-foot red fireboat and the EOD guys on their little 25-foot ocean-camouflaged, rubber inflatable, which carries more high-tech goodies than you can imagine. The EOD leads us to the first site plotted by the sonar ships. They put their first diver over. These guys free dive, with no tether line to the surface or even an anchor line to follow down. The diver carries a "pelican" float that he can attach to the debris pile and release to float to the surface to mark the spot if he finds anything. Then we can send down our teams of divers to begin the recovery. The first diver seems to be down a long time, with no sign of a pelican or anything else. We have to stand off about 300 yards from the divers' descent point, since in the open ocean we have to keep our screws turning to keep from being carried away by wind and tide, and we don't want to risk being in the diver's way as he ascends. Finally, after what seems like an hour, the EOD crew gives us the thumbs down. The site was clean of debris (fig. 25-3).

We move off several hundred yards and repeat the process. Again, the bottom search reveals the "blips" on the sonar screen to be nothing related to the plane crash. The south shore of Long Island is littered with relics of shipwrecks dating back to the 1700s, and some of these are what show up on the sonar. The day continues in this fashion until late afternoon, when the EOD is down to one diver. This "hit" on the sonar was not particularly strong, so the command staff has left it for last to investigate. The Navy diver is just getting ready to go over the side when the command post radios us with word of a strong thunderstorm approaching. They want to know how long it will take to check out this site and get back in to shelter. I check with EOD and they say 15 minutes of descent and ascent just to hit bottom, plus whatever it takes on the bottom. Command radios to "can the rest of the operation and head for shore." The first bolt of lightning on the horizon convinces us all that the decision is correct and there's no time to lose. The EOD crew takes off and leaves us in their wake. The *Kane* can do about 20 knots in the open water, but the inflatable raft with its twin 155 hp outboards must be doing 50 knots! Well, they have no cover. The *Kane*, at least, has a cabin that we can squeeze everyone into in a pinch.

By the time we make it to the Moriches Inlet, it's pouring buckets. This inlet is, as I said, very difficult to navigate. The pilot wants some help in navigating, since even with the windshield wipers going, the rain makes visibility very limited. I don an exposure suit and take a position up near the bow with Cliff Stabner, who is dressed in his dry suit. We act as lookouts, giving

Figure 25-3. Richie Euler and Cliff Stabner aboard the fireboat *Kevin Kane* after a disappointing day searching the Atlantic for Flight 800 wreckage.

hand signals to the pilot. Even with the cold-water exposure suit on, I am getting soaked. I imagine the old salts up in the "crow's nests" of sailing ships in times past, as they must have fought through storms to reach safe harbor. No wonder there are so many shipwrecks along this beach!

After a week of this, the Fire Department is phased out of operations as more and more Navy resources arrive. I am relieved by Terry Hatton and Phil Ruvolo, newly promoted captains who are surplus, not regularly assigned to a company. I have Rescue 1 to run, so I am taken off the crash detail and returned to my regular assignment. That's just fine with me. I've seen enough of what plane wrecks can do. I am more than happy to return to work in Manhattan, where the more mundane disasters await. There is no way that I could have imagined that that is where my next airplane encounter would come some five years later at the World Trade Center.

My final (hopefully) airplane disaster occurred two months after September 11th. On November 12th, 2001, I had been designated as the Chief of Special Operations for nearly the entire two months, and as such I had been chosen to represent the FDNY at a conference concerning the World Trade Center disaster, which is being held at Columbia University. The conference had started at 9:00 a.m., and as I was sitting in the front row, waiting for my turn to speak, my pager went off, announcing a 5th alarm for Box 1407 in the Belle Harbor section of Queens, on the Rockaway Peninsula. My initial reaction was "What the hell is wrong with this pager? Why did I not receive the notifications about the 2nd, 3rd, and 4th alarms?" Well, there was nothing wrong with the pager. Chief Howie Carlson of Battalion 47 had immediately sent the 5th alarm on his approaching the scene of yet another disastrous plane crash; American Airlines flight 587 had suffered a catastrophic mid-air failure while taking off from JFK, and plunged nearly nose first into the middle of a quiet residential neighborhood. The crash instantly flattened three homes, killing all 260 aboard the plane and six on the ground.

The ensuing fireball ignites about a dozen homes on three sides of the large intersection near the main impact site, and the engines, which had snapped off in midair and landed in two separate locations several blocks away, ignited several other large three-story wood frame homes. The Rockaway Peninsula is approximately nine miles long, projecting to the west from its border with Nassau County, including my hometown of Inwood. Fire protection on the peninsula consists of seven Engine Companies, three Ladders, and the Chief of the 47 Battalion. All additional assistance comes from many miles away, either across one of two bridges, or by going around JFK airport and traveling down Rockaway Boulevard through Nassau. Chief Carlson knows he is in a desperate situation: there is heavy fire in several locations, with tightly packed wood frame dwellings throughout the area. The prevailing winds come off the Atlantic Ocean, threatening to spread fire rapidly. The area has seen previous conflagrations, one of which burned out over 13 densely packed acres, with more than 140 structures destroyed. The stage was set for yet another such inferno.

Chief Carlson did a magnificent job deploying his scarce units, augmenting their personnel with many off-duty firefighters who live on the peninsula, and immediately reported for duty at the scene. By positioning units with master streams at critical locations, fire spread was limited to only a few buildings beyond those that were already involved on arrival. By the time I arrived, the fires were largely under control. Now the job became one of tremendous difficulty: digging through the smoldering remnants of the plane and the homes to recover the bodies of all the missing. It would take two days of heartache and sorrow (fig. 25-4).

Figure 25-4. Firefighters go about the grim task of recovering the dead from the wreckage of Flight 587 in November 2001, impacting a residential neighborhood that shook many New Yorkers still reeling from the collapse of the Twin Towers only two months earlier, and sparking renewed fears of terrorism.

CHAPTER 26

When You've Got a Great Act, Take It to Broadway!

My time in Rescue Company 2 was brought to a close with my promotion to captain in September 1993. The four and a half years I spent on Bergen Street were some of the greatest in my life, but as the saying goes, all good things must come to an end. Newly promoted officers of all ranks are expected to "bounce" around from vacancy to vacancy wherever the need for an officer of that rank exists. As a new captain, I was assigned to the 15th Division, covering the southeastern third of Brooklyn.

The companies in this area are all good companies, including my first unit, Engine 290, and while I enjoyed my times back on Sheffield Avenue, another old saying is equally true: you can't go home again. It is never the same as it was the first time. When the chief of the Special Operations Command (SOC) called to ask if I was interested in a position with the Rescue Liaison Unit, I jumped at it. The Liaison Unit was established several years before to coordinate the unique skills of the Special Operations Units with activities of other fire department units and other city agencies at incidents such as building collapses, pier fires, and train wrecks, as well as at the scene of multiple alarm fires. Unfortunately, about one month after I was assigned to the Liaison Unit, it was disbanded (it would be reincarnated four years later as the SOC Battalion, with chief officers replacing the Liaison's captains). Fortunately for me, there were several promotions and retirements among the five rescue companies at about that time. Brian O'Flaherty in Rescue 1 and Ray Downey from Rescue 2 were promoted to battalion chief, while Charlie Driscoll in Rescue 5 retired. This created several vacancies at just the right time.

I was assigned as the relief group captain covering in SOC, working one day and one night in each of the five rescues and in Squad 1, on a rotating basis. The relief group has always been called "amateur night" in the rescues because it was always filled by a new officer who was waiting for a permanent assignment. Some officers hate the relief group since it means bouncing from house to house every tour. You live out of your car, schlepping your turnout gear and uniforms with you every tour. Also, there are no 24-hour mutual exchanges available, which give a few days off in a row. The relief group works the straight work schedule, meaning it's much more difficult to get time off for family events like weddings and birthday parties. There is also a lot more travel involved, since in Special Operations, the units are spread out all over the city. The schedule had you working one night in Brooklyn, the next in Manhattan, then the Bronx, Brooklyn again, then Queens, and then Staten Island, before starting all over again. I liked it! I liked the variety of buildings and types of incidents it offered. I liked working in the heart of the midtown high-rise area one night and the heart of the poorest neighborhoods the next, going from a "yuppiefied" neighborhood to near suburbia and back to the epitome of urban blight. You never know what type of problem you could face when you walk into the firehouse. It's totally unpredictable. My favorite saying was, "If I can't be the captain of one of the rescues, let me be a captain in all of them!"

I did that routine for a year, until it too sadly came to an end—sadly, because of what created an opening for me to be permanently assigned. Chief of Rescue Services Ray Downey called me in October of 1994 and told me he was putting me into Rescue 1 UFO (until further orders). While I had worked in Rescue 1 many times as a lieutenant, working overtime to replace a lieutenant who was on vacation and as a covering captain, the sudden realization that I could now be *The Captain* of Rescue Company One came as a shock. There were so many great candidates who would love this spot: Patty Brown in Ladder 3, John Vigiano in Ladder 176, and Jimmy Ellson in SOC, to name just a few. I spoke with some of my best friends/mentors: Jimmy Curran, a former lieutenant in Rescue 1 whom

I had worked with closely starting the fledgling Rescue School, and Steve Casani, who was a current lieutenant there. Both asked the same question: "Do you really want it?" If someone had told me it was possible 10 years earlier, I would not have believed them. Jimmy's advice was simple: "If you've got a great act, you have to take it to Broadway!" Steve wholeheartedly concurred. I still wasn't sure.

I had worked my day and night tours in Rescue 1 as part of the relief cycle just two weeks earlier and had spoken with Jimmy Rogers, the captain at that time. Now Downey told me that Jimmy was hospitalized with cancer, and that it looked like it had spread to his stomach and lymph system. Jimmy had shown no sign of being that sick. I don't know why it is always the good guys that get hit with stuff like this. Mercifully, it was over quickly. In a matter of weeks, Jim went to that great firehouse in the sky, where he could break in a whole new crew. Before he died, Jimmy was able to give me his blessing and advice and guidance on how to handle my new assignment.

Rescue Company 1 is very different from Rescue 2 where I was a lieutenant, and Rescue 3 where I had been a firefighter. Rescue 1 does far more technical rescue work but less everyday fire duty than either Rescue 2 or Rescue 3. But when they do have a fire, they are very often high-rise office buildings or apartments, which have unique challenges. Because of its location in the heart of the Manhattan high-rise and theater district, Rescue 1 has a greater degree of interaction with the news media, city officials, and executives of some of the nation's largest corporations. The work might see you walking right into the corporate boardroom of a major company to rescue a window washer trapped on a faulty scaffold on the 50th floor or provide standby crash rescue protection at the helicopter landing zone for the President of the United States. The officers of Rescue 1 still wear neck ties while on duty during day tours, while none of the other Rescue officers do.

The high profile nature of the unit has earned it the nickname the "Broadway Rescue." Since the headquarters of all of the major television networks are within a mile of Rescue 1's quarters, they tend to get a lot of air time. A fire in midtown Manhattan often has TV coverage from the ground as well as the air, as traffic helicopters and the ubiquitous skycams mounted atop a number of high-rise buildings zoom right in on even a one-alarm mattress fire. Meanwhile, out in the other four boroughs of the city, you could have a raging five-alarm fire in an orphanage with 50 rescues made, and that gets no film coverage—if they're lucky and it's a slow news day, maybe it gets a brief mention on the 11:00 news. Ah, the pressures of working in midtown! Nobody ever said life was fair.

One source of Rescue 1's many responses is a city ordinance, Local Law 11 of 1990, which instituted a requirement that the exterior facades of all high-rise buildings be physically examined for structural defects at least every five years. The law was passed after a large piece of a stone façade came loose and fell to the street, killing a passing college coed. Of course, when the law was first passed, it was practically ignored. After all, building owners had five years to comply! By 1994, however, the building department was cracking down on the owners and a tremendous boom in inspections was developing. This had two effects—one was to cover the exterior of nearly every high-rise building with suspended scaffolds, the kind that are anchored on the roof and move up and down along the façade along ropes or cables; and the second effect was to deplete the number of experienced scaffold workers. There was more demand than there was supply. As a result, a large number of unskilled laborers were being hired to work on these scaffolds. Very often this included recent immigrants who spoke little or no English and had difficulty understanding instructions. The combination of these two effects resulted in a dramatic increase in the number of workers stranded on the outside of high-rise buildings on malfunctioning scaffolds, workers injured when these malfunctions occurred, and a number of mishaps that involved scaffolds or debris falling from them to the street below. All this kept Rescue 1 quite busy! By late December, we were responding to several of these episodes a day. One of the busiest days was actually the Sunday that Rescue 1 had planned our families' Christmas party. All the kids had gathered on the apparatus floor waiting patiently for Santa to arrive, but Santa was busily running from one scaffold to the next.

The first call is for an unattended scaffold in lower Manhattan which is being blown around by high winds. The heavy scaffold is being slammed into the walls of the building with such force that it is knocking bricks loose from the 23rd floor. The wind is so high that it has spun the scaffold around the corner of the building. Since there is no life hazard on the scaffold, we will be extremely careful in securing it. We don't want anybody getting hurt in the process. We send a team up to the 24th floor with some Halligan hooks. The trick is to lasso the swinging scaffold and tie it down securely. Unfortunately, the way the thing is moving with the wind, throwing the ropes around, it takes several attempts. Even with hooks to extend their reach, it takes time to pull the ropes in, since the wind pulls the scaffold back out of our hands. The last thing we want is someone to be pulled out a window by this thing. Everyone in proximity to the windows is tied off with a rope. After about 20 minutes, it's secured.

The trip back to the firehouse has a bit of comedy. Santa is trying to get dressed for his appearance when we get a 10-75 up in lower Harlem. Off comes the Santa suit and back on goes the bunker gear. We get to 96th Street before the 11th Battalion tells us we are not needed. Off comes the bunker gear, back on goes the Santa suit. As we turn into 43rd Street, we get another scaffold emergency, this time way downtown on Greenwich Street. It too is unoccupied and being blown around by the wind. This one is a relatively small aluminum model, but we used up most of our utility ropes at the first incident, so we decide to dismantle this one and bring it into the building. The kids wait another hour for Santa. On the way back, we get sent to a Class 3 automatic alarm at the Empire State Building, and then a gas leak on East 13th Street. Santa is really late! Finally, we head back to quarters. We pull up just as the *tooo-dooo* sounds inside. Fortunately, Santa is ready for his show and Joey Angelini is ready to replace Gary "Santa" Geidel on the rig. This run is another scaffold job, but this one is a lot more serious. Three workers are stranded on a scaffold 47 stories above 57th Street. Normally, a stalled scaffold is not a life-threatening emergency. The workers are not usually in imminent danger and can wait for a mechanic to solve the problem with their electric hoists. If things are more serious, it is usually a relatively simple matter for us to remove a pane of window glass and bring the stranded workers inside. This is not one of those cases. This one is serious.

The high winds are really rocking this scaffold, swinging it out from the building several feet, and then slamming it back against the wall. The workers are on the verge of panic. Darkness is rapidly descending, and the temperature is plummeting. To complicate matters, the workers are in a section of wall that is nearly solid granite panels, three stories below the roof level. We have only a few options. For one, we can set up a hauling system, lower a rope over the edge of the roof to the workers, and haul them up to the roof. The other is a bit more work but may be a little safer. Just off to the side of the scaffold, about three feet away, is a fresh air intake for the building. This is a four-foot by four-foot grating into which outside air is drawn to provide fresh air for the building's heating, ventilating, and air conditioning (HVAC) system. We send a team from Rescue 1 and Ladder 2 down to try to locate that grate, to see if we might be able to use it to bring the workers in through. After a few minutes, the scouts locate what they believe is the duct-work leading from the grating. We try to get the workers to confirm that this is the duct by having our members bang on the steel with tools, while on the roof we ask the workers if they hear the banging. There is no reply from the three men. I don't know if they can hear our shouted questions. They're about 40 feet below us, and the wind is really howling. We lower one of our radios down to them and repeat the process. Still no answer. We lower three separate safety lines to them and tell them to hook into their harnesses, but they just look at the ropes and back up at us like they don't have a clue, which they don't. Just then a representative of the inspection contractor shows up. He tells me the men are all Polish immigrants who speak no English. He has a translator on the way. Ladder 2's officer is convinced that he must be in the right place, so he starts cutting a hole in the gigantic ductwork, but first he has the HVAC system shut down. The reciprocating saw slices through the steel easily enough. But once the panel is removed, the grate is exposed and the next challenge is revealed. They're in the right place all right, but the grate is a heavy stainless steel mesh. The Sawzall can't cut through this heavy steel as easily, so another plan must be developed. We send a guy back down to the street 50 stories below for the metal cutting K–12 saw.

The translator comes up in the elevator along with the firefighter with the saw. He radios to the workers what is happening. Two of the men are fine with the plan, but one of them wants no part of it. He doesn't like the idea of stepping off the platform to reach the opening with 47 stories of air underneath him. I tell the translator that it will be perfectly safe. We'll have a safety rope on him from the top, and another rope attached to him from inside the grating, so that he can't go anywhere. The man is still unconvinced and has a white-knuckle grip on the scaffold rail. It takes a particularly nasty gust of wind, which pulls one side of the scaffold away from the building and then slams it back into it, to make the man see the wisdom of our plan. Ladder 2 starts cutting the grate with the saw, which sends a shower of sparks out into the darkened sky, lending an eerie appearance to the scene. It just looks spectacular. To ensure there is no danger to anyone, the area below has been cordoned off by the police in case anything falls from the scaffold during the effort. In just a few minutes, the grate is removed and the workers are secured with their safety ropes. The transfer from swinging scaffold to safety takes barely two more minutes, but it's not the end of our problem.

We still have the matter of the scaffold itself to deal with. Now it is swinging even more violently, since the weight of the workers has been removed. This one is too large to dismantle and since there are no windows above and below it, we can't simply tie it down. We will have to haul it back up to the roof to keep it from causing further damage. The normal hoist system has malfunctioned, which is what caused the problem in the first place, but the davits that support it (the crane-like structures that extend out over the side of the building which support the scaffold) are still in place. We can use them as the anchors and attach our rigging to the existing wires and "inchworm" it back to roof level. We send for our griphoists from street level and begin the painfully slow process of hauling the offending scaffold out of trouble. The griphoist is a very safe, reliable mechanical hoist that is operated by a hand lever. It can support loads of over eight tons, but it hauls very slowly only about two inches per stroke of the lever. We have to haul 40 feet!

By the time the scaffold is hauled up and secured, and all of our tools are brought down to the rig, it is almost 8:00 p.m., well past the end of our shift, and also the end of the kids' Christmas party. All of the families have left, and the incoming night tour guys have cleaned the mass of wrapping paper from the apparatus floor. We turn on NY1, the local cable news channel, to see if any of the day's events have made the news, but not a word is mentioned.

Who would have guessed that would happen in mid-town? I guess all the news folks were at their own Christmas parties.

Little did I know that that day would mark the start of a run of incidents that I call our "griphoist exposition." As I said, the tool is very reliable and powerful, but it is not very commonly used. In the four years that I was a Lieutenant in Rescue 2, I only used it once at a real incident, although we trained with it and all the other tools regularly. Now at Rescue 1, I was going to use it a lot!

The next opportunity came less than a month later, January 16, 1995. A very large movie theater was being demolished in Jersey City, New Jersey that afternoon when something went awry. A tremendously large section of two-foot thick brick wall broke loose at the sixth floor level and fell outward, right onto a two-story commercial building next door. The section of wall smashed its way right down through the wooden roof and floors of this structure, crashing down through a secretarial school and a Woolworth's department store, finally coming to rest on the concrete floor in the employees' locker room in the cellar of the store. The Jersey City Fire Department responded and began initial operations. They examined the site for surface victims, identified and secured site hazards like live electric and gas lines in the area, but soon realized the scope of the problem was greater than they could handle alone. Located right across the Hudson River from Manhattan, Jersey City is only minutes away from the Holland Tunnel and a virtual cornucopia of collapse expertise and equipment. The chief in charge wisely decided to ask for help.

The initial mutual aid request dispatched Rescue 3 and the Collapse Utility Unit, as well as BC Charlie Kasper in the Special Operations Battalion. (Kasper had been the captain of Rescue 1 for a time in the early 1990s, after Captain Brian O'Flaherty had gotten promoted to Battalion Chief, then Kasper himself was also promoted to BC.) Rescue 3 worked with Jersey City firefighters (probably for the first time ever) to secure the remaining large sections of free-standing wall overhead and began shoring the edges of the roof and floors that remained, allowing them to work near the edges. The examination of the rubble did not reveal any likely void areas, so operations focused on general debris removal. FDNY and JCFD members worked side-by-side, cutting and pulling debris out of the mountain of brick and splintered timber by hand, but the task was enormous. The debris pile was about 10 feet high, 50 feet long, and 20 feet deep. Chief Kasper requested additional reinforcements from the FDNY. Rescue 1 and Battalion 6 responded. BC Brian O'Flaherty was working in Battalion 6 that night, a fact Kasper knew from consulting his Chief Officers Assignment Sheet, which meant that three present or former Rescue 1 Captains would be working together: rare enough in New York, but rarer still in New Jersey!

On our arrival, Charlie briefs us on conditions. The loose brick and timber is being dealt with OK, but there are still several very large pieces of what used to be the wall pilasters (thickened sections of brick that are built into the wall, like columns, that act to give the wall greater stiffness). These pilasters have remained intact and are impeding the search. We have a couple of options. We can bring out the jackhammers and break them into smaller, more manageable size pieces, or we can try to move them whole. I suggest the latter. We have four of the large T-32 griphoists between the three New York City rigs, plus all the cable and slings we could need. It will be much faster to sling them and move them than it will be to break them and then move them. We set to work, and within a few minutes the first of these massive blocks is on its way down the debris pile. At the bottom, we have laid a bed of two-inch pipe that we carry for use with the trench jacks. We pull the blocks onto the pipes, and then it's a simple matter for two firefighters to push them over into a corner out of the way, much the same way as the ancient Egyptians built the pyramids. As each of the massive pieces are moved out of the way, a group of firefighters moves in, removing the loose debris that remains behind, until they encounter another immovable piece. Then the process begins again. After 14 hours, the pile has been cleared down to the foundation, all by hand and with the help of the griphoists. Miraculously, not one person was in the collapsed area. This was due largely to the date of the event. The collapse occurred on the day that Martin Luther King Jr.'s birthday is celebrated. The secretarial school was closed for the holiday; otherwise it's likely that 30 to 40 people would have been killed.

A few months later, we respond to yet another unusual "griphoist event." About 7:00 a.m., an oil truck attempts to back into a Manhattan auto repair garage, to pump out the used motor oil storage tank for recycling. No sooner has the truck backed in when the concrete floor beneath it collapses, dropping the truck into the cellar. The 7th Battalion, with Chief Tom Healy in command, arrives shortly before Rescue 1. He transmits his preliminary report describing the condition and concludes by telling the dispatcher to "have Rescue 3 and the collapse rig continue in." A search of the area reveals no one trapped and no serious threat of fire or secondary collapse. The question now is, how do we get this truck up out of the cellar? I tell Chief Healy, who is a former Rescue 2 man himself, that I have a plan, but it will take a lot of lumber. We'll need Rescue 3's collapse rig for its 6' × 6' and 4' × 6' timbers and lots of box cribbing.

I propose to crib up under the frame and insert a set of air bags, lift the truck enough to set the 6' × 6' timbers beneath the wheels (spanning from bearing wall to a steel column line and beam below) then drive the truck out on this makeshift deck. Chief Healy is not too sure of this, but I tell him this is exactly the way that Rescue 2 and I had lifted an engine company pumper that had fallen through the boardwalk in Coney Island about four years earlier. Rescue 3 arrives and we lay out the plan to Lt. Gerry Murtha and his crew. Gerry is not sure if they have enough lumber for this and proposes another plan. He suggests rigging two of the large griphoists to a steel I-beam just to the rear of the truck at the first floor ceiling, and hooking the Rescue apparatus winch to the oil truck's front axle, which is still on the sidewalk in front of the garage. We will use the two griphoists, first to raise the truck up out of the hole, and then as the winch pulls the truck forward, the griphoists will be operated in reverse, letting out on the cable at the same rate as the vehicle is being pulled forward.

It sounds pretty complex, but with a little coordination it should work. We set up the rigging. Patty O'Keefe and Warren Forsyth set a pair of slings around the front axle while Nicky Giordano, Bob Athanas, Ray Meisenheimer, and Chris Blackwell rig the griphoists to the rear frame of the truck. The street that the truck faces onto, 27th Street, is too narrow to get the rescue rig in with the winch in line for a straight pull. Patty O' and Warren are working on a directional pulley using a snatch block, which will allow us to keep the rig straight and use the winch to "pull around a corner," when a heavy duty tow truck arrives, summoned by the oil truck's owner. This rig is only about half the length of our rig, so we ditch the snatch block idea and put the tow truck in place of the rescue rig. We explain the plan to the tow truck driver and run his winch through a snatch block and anchor it off back to his own frame, then connect it to the slings on the front of the oil truck. This will slow the speed of his winch by half, allowing Ray Meisenheimer and Chris Blackwell to keep up with his pull with their manual operation of the griphoists.

Finally, all is set. Gerry and I have calculated the load of the truck and its contents to be about 24,000 pounds, which is well within the capacity of the two T-32s (34,000 lbs.), especially since we don't have to actually lift the front of the truck; the front wheels are on the sidewalk, taking almost half the load. The only thing that concerns me is what effect the eccentric load will have on the steel I-beam at the roof level that we have rigged the griphoists to.

Hauling the truck up was no problem. The load was nearly vertical down the column line. When we start pulling the truck forward, though, the griphoists are pulling sideways on the beam. To be sure we don't cause a roof collapse, we add some cable slings to the beam, tying it back to the next line of columns. We order all nonessential personnel out from under the area anyway, just to be on the safe side. The operation goes off without a hitch. Then, just as the rear wheels touch down on the sidewalk, the radio calls for Rescue 1 to respond to a 10-76: a high-rise office building fire. We take off in a hurry, leaving Rescue 3 to finish up, glad that we decided to let the tow truck take the job of winching the truck forward. Otherwise, we'd be stuck here for another 5–10 minutes while we rewound the cable on the winch and stowed all the rest of the rigging equipment. That decision freed us up to be available to take in a working fire in a high-rise office building on Williams Street, way down in lower Manhattan. Sometimes, things do go our way (fig. 26–1).

The next week brought a series of unique responses that had us all commenting on the unusual nature of the events and how "bad things come in threes." The first incident involved a guy who had fallen down a narrow shaft in the rear of a commercial building on 30th Street late one Friday night. We responded to a report of a "man trapped" with no other details.

Figure 26–1. Rescue 3 manually hoists a fully loaded oil truck up from the cellar into which it had crashed. From left to right are Bob Athanas, Ray Meisenheimer, Chris Blackwell, and Nick Giordano.

At the scene, a group of people direct us to the rear of a parking lot adjacent to a hip-hop club. The pounding bass emanating from the club makes it difficult to hear in the street, but the people insist that they were passing by the parking lot, "minding their own business," when they heard a person calling for help. Ladder 24 puts its aerial up to the roof of a one-story extension at the rear of the lot. Sure enough, they call down to us that there is a guy wedged into the bottom of a shaft about 20 feet below roof level. We take our high angle and confined space gear bags up the aerial ladder and try to figure out how to get to a high enough point to lower a rescuer down to him and then haul them up. Our tripod for this purpose won't fit on the edges of the walls that surround the shaft, and we can't get the aerial ladder directly over the hole because of a projection of the building line. We end up setting a portable ladder across the gap and leaning it against the building across the shaft. Warren Forsyth is the "meter man" for the tour and lowers a probe down into the shaft to test the air for any chemical hazards.

It's an open-topped shaft, so that is unlikely, but it pays to follow procedures every time.

Joey Angelini has the designation on this tour as our first entrant for confined spaces: part of our assignments issued at roll call at the start of each shift. We designate first and second entrants: a rigger who is responsible for setting up the rope systems for hauling and lowering, and a meter man to monitor the atmosphere inside a space. Lloyd Infanzon is the rigger tonight and has suspended a four-to-one mechanical advantage system that will carry our entrant. Lloyd gives the signal to the firefighters of Ladder 24 who are manning the rope to "prepare to lower." Joey gives me the nod and steps off the wall to begin his descent as I signal 24 to lower. The members of 24 pass the rope hand-over-hand, controlling his descent. The first thing Joey does on reaching the victim is evaluate his ABC status: airway, breathing, and circulation. The guy is lying nearly face down, almost doubled up, wedged between the walls in a narrow space. I often designate Joey as our primary entrant for this particular reason. The small, wiry man can get into an amazing variety of positions. The "bruisers" like Lloyd and Warren provide tremendous strength but could never fit into many of the spaces we have to enter. It takes a team!

To reach this trapped man's head, Joey must now turn himself totally upside down inside this tight space while suspended at the end of the rope. The man's breathing is extremely labored, but Joey is not sure if it's from the injuries resulting from his fall, his extremely contorted inverted position, or landing face down in decades worth of accumulations of dust and pigeon shit. Joey calls for an oxygen mask to be lowered and quickly places the lifesaving facepiece on the trapped man. He checks next for any obvious severe bleeding, and finds the man's back has a severe puncture wound. The man is barely conscious and can't really offer much information. We must treat him for spinal injuries. Tim Kelly lowers a half-back, a spine immobilization device, down to Joey. We don't want to kill this poor guy by twisting his neck as we extricate him. Usually, it takes two people to place this device on a victim. The extremely cramped space means that Joey will have to do it by himself, and the victim's deteriorating condition means he will have to do it quickly!

Joey is a man of very few words, but from the top of the shaft we can hear him talking to the nearly comatose man beneath him. "OK. Listen, guy, I ain't trying to be fresh or anything like that," I hear Joey say as he passes the nylon straps between the man's legs and up through his crotch. I have to smile to myself at the image of it. Joey is devoutly Catholic, and the mere suggestion of sexual contact between him and another man is almost enough to send him into shock. After another two minutes, Joe has the leg strap, as well as the chest straps, fastened, and the man's head anchored to the back plate. He wraps the entire ensemble in a triangular canvas "diaper" and attaches the main hauling line to it. He gives the signal to haul and the members of Ladder 24, directed by Lloyd, start to haul.

We get the man up to roof level and lay him on a long backboard to immobilize his neck and spine. Now we have to get him to the ground. The long board is laid in a stokes stretcher, to which Lloyd and Warren have already secured another rope for lowering, and the stokes is slid down Ladder 24's aerial to waiting paramedics. I tell the 7th Battalion that we'll be out of service a little while longer after we wrap up our equipment here. I want to follow the ambulance to St. Vincent's Hospital to retrieve the half-back. The device costs over $1,200 and the last time I let it go into an emergency room with a victim without an escort, a doctor cut all the straps off of the device, destroying it. At St. Vincent's, the ER staff doesn't want to take it off the guy until they've got a full spinal work-up done, x-rays and all. We accompany him up through the entire process. It turns out that the man's spine is fine, but later an infection from the puncture wound damn near kills him. The pigeon shit in the bottom of that shaft was nearly deadly. He was in the hospital for about six weeks.

Another unusual incident happened a few weeks later. A 30-something architect was supervising rehabilitation work on the Croatian Orthodox Cathedral of St. Salva, on West 27th Street, only a few blocks from the guy down the shaft. This time, the problem is the opposite of the last. This architect was inspecting stone repairs on the very top of the spire when he blew his knee out stepping onto a staged scaffold. He has to be brought down, not up, but the challenges are the same. Rescue 1 arrives at the same time as a two-man police emergency service squad. The cops, seeing that we are going to go to work to get this guy, jump out of their rig and run up the series of ladders that lead the 100 feet or so to the injured man. They want to beat Rescue 1 to the victim. The game they play is called "staking a claim." Because they got to the victim first, they feel that they should have "owner-ship" of the rescue.

The fact that they ran up empty-handed (I mean absolutely nothing with them to solve the problem) does not bother them at all. Meanwhile, Rescue 1 is bringing up all the tools: bags of ropes and rope hardware, stokes basket, trauma kit, the works. The access to the victim's location is via a single 40-foot long aluminum ladder that is laid along the church's steep slate roof. I reach the top to find the two cops playing their game. "Oh, we got it, Cap. We'll take it from here." When I point out their obvious shortcom-ings in equipment and personnel for what I envision to be a fairly complex operation, they give me a story about "more" help coming. I put an end to it by being blunt. "Listen, guys, we're here and ready to take him down now. If you want to help, fine. You can do the victim packaging while we do the rigging. This is going to be a real project to get him down." They go with the obvious. The guy is conscious. If they make a fuss or try to stall, the victim is not going to be happy. There is also the ubiquitous media everywhere to consider. We pass the stokes basket over to them.

I send Ed Myslinski and Lloyd Infanzon around to the opposite side of the church from the ladder to look for anchor points that will be strong enough for us to tie off to. We can't get an aerial ladder up to the peak of the roof; it's too high and set too far back from the street. We need something very substantial to tie our ropes off to so we can rig a lowering system. Being at the highest point of the roof complicates things. There's virtually nothing substantial that is higher than where we are working. There is a small stone projection, right at the edge of the roof, that we consider for a moment, but then the victim himself tells us not to use it. That projection is why he was up there in the first place, inspecting it for Local Law 11, and he found that it was unstable! What we could use right now is the legendary "skyhook," an imaginary high-point anchor that riggers wish they could just put wherever they need an anchor. We'll have to improvise. Eddie and Lloyd radio back that the only thing that's close to being in line with our operation is a round bell tower about 20 feet in diameter. They will have to wrap the entire tower to create a substantial anchor, then throw the ropes up over the peak of the roof to our side. There's one slight problem. They'll have to do it 60 feet in the air. I send Brian Foy to give them a hand. They have to spread themselves out around the perimeter of the tower on segments of scaffold and toss the ends of the rope over from one to the other. Eventually, they make two complete loops around the tower.

In the meantime, Joey and Jerry Nevins are rigging a lowering line, a second safety line, and a tag line to guide the stretcher. I have lowered one end of another rope to Lloyd, who secures it to the anchor rope around the tower. When Lloyd's end is secure, I can tension the line by putting my weight on it to take all the slack out. Once the slack is out, I can tell where the rope will cross over the peak of the roof. Right at that point, I tie a "butterfly" knot to which we will secure the main lowering line. Once all the rigging is complete, we attach the two lines to the stokes and directly to a harness secured to the victim. Now the trick is to pass the basket across a five-foot wide gap at the outside of our access ladder, turn him 90 degrees, and lower him without any undue impact or shock. Jerry has secured the free end of the safety line to the foot of the stretcher and I pass it through a carabiner attached to the butterfly. Now with the police officers on the head holding back, Joey and Jerry can pull out on the foot end and winch the basket slowly outward across the gap. Once the basket is out and in position to go vertical, it's a simple matter of removing this line from the feet and letting the main lowering line control the descent. Gravity takes over from there. But there is one surprise left in this for me. As the victim comes to the vertical position right in front of my position on the top of the ladder, I see that one of the cops, in a transparent effort to make it look like the police had contributed something to this very complicated effort, had removed his jacket, turned it around, and placed it with its back across the victim's chest with the big patch "NYPD Emergency Services," facing out so that the ever-present Manhattan media would have no choice but to photograph it. Sheer childish behavior, but you have to give them the "Benny Goodman trophy." They never miss a chance to blow their own horn! The next day, there it was, all over the front page of the newspapers in living color.

A few months later, we go to another incident in the Brooklyn Navy Yard, where this same childish game almost has tragic consequences. We responded as part of the standard FDNY confined space response: two engines, two ladders, two rescues, one squad company, hazmat company, and a battalion chief, to the MV (motor vessel) *Cape Lambert*, in dry dock for repairs in the Brooklyn Navy Yard. This large an assignment might seem like a lot until you have been to a few of these incidents. The 45 people are quickly used up in the myriad tasks required. On a ship, it gets even worse due to the difficulties in radio transmissions inside the steel hull.

An immigrant worker with little experience had been hired to cut steel partitions out of the ship's double bottom using an oxy-acetylene torch. The double bottom is a space less than four feet high that runs the length of the ship. It is subdivided by steel bulkheads or "frames" every three feet in all directions. The bulkheads have openings in them that are just about sufficient to let a person squeeze through: 18-inch wide × 24-inch high ovals. The access to this space is through one 16-inch diameter manhole-like hatch about every 100 feet. This poor, untrained worker is given his oxy-acetylene cutting torch and a paper dust mask and brief directions: "Go into the hole and go 10 frames to your left and start cutting out all the frames in that area." These would later be replaced by new steel. Unfortunately, this guy must not have known his left from his right, because he goes to his right, not his left. When he reaches what he assumed to be the designated spot, he lights up the 5,000 degree oxy-acetylene torch and starts burning away steel, cutting right through the bottom of a 90,000-gallon, number two diesel fuel tank! The resulting spray of diesel fuel is instantly ignited by the torch, causing the man severe burns, but not killing him. The man's partner at the frame below the manhole sees what's happening and calls for help, shouting that the worker is hurt. The resulting 911 call initially indicates a relatively simple medical emergency, to which Engine 210 is assigned as part of the CFR-D program. Boy, are they in for a surprise!

I have always believed that God watches out for firefighters, even if he sometimes slaps us around to teach us a lesson. More than once, I've felt that divine intervention has saved us from catastrophe. This is one of those cases. Up to this point, the similarities between this incident and another shipboard disaster right here in the Brooklyn Navy Yard are scary. In 1961, the aircraft carrier USS *Constellation* was under construction in a similar dry-dock, when a much smaller tank of diesel fuel on the hanger deck was punctured, leaked, and was ignited by a welder or cutting torch. The resulting fire killed 50 shipyard workers and took 17 hours to bring under control. And that tank only held about 1,000 gallons, not 90,000! The only good thing that happened onboard the *Cape Lambert* is the fact that the burst of flame was so intense from the atomizing effect of the spray through the initial hole, that the fire consumed all the oxygen in the under-ventilated space, and basically snuffed itself out. Thank you, Lord! Unfortunately, if there isn't enough oxygen in a space to support combustion, there's barely enough to support human life. The two workers in the double bottom are overcome before they can escape. Ships' personnel responding to the initial call for help now recognize the seriousness of the situation and contact 911 with additional information, which results in our turning out to the full confined-space response.

Squad Company 1, located about one mile from the Navy Yard, is the first Special Operations Unit to arrive, followed by Rescue 2 and then Rescue 1. As Squad 1 with Captain Terry Hatton in command arrives, two emergency service unit cops run past them, again empty-handed, up the gangway to start "the game." Captain Hatton is very knowledgeable and knows that the steel construction of the ship hampers communications. He also knows the great distances to be traveled aboard the ship, from the gangway, which is located at the stern of the ship, to an operation site nearly anywhere aboard; this is going to slow the arrival of any equipment that you might discover you need later. He tells his crew to bring everything they can, not knowing exactly what he'll need. The squad guys load up ropes, hardware, breathing apparatus, lights, air monitoring equipment, victim handling equipment, half-back, stokes: the whole works. It takes a little longer to get ready, but when you get to the scene, you're ready to go to work (fig. 26–2).

When they finally cross the gangway and make their way through the passageways down the six ladders to the bottom deck, Terry finds an ESU cop standing in the manhole and asks, "What have you got?"

"Aah, not much," the cop replies. "One guy's down inside there, but his buddy's in there with him buddy-breathing from a mask. We got it."

Figure 26–2. A confined space rescue in the double bottom of the Motor Vessel *Cape Lambert* killed one worker and nearly resulted in a catastrophic fire. The logistics of conducting a complex rescue in this environment are extraordinary.

Terry tells the cop he wants to take a look and the cop starts to give him a hard time. The cop is out of his league. Mayor Rudolph Giuliani has issued an executive order placing the Fire Department in exclusive command at confined space incidents, just to prevent situations like this ESU cop is trying to create now. Terry knows this directive like the back of his hand and cites it now verbatim. Realizing he's not going to bullshit Terry, the officer reluctantly climbs out of the way. Terry takes his place in the opening and as soon as his head clears the underside of the deck, he is absolutely shocked at what he sees.

There's no worker "buddy-breathing" with his partner. Neither one of them have an SCBA or any other source of air. Their only protection is the paper dust masks they wear! The worker nearest the manhole is semi-conscious, three frames in from the opening. Apparently, he tried to reach his co-worker after the initial fire but was overcome by the lack of oxygen and high carbon monoxide level present. Captain Hatton immediately orders his members to prepare for a very involved confined space entry.

Air monitoring is begun and it indicates a life-threatening situation. Oxygen was depleted to below 17%, and over 300 parts per million of carbon monoxide and dangerous levels of sulfur dioxide are found. Squad 1 members wearing SCBA are able to reach the first victim and hoist him out of the hatchway.

By this point, Rescue 2 arrives with their air source cart and additional air hose, and all their members are wearing surface-supplied air respirators (SAR), which take their primary air supply via hose from the cart. Lt. Tony Errico of Rescue 2 directs Firefighter Sal Civitello to enter the hatchway, carrying a spare mask for the victim, and make his way in to perform victim assessment. The second worker's location cannot be seen from the hatchway, but the job foreman swears he can't be more than 50 feet in; that's how much hose for the torch was in there before he pulled it out. Lt. Errico orders Sal to make the initial entry wearing a one-hour bottle in the SAR's back plate as his "escape air." While the air cart provides virtually unlimited air through a hose to the wearer inside the hole, it does absolutely no good if the hose gets cut through, burned, or even just kinked. The cylinder on the wearer's back is there to ensure the rescuer gets out alive, just in case "Murphy" shows up.

The problem with this plan is fitting nine inches of air bottle and nine inches of Sal through a 16-inch-wide opening. Sal will have to take the mask off to fit through the opening, while keeping his facepiece in place, a move he has trained for extensively but only rarely used. He then will have to carry his own mask, as well as the victim's, through the series of frames, successively passing first the victim's, then his own, through the 18-inch × 24-inch openings, before squeezing through himself and repeating the whole process all over again nine times. All the while, he is sloshing through diesel fuel that sprays at about a 20-gallon per minute rate from the hole in the overhead. At the 10th frame, Sal finds the burned, unconscious but still living, victim. He quickly places the mask facepiece on the man and feels for a pulse. It is there, but very weak. Sal attempts to move the man, but the burned, oil-soaked skin repeatedly slips from his grip. He makes his way back to the manhole and briefs us on conditions (fig. 26–3).

Rescue 1 had arrived just prior to Sal's exiting the double bottom. Lt. Errico questions Sal about distances and obstacles, and a plan is developed.

Figure 26–3. An investigator examines the spot inside the double bottom where a worker was burned. He had to be removed through the series of narrow holes in these baffles.

Two new rescuers enter with some specialized victim handling gear. Based on Sal's assessment, these guys—Steve Brown from Rescue 2 and Kevin Kroth from Rescue 1—will be able to enter without the extremely heavy and cumbersome one-hour bottles. They will wear 10-minute escape cylinders. We call them "gunslingers," because they strap along the wearer's thigh like an old west pistol holster. That will free both hands, narrow their profile, and still leave them enough air to get out, if needed. In they go. In the meantime, Chief Ed Collins of the 31st Battalion orders me to take the rest of Rescue 1 and Engine 207 back up through the ship, across the gangway and then down the 50-foot ladder to the bottom of the dry-dock. The ship's master reports that there is a hatch cut into the bottom about 20 feet away from the manhole. It might provide an easier access point for victim removal.

By the time we make our way up, out, and back down to the bottom of the dry-dock, Steve and Kevin have already reached the second victim and begun removal. Standing on a milk crate under the massive ship, I can stick my head up through a 24-inch × 24-inch hole that had been cut in the ship's outer bottom to allow the steel that had been cut to be passed out to the outside. I still can't see the operation, though, because the bottom of the oval holes in the baffles is raised up 12 inches off the bottom of the ship. It takes two milk crates to let me clear this obstacle. That 12 inches is making life miserable for Kevin and Steve and is making life impossible for the man they are trying to maneuver the 35 feet back to the manway. An unconscious person is absolutely flaccid, no rigidity. You pick one end up, the other end sags. You pick both ends up, the middle sags. The men have tied loops of nylon webbing in girth hitches around the man's chest and legs, which tighten down and prevent slipping, but now they must lift his dead weight up and over the 12-inch steel dividers every three feet. They have to do this while crouching down, since the headroom is only 44 inches high. They're busting their asses!

From my vantage point, I think it might be easier if the men bring the worker out that way instead of trying to squeeze him up through the 16-inch manhole. I radio them and wag my light toward them to let them see where the opening is. They are nearing the manhole and must make a decision about which way to go. There is no discussion. Kevin looks at Steve, both look at the five frames separating them from this hole that I'm in, and Kevin calls back, "No way, Cap. He's going up right here, right now." They pass the webbing up to the other members inside the ship and guide the badly injured worker up through the narrow opening. The man is carried up the multiple shipboard ladders and across the gangway to waiting paramedics. Unfortunately, after a long hospital struggle, he would not survive his burns.

The operation is far from complete: all three rescuers and all their equipment are saturated with diesel fuel, and most of the equipment will have to be discarded. It is a small price to pay to save two lives. But it's time for that nasty "D" word again—decontamination. As Steve and Kevin climb out of the narrow manway, they are drenched in sweat from their exertions, even though it's early March and only 40 or so degrees in that steel chamber. Showers are put off at the scene, since the contaminant is known to be diesel fuel; we take the boys the three short blocks to Engine 210's firehouse where real hot water showers, shampoo, towels, and real clothes are available. So much for a nice quiet Saturday. It makes for an interesting way to end the week! Just one more incident that makes life in a rescue company so unpredictable.

CHAPTER 27

30 Rock

One of the drawbacks to working in midtown is the abundance of automatic fire alarms in the area's buildings. Midtown is the high-value district, and the buildings are usually better protected from fire than most other buildings. All of the high-rise hotels and office buildings have automatic fire detection and/or sprinkler systems. They create two concerns for firefighters. The first is the large number of false alarms the smoke detectors generate: it is very common for some companies to do 10 or more false alarms every day due to these devices. At times, units go back to the same building three or four times a day. This "crying wolf" syndrome breeds complacency in the firefighters. Guys turn out slower, and they are not expecting to go to work when they get there. Things were getting so bad that the two ladder or "truck companies" in the heart of midtown, Ladder 2 and Ladder 4, were doing close to 6,000 runs a year, earning the area the nickname of "The Electronic Ghetto," since normally only fire companies in the heart of the poorest neighborhoods with the most fires would respond to that staggering number of runs. The 2 Truck and 4 Truck rarely have a moment's rest.

Eventually, the department was forced to take action, and legislation was passed that authorized fines and other penalties for building owners who continued to have repeated, unneeded alarms. This program has had some success, but the responses continued to escalate. Eventually the department decided to reduce the response to these alarms, from one engine, one ladder, and one battalion chief, to *either* one engine *or* one ladder. Still, the midtown units hate the "Class E" alarms (so named because the office buildings are occupancy group E in the building code). Rescue Company 1 is assigned on automatic alarms in the hotels in its response area (34th Street to 57th Street, from 5th Avenue to the Hudson River, 12th Avenue) and we typically go on Class E's only at night, when the life hazard is greatest due to sleeping occupants.

Thus, when the alarm came in for 30 Rockefeller Center at 4:00 a.m. one morning, it was not anything that would make you particularly wary. On the way across 50th Street, however, the situation started to sound somewhat more serious. The dispatcher relays to the 9th Battalion that they received a call from Rockefeller Center security reporting heavy smoke on the seventh floor. I yell into the back to start putting on the hour bottles, that it sounds like we might really have something. (At the time, fire companies were equipped primarily with 30-minute cylinders, which are barely adequate for a relatively small building, such as a house or apartment. The 60-minute cylinders are much larger and heavier, and are saved for fires in subway tunnels, high-rise buildings, and large warehouses. We have to remove the smaller cylinder and insert the new larger cylinders, which can take a minute or two.)

Rockefeller Center is a huge complex: about six square blocks, with a tremendous underground shopping arcade, numerous low- to mid-rise buildings (6–14 stories) and at the center, 30 Rockefeller Center—a 66-story, 250 ft × 500 ft office tower that is the home of NBC, the National Broadcasting Corporation. Several of its programs are broadcast live from the building. Thousands of workers and audience members visit it every day, and even overnight there are dozens of technicians and maintenance personnel present.

We get off the rig and I remind everybody (quite unnecessarily) to be careful, to stay together as a unit, and to make sure we have the search ropes. We report in at the fire command station to Battalion Chief Joe Grosso of the 9th Battalion. Chief Grosso is nicknamed the "High-Rise Drifter," after Clint Eastwood's character in "High Plains Drifter," due to Grosso's aura of total cool and calm even under the most stressful circumstances. The chief tells us that the fire alarm system had been taken off line (so the building didn't risk false alarms that could generate a fine), so he has no indications from the alarm panel of where the seat of the fire is, but he has a report from the security guards of heavy smoke on seven and now another report of smoke on five. He says, "Two Truck is up on five, and four is up on seven. Get up there and give them a hand and let me know what you've got." I call out to the chauffeur, Hank Molle, to bring the thermal imaging camera with him. We don't like the idea of taking the elevator with smoke on two floors, so we hoof it up the stairs to the seventh floor.

Coming out of the stairway on the east end of the building, we find the hallway is clear, and I momentarily think this must be the wrong place, but we are in one of the north-south corridors that runs parallel to the short (200 ft) sides of the building. We proceed south to one of the main corridors that run the long axis of the building, and I can see smoke down the hall, toward the far end. "OK, this looks like it might be the place. Let's find it," I say as we set out down the hall. About 200 feet down the hall, it's like walking into a wall of smoke. The lights on this floor are on, and the smoke is a light white color so visibility is not that bad, but the nature of the smoke makes use of the mask mandatory. This is a really "biting" irritating smoke: the kind of smoke we smell at car fires where all the plastic is burning. That tells us this one is electrical in nature, and probably wire insulation burning. In a building like this, the wire is usually hidden above the ceiling, or beneath a raised floor. We move into the smoke and meet Ladder 4 Lieutenant Ray Ziegler who tells me they haven't found any indication of heat, and they've been down all the hallways. I tell him we might be above the fire, but Two Truck is checking on the fifth floor. I tell him we'll have to start forcing some of the locked doors. We spend the next five minutes lifting ceiling tiles and opening doors looking for flames while the smoke just keeps getting thicker.

Finally, someone opens a door into another stairway at the northwest corner and the smoke pours out onto the floor. "Well, this has some promise." We move into the stair but the smoke, instead of rising as expected, is pumping *down* from above. "What the hell is this?" We start up the stairs, where the smoke is definitely getting heavier. At the 10th floor, the staircase suddenly stops at a pair of doors with a very small landing. "Well, this must be the place." Warren "Woody" Forsyth forces the right door with the adz of the Halligan. There is definitely heat, and we hear loud crackling and popping noises. We definitely have fire here. Engine 54 is soon there with a hoseline they have connected to the standpipe on the ninth floor. Warren moves out of the way to let them up and then he forces the remaining door off to the right. I tell him to take Billy Henry and Hank Molle with him into the other door and to be careful; this seems to be a mechanical equipment floor. I also try to contact Chief Grosso at the fire command station via the handie-talkie, to tell him we have located the fire on the 10th floor and are about to begin the attack. The handie-talkie is full of static; I hear a lot of broken conversations, but I can't tell if any of them are acknowledgments from the command post. I try it again but then Engine 54 is ready to go with their line. There is a glow off to the left and ahead as I move in behind 54's line. My suspicions about the nature of the floor are confirmed once inside the door.

The floor is a gray painted concrete, and off to my right the wall is lined with painted gray metal cabinets, typical of electrical equipment panels. Fifty-four opens their line and directs their stream at the fire. After about 15 seconds of this, there is a loud explosion and a shower of white sparks. There is a collective yell of, "Holy Shit!" as 54's nozzle operator shuts down the stream. I suggest we all back up to the stairway and close the door, but the brothers are way ahead of me. I make a quick check to be sure no one is left behind, and I too scoot into the now crowded stairway. I call Woody and Hank on the handie-talkie and tell them to get back into the stairs but they reply in person. The blast got their attention also and they didn't need orders to know what to do. I tell Ray Ziegler and my guys to drop down to the other floors and round up all of the dry chemical or CO-2 extinguishers they can find. This is an electrical fire in high voltage equipment, and water is a terrible thing to use on it while it is still energized. Then I try raising the command post on the radio. Still no reply. I grab Billy Henry and try his radio, thinking mine might have a low battery, but still no answer. The concrete floors and limestone walls and steel columns and beams of the massive structure are preventing radio transmissions from penetrating the nine floors to the lobby.

I know by now Chief Grosso must be pissed off not getting any reports, so I decide I had better get him a preliminary report and ask him to get the electricity turned off. I drop down to the ninth floor to look for one of the "Floor Warden Stations," which are hard-wired intercom systems with two-way communications to the fire command station. These are required parts of the Class E system. I find it in the middle of one of the long hallways. It is dead. No power. "Aw, shit!" I move into one of the offices, planning to use a telephone to call the command post. They're dead. "Aw, *double* shit!!" I search for a pay phone and I find one on the eighth floor, but I don't have a quarter. No problem. I dial 911 and ask for Manhattan Fire. The dispatcher comes on the line shortly and asks, "Where's the fire?" I quickly explain who I am and where, and tell him to relay to the Ninth Battalion a quick size-up and request to have all power to the 10th floor shut down. I then tell him to transmit the second alarm on my orders. He replies, "You're too late, pal. Chief Grosso sent the second about five minutes ago, said he had fire on the fifth floor." "Aw, *triple* shit! Well, then send the third alarm, and make sure the command post knows we need help on the 10th floor!"

As I get back to the mechanical room on the 10th floor, Ladder 4 and Engine 54 are exiting the room, dragging a bunch of empty fire extinguishers. They say they dumped them all and it didn't make a dent in the fire. Just then Hank calls me. He says that he's down on the south side of the 10th floor near the "Z" stair. While he was looking for more extinguishers, he found another electrical closet over there and it too is blazing to beat the band. Then Billy Henry calls. He's on the north side hallway, about 150 ft east of where I'm located and he's got heavy fire over there, also in an electrical closet. "What the hell is going on here?" I ask myself. I drop down to eight and call the dispatcher again. I tell him to relay our situation to the command post and that we need more help. Our vibralerts are starting to sound on the masks. I cross over to the "Z" stair and find Hank, who has gone down to try to get help. He has one engine company with him, which is better than nothing, but it's nowhere near what I'm hoping for. He shows me the electric closet and now the smoke has lightened up enough to see that there is fire blazing away, roaring up a shaft toward the 11th floor.

Racing up there, we find there is no closet in that location on 11, just a solid blank wall. We go up to 12, where there's a closet. The fire there is just coming through the floor in the openings around the four-inch electrical conduits. There is not supposed to be any storage of combustibles in an electric closet but, of course, here we find several large cardboard boxes full of paper goods. We move them out into the hall, hoping to delay the fire's extension to this floor, or at least reduce the fuel load. Out in the hall, I meet Deputy Chief Jim Bullock from the First Division, who responded as the second Deputy Chief on a multiple alarm in a high-rise. He is now the operations chief for this fire and he wants to know what's going on.

Chief Bullock used to be the commander of the Special Operations Command that oversees the Rescue companies. When I tell him what we have so far, he too gives an "Oh shit." We both agree that we are going to have to use water on this; there is just way too much fire for anything else. We will have to stand back from the panels and try to deflect the water in a spray off the ceiling and other surfaces, to reduce the risk of electrical shock to the nozzle operators. He tells me, "John, you and your men are all the help I have up here on this side right now. The Third Division has sent the fifth alarm already, and I'll get more help, but you've got to try to cut this thing off here." I tell him we're about out of air and need fresh cylinders. As he heads down the stairs, he's already relaying this into his cellular phone to the command post and he gives me a nod and a wave. "Try to get above it and cut it off," he says. Hank and I make our way up to 13, where there is only a little fire in the electrical shaft, just the insulation on the wires in two conduits, and a few cardboard boxes of copy paper, so we grab a house line off a standpipe and flake it out. The hose on these buildings' house lines is notorious for leaking, if not bursting, since it almost never gets used or tested, and the nozzle is a smooth bore "suicide pipe": so-called because it has no built-in shut off. Once it's charged and in operation, the operator has no way to shut off the flow by himself in case they want to retreat. It is either put the fire out or else, because if you try to let go of the flowing hose, it will whip around and club you to death. It's better than nothing in this situation.

Hank has the nozzle at the door to the room and I back him up. Paul Baldwin is out in the staircase, where the air is a little better. He's out of air entirely. I tell him to open the valve slowly; we don't want to go for a ride from too much pressure. In the back of my mind I am seeing reruns of *The Towering Inferno* with Paul Newman and Steve McQueen: a movie about an electrical fire that destroys a high-rise building in San Francisco. I know that we have fire in at least three widely separated locations on at least three different floors, and it seems to be involving the entire electrical system. We're only on the 13th floor; there are 55 stories above us. I wonder what conditions are like in the rest of the building. I jokingly yell to Hank, "Listen, if this doesn't work, we might have to blow the roof tanks!" Hank knows exactly what I'm talking about and doesn't miss a beat. "Well, fuck you! I ain't walking up 50 stories, and I don't like explosives, so it's all yours, Steve McQueen!" I nearly cough up a lung from laughing so hard. To try to conserve my remaining air, I've been taking the facepiece off for a few moments at a time, and this PVC smoke has gotten to me. I'm OK as long as I take shallow breaths, but to laugh was just too much and now I can't stop coughing. Fortunately, at about that time an engine company shows up to relieve us on the hoseline. I grab Hank and tell him, "Give them the line and let's go find some fresh air." We stumble into the staircase, where it's a little better, and head down.

I've been calling the rest of the guys on the radio, but the building prevents the radio waves from penetrating more than one or two floors. As we get to the 10th floor, I finally get through to Billy Henry. "Where are you Billy, and are Warren and Kevin (Kroth) with you?" He replies, "Yea, they're both here. We're in the fire tower over on the north side, near that electric closet that Woody and me showed you before." The smoke is still bad here, but at least you can taste the fresh air mixed in with it, because it doesn't burn your throat as much. We find the doorway leading into the open balcony that separates the enclosed fire tower stairs from the floor area. The fresh air coming up the shaft is a godsend. We crawl into the staircase and flop down for a rest. I visually size everybody up. These boys are whipped.

"Get your gear off and take a break," I tell them, but they're too tired to comply. Woody in particular looks dazed. That's scary. He's the biggest guy in the company and as strong as an ox. If Woody's in bad shape, we're in trouble. I ask him, "Warren, are you all right?" He's shaking his head, not a "yes" or a "no," just shaking. "Warren, what's the matter?" Finally, "Hey, Cap, I got hit in the head with a line, right in my ear. I can't hear too good, and I'm a little shaken up." I ask if he can walk and he says, "Yeah, I'm OK." I tell

Billy and Paul to walk him down to the street. Just then one of the dozen or so special-called chiefs pokes his head into the stairway and says, "Oh, there you are, Rescue. The command post has been trying to locate you." I tell him, "Yeah, Chief, the radios don't reach the command post." "Well, get your guys together and bring your thermal imaging camera out here, we want to trace the run of all these conduits to see if we have any more fire hiding on us." "Sorry, Chief, no can do. We need new air bottles and we need a blow. We're whipped. How about calling another Rescue?" "Rescue 4 is already operating downstairs on five. We need you." "We'll have to get new bottles," I say, hoping to dissuade him. "Fine," he says, "go get them and come back up." While this is not the reply I'm hoping for, it will do. It's better than if he said, "Wait here, I'll have somebody bring them to you."

When we get to the lobby, there are about 20 fresh companies milling around, some of them laid out against walls waiting for an assignment. I make my way through the throng and catch Chief Grosso's eye. There are deputy chiefs and staff chiefs up the ying-yang here at the command post, so Grosso has time to make his way over to me. "How are your guys, John?" he asks. I tell him honestly, "We need a blow, Chief. We went through our hour bottles and kept at it even after they were empty, and I've got one guy on his way out to EMS." "Who is it? Is he all right?" (That's "High-Rise Joe," always thinking of the people!) I tell him what I know about Woody, which isn't much to start, and he says he'll speak to Chief of Department Cruthers and see if they can't do without us for a while. Now he wants to know what happened, where we were, and how it looked when we left the thirteenth floor. I give him my rundown and ask him about reports from the upper floors. He tells me that so far things seem all right once you get above 12. According to the building engineering staff, one electrical grid serves floors five through 15, with different grids serving other groups of floors. It seems that the power supply room on 10 was the main distribution room for this grid. When a voltage surge or an arc occurred there, it ignited numerous fires along the entire high-voltage network between the fifth and 15th floors.

All around us, companies are being assigned to different floors to assist in the extinguishment and search efforts, as more and more exhausted crews make their way down from upstairs. Grosso makes his way back to the fire command station, and after a moment, talks with Chief Cruthers, who gives me a look up and down and then gives Grosso a few words. In a moment, he comes over and tells me to have the guys go out and get fresh bottles, but just stand by. Chief Bullock is telling the command post that he thinks they have all the main pockets of fire under control and he has all the other rescue companies up there with their thermal imaging cameras. He thinks that he will be able to finish this mess without us and our camera.

It is daylight when we walk outside, and the streets are choked with fire apparatus. This is going to make some mess out of rush hour traffic in mid-town. We all strip off our turnout gear and try to cool off and unwind. I find out that EMS wants to take Woody to New York Eye and Ear Hospital. They think he's got a ruptured eardrum and it's affecting his equilibrium. I tell him we'll pick him up when we get done, but he says he'll probably be a while in the ER and he'll call us when he's ready to be picked up. The rest of us are sitting on the back of the rig drinking Gatorade when we get the first inkling of more trouble coming our way. Over the handie-talkie we hear the command post talking about PCBs in a transformer on the 10th floor. I look at Hank and Paul and their expressions match mine. "Aaaaaw, shit!" everyone moans. We all know what's going to happen next.

Polychlorinated biphenyls, or PCBs, are a type of cooling oil used in electrical transformers and capacitors. It has good electrical insulating qualities and a high ignition temperature, making it highly desirable as a cooling fluid. Unfortunately, it also causes cancer. This wasn't discovered until the mid-1970s, when its use was banned in new installations in the United States. By then, however, there were millions of electrical devices filled with the stuff scattered all over the country. Beginning in the 1980s, the EPA mandated a replacement program to begin to get rid of the PCB oil, but it would take years to accomplish. In the meantime, there are all these transformers and such sitting in manholes, vaults, and in buildings, just waiting for something to go wrong. To make matters worse, as bad as PCBs are, they are positively mild when compared to what happens when PCBs are exposed to high heat—say 800 degrees Fahrenheit. At that point, PCBs break down into dioxin, one of the most deadly compounds around. And you know who was up there crawling all around on the 10th floor!

Now the crowds that are gathered to watch the fire, the same ones that usually gather in this spot to watch the live broadcast of the *Today Show* from Studio Two-A, are in for a real show. Everyone who was anywhere near that transformer is going to have to be decontaminated (decon, for short.) Decon is not the most pleasant experience to undergo. First of all it means we're not going to get out of here for hours. A special tractor trailer for this purpose has to be called from Ladder 15's quarters in lower Manhattan. Then it has to find a suitable location and get set up. Once this is done, the actual process can begin.

First, all the fire tools that were used in the area, including our $17,000 thermal camera, get bagged in large plastic bags and deposited in one pile for later cleaning. Then all the members' outer clothing is removed and bagged. Next, the masks are removed and bagged and placed into yet another pile. Then the members climb a set of stairs into the back of the decon trailer, where they strip naked and place their uniforms and underwear in still more bags. Everything you have on you: watches, wallets, and jewelry included, are bagged for cleaning, or in the case of porous items like cash, disposal in a hazardous waste site. Next, the naked members proceed into a series of showers with a harsh soap and shampoo under the watchful eye of EMS personnel who are dressed in Level A protective clothing, which resembles an astronaut's suit. (What gets to me is that at least some of these people watching us are women. Can you imagine the holy hell that would break out if a bunch of male firefighters ordered a bunch of female paramedics to strip and shower in front of them? Not that I'm a prude, mind you, it's just that I have the Irish curse.) After showering, you step out and towel off and then bag the towel. Then you step forward and you yourself are bagged; being issued a disposable one-piece jumpsuit-like garment made of a special space age fabric which promptly rips right in the crotch as soon as the wearer sits down—even in spite of having the Irish curse! To make sure that the crotch rips right out of everyone's assigned "bag," the next stop is a chair where another EMS person takes and records your vital signs. I swear this chair has a special edge on it, designed to rip the crotches out of these suits, but I haven't been able to find it. Anyway, once you're done here, the next stop is out the door into the public eye with your ripped crotch and no shoes. This is fine attire for the beach at Fire Island in August, but not so good if it happens to be Manhattan in February, or even May at dawn.

By the time we get done with this, it's approaching 11:00 a.m. Billy Henry is working his straight tours, meaning he was supposed to finish work at 9:00 a.m. Even though he is receiving overtime pay for the extra two hours, he's pissed. He was supposed to pick up his mother at 11:00 a.m. for a visit to the doctor out in Queens, and now here he is dressed up in a funny plastic suit with the crotch ripped out (and he's *not* Irish) in the middle of Rockefeller Center, with no way of even communicating with the dear woman to let her know that he's okay but will be late. Of course, the fire and the reports of injured firefighters and the pictures of the lines of firefighters going in one end of a trailer labeled "Hazardous Materials Decontamination" and coming out the other end in their funny suits with the ripped crotches are all over all the morning news shows (or at least all the morning shows except NBC, which has been knocked off the air by the fire). So Billy's reaction is actually to be expected—even so, it only adds to the comic, almost circus-like atmosphere that now surrounds the entire area. Just another day in the Big Apple (fig. 27-1)!

Figure 27–1. Rescue 1 takes a break during a month-long operation atop the 750-ft tall 4 Times Square building, where a large scaffold collapse required around-the-clock rope rescue capabilities. From left to right are Thor Johannesen, Paul Baldwin, Lloyd Infanzon, Billy Henry, Capt. Norman, and Tom Sullivan.

CHAPTER 28

Call the Veterinarians

While most of the lifesaving efforts of firefighters are directed at humans, we also have our share of interactions with our four-legged friends. Almost everyone has heard stories of firefighters called to rescue a cat from a tree. While this may make for a good public relations story, it is a practice that I personally frown upon. The last thing most cats who have been chased up a tree need is a strange human riding the tip of a very noisy aerial or tower ladder coming to retrieve them, usually breaking a few branches in the process of trying to reach them, and invariably further exciting an already frightened feline.

What usually happens in these cases is the traumatized cat lashes out at the would-be rescuer with all claws unsheathed. The firefighter ends up cut to ribbons and the cat slips from their grasp and falls to the ground, magically landing on all four feet and running away, leaving the fire officer to stop the bleeding of the sliced and diced member, and explain to the chief why their company is now one person short for the next few hours. In one case I know of, a firefighter nearly lost his eye when the cat turned Tasmanian devil and clawed the member's face before escaping. No, rescuing cats from trees is not a good use of fire department resources. All that's really needed is to chase away any dogs in the area and disperse the crowd, then let the owner coax the little darling down with food or catnip. They all come down when they get hungry enough; otherwise, the trees all over the country would be littered with the skeletons of cats that have died of starvation. They're not! If he's still there after three days and nights, then call me and we'll see what we can do.

While cats, dogs, and other domestic animals are routine players at many fires in every city, many people would think that in New York City there would not be much else in the way of animals for the fire department to deal with. It, therefore, comes as a shock to many people to find out that we have had quite a few dealings with other animals like bats and horses. Bats are usually handled by the ASPCA, a private exterminator, or even the police department's Emergency Services Unit. On occasion, however, various fire department units have been called for use of their carbon dioxide extinguishers, which have been used to chill a bat and put it to sleep. I have not had a lot of luck with this technique. The noise of the discharge (or maybe it's the moving flecks of solid carbon dioxide that resemble flies) sets the bat in motion before the gas has the desired effect. I prefer the large net approach to bats.

At the larger end of the spectrum, animal-wise, have been the horses. In New York City, there are many stables. Some are located in parks like Van Cortlandt Park in the Bronx, or private stables like the Jamaica Bay Riding Academy in Brooklyn, but the largest collection of stables in the city is actually located in the high-rise building capital of the United States: the west side of Manhattan, which is Rescue 1's first alarm area. The horses here pull the hansom cabs that take tourists on their scenic journeys through Central Park. Having the horses traveling the heavily congested streets of midtown Manhattan to reach the park is bound to produce some unusual situations, usually involving the horses and vehicles, although horses and pedestrians, and horses and stationary objects also interact (horses win vs. pedestrians, but they usually are the losers in confrontations with cars, trucks, or buildings). Rescue 1 has had to free horses from some strange places—legs caught between the bumper and fender of a cab, hooves caught in between Belgian block paving stones, or a head stuck between the bars of a fence. But by far the most complicated removal has to have been the one inside its own stable.

The value of land in midtown had the same effect on stables as it did on office buildings: they built up instead of sideways. (In 2015, Mayor Bill

DeBlasio tried to ban the horses from Manhattan, and many suspected it was a ploy to turn over the stable properties to wealthy developer friends.) As a result of the high prices, the multi-story stables are common, which incorporate some features that are unfamiliar to the average horse: like manure chutes (holes in the floors into which the stable hands would shovel the droppings instead of having to carry them down from the second or third floors). One afternoon in 1995, one of these chutes was left uncovered while a horse was walking through the area and, you guessed it, the horse stepped right into it. Both front feet plunged in until the horse's chest met the floor. The horse went wild. Its trainers and the stable hands tried to free the animal, but to no avail.

After about 15 minutes of frustration, with the horse struggling mightily, a call was placed to 911, bringing Engine 34, Ladder 21, Rescue 1, and Battalion 9 to the scene. The firefighters are greeted by the sight of the horse's forelegs flailing like two windmill blades in midair, accompanied by frantic neighing as they make their way up the hay-strewn ramp to the second floor. Lt. Dennis Mojica and Chief Joe Grosso confer with the stable personnel and try to work out a plan to lift the now frantic beast. The horse's legs being wedged in the chute and his body flattened against the floor prevent anything from being slid under him.

Firefighters build up a gigantic box crib platform below the chute. They build the platform up until it is just under the horse's front hooves, projecting down from the manure chute. On top of this platform they insert two air bags which are sort of like inflatable pillows. The air bags expand when inflated and can lift several tons. The firefighters then top the bags with a final layer of plywood, which just touches the horse's hooves. As soon as the horse senses something solid under his front legs, he resumes his frantic thrashing. A veterinarian gives the horse a mild sedative, not enough to knock him out, just enough to calm him down so the firefighters can get close to the panicky animal. When the air bags are inflated, it raises the horse enough so that a pair of nylon slings can be slipped underneath him. These are secured to two Griphoists, that have been slung from the third-floor beams.

The signal to hoist is given and very slowly and very gently, the legs begin to emerge from the confining chute. Inch by inch, the now calmed animal is hoisted until both forelegs are clear of the opening. As the horse is slowly lowered to let his front feet again touch solid ground for the first time in over an hour, the animal, unaccustomed to his new-found mobility, takes two steps forward. His entire rear end now plunges into the chute. It was at about this moment that a camera crew from one of the television networks decided to interview Chief Grosso about the operation. The usually unflappable Grosso was asked to describe the ongoing effort, how long it had been underway and what actions were being taken—another human interest story. Concluding the interview, the reporter asked, "By the way, Chief, what's the horse's name?" Without so much as a blink of an eye, the exasperated Grosso replied, "Stupid!"

This time, the operation only took another 10 minutes. The horse was still drugged, most of the rigging and cribbing was already in place, and everyone knew what to expect. It was a relatively simple matter to slide the slings back under the horse's rump and midsection and winch him upward. As soon as he was clear this time though, a newly fabricated hatch cover was placed over the opening, before the slings were released. The entire operation took just about an hour and a half. It took a solid week to rid the firehouse and the apparatus of the stench of horse manure from all the cribbing slings, bags, and other tools, however not to mention the members and their clothing. I think the horse won that one!

Another animal encounter involving Rescue 2 has become a classic firehouse story. While I didn't witness any of it, I came in to work the following morning to find the brothers still howling at it every time they retold it. It's still one of my favorites.

Shortly after 2:00 a.m., Rescue 2 gets a special call to "assist civilians with a dog." (This is not a typical call, but neither was the run I got one morning for a report of "a man who fell into the shark tank" at the New York Aquarium in Coney Island. You never know what to expect.) The company turns out to find Ladder 102 and Battalion 34 in front of the building awaiting them. They had gotten an alarm from the street alarm box and had been directed into a ground floor apartment in a brownstone building by neighbors who reported "some people being eaten by a dog." It seems the occupant of the apartment had been training this dog to fight other dogs. To make the animal more vicious, he had taken to administering almost continuous beatings, until this last one. The dog had had enough and went after the man and his girlfriend.

The firefighters cautiously entered the front door and made their way back through the hall until they located a doorway from which terrifying screams, as well as horrendous snarling noises, could be heard. Ladder 102 forced open the door and was confronted by a gruesome sight and a ferocious pit bull. The walls were splattered with blood, and at least two people could be heard screaming from inside, but no one could be seen. When the firefighters attempted to enter, the snarling pit bull attacked them and they were forced to beat a hasty retreat to the hall, closing the apartment door just in time to prevent the dog from sinking its fangs into one member's arm. At this point, the chief called for assistance. The police department's Emergency Services Unit (ESU) is equipped with tranquilizer guns for this eventuality and he requested them, as well as the ASPCA. He also directed the dispatcher to request an ETA (estimated time of arrival) from each so that he could evaluate his options. The police department simply refuses to give an ETA. The ASPCA gives theirs: 45 minutes to an hour. The chief knows he can't wait that long, so he calls for Rescue 2, less than 10 minutes away. A police officer from the local precinct arrives and is briefed on the situation. ESU will not give him an ETA either. The firefighters ask the police officer to shoot the dog so they can enter and locate the victims, but the officer can't safely fire. He doesn't know the location of the two people. If his bullet misses or ricochets and hits a person, he could be in deep shit. The people will have to wait. At least they're still screaming, so they're still conscious.

On arrival of Rescue 2, Lt. Artie Connelly is briefed by the 34th Battalion. Lt. Connelly formulates his plan. He tells Timmy Higgins to bring the dog noose, a six-foot long pole with a loop of ⅜-inch rope at one end that can be slipped over a dog's head from (hopefully) a safe distance, and then tightened by pulling on the other end of the rope that runs up the middle of the hollow pole to the operator's end. We occasionally have to use the device to restrain guard dogs that we encounter at fires in factories, junkyards, and other commercial occupancies. Usually in those incidents lives are not at stake as they are here. The noose is often very difficult to slip over the head of a moving animal, especially one who is coming directly at you with the intention of ripping your throat out. As a backup, Lt. Connelly tells John Kiernan to bring the Lyle gun.

The Lyle gun is a .45-70 caliber, single-shot rifle with an 18½-inch barrel. It has been around since the mid-to-late 1800s. The gun has been associated with lifesaving and rescue for nearly as long. In fact, the Lyle gun is prominently figured on the badges that symbolize Rescue company officers. The gun is used to shoot a rope across an open space, such as from one side of a river to the other, or from a ship to the shore, in order to pull a much larger rope across the divide. Such methods were used by the U.S. Lifesaving Service for years to rescue people who were stranded on vessels that had run aground. They are still used to rescue people from flood-swollen rivers and streams. The rescue companies carry them for this purpose and could use them to span from one building to another, although I have never actually used one for that purpose. In fact, the only time I have ever actually shot one was in training.

The gun uses blank .45-70 cartridges (a really large shell) to propel an 11-inch long by $^{45}/_{100}$" diameter brass rod toward its objective. A high strength nylon string of about 600-pound test strength is attached to the brass rod. Depending on the strength of the cartridge used, and the weight of the line attached, the gun is capable of shooting the bolt up to 600 feet. The rod will actually go much farther without the string attached, which tends to act as a brake on the projectile. (Don't ask how I know that, but it is a verified fact!) With the huge shell and the short barrel, the gun makes a tremendous bang when fired. That was what Lt. Connelly had in mind, an improvised "flash-bang grenade" that might be used to stun the dog without delivering any serious harm. John Kiernan, on the other hand, was all for using the gun for real, with the brass rod as the bullet. Lt. Connelly, of course, would have none of that, but John at least had to ask!

At the end of the narrow hallway is the door to the apartment. Just inside this doorway is a partition that runs perpendicular to the doorway, which means you have to get the door open about six inches in order to slip the dog noose in through the opening. Lt. Connelly directs Tim Higgins to quietly open the door and position the noose, before the pit bull realizes they're there. But the dog is waiting for them. As they open the door, the dog is right there on top of them, trying to get out into the hall. Connelly is trying to hold the door closed enough to keep the dog in, but open enough to let Timmy get the pole and noose in place. The dog wants out at them. He is snarling ferociously and is snapping at anything that comes within reach of his teeth. Timmy is trying to maneuver the noose around the dog's head, but the tight quarters in the narrow hallway and the odd angle created between the door and the wall prevent him from being able to use the pole to advantage. He now has to get up real close and personal with the four-legged eating machine and use his hands to try to maneuver the noose over the animal's head. "Make sure you hold that door closed, Lieu!" Timmy calls. "Hurry up out there!" calls a female voice from within the apartment. A male voice also calls out from somewhere within the apartment, seemingly muffled, "Just open the door and let the dog out, man! He won't bite!" "Kiss my rosy red Irish ass!" calls Timmy.

The dog has the noose in his jaw now, preventing Timmy from getting it around his head. The carnivorous canine seems possessed, like "Cujo," the vicious St. Bernard of the horror flick of that name. He also has grabbed the edge of the door in his powerful jaws and is now backing up, pulling the door open, even as Lt. Connelly and Timmy struggle to maintain it in the nearly closed position. The dog is powerful, and it is pissed off, and it's winning. "Shoot the dog!" orders Lt. Connelly. "With the rod?" asks John Kiernan, hopefully. "No, John, just the blank, and hurry!" John puts the muzzle to the dog's head and fires.

The roar of the monstrous cartridge going off in such a small space sounds like a cannon, not a rifle. A tongue of flame about three feet long blasts out of the barrel, singeing all the hair off the animal's head and setting the rope of the noose on fire! The blast sends the animal skittering backward, tumbling along the floor until he bounces off the far wall, stunned and smoking. In spite of being somewhat shell-shocked themselves, Timmy and the others open the apartment door and call in, "C'mon out! The dog's knocked out!" A hysterical female suddenly appears from around the corner of the partition wall, screaming for her boyfriend. At that, a large cardboard carton in the corner of the sparsely furnished apartment begins to erupt. Like a volcano spewing fragments of cloth, a figure suddenly pops up out of the washing machine carton, covered in dozens of rags. Timmy pulls the woman to the safety of the hallway and passes her to others for treatment of several very nasty looking bite wounds of the hands, arms, and legs. Timmy calls out to the human jack-in-the-box, "Let's go, pal!" As he sees the still groggy pit bull getting back on its feet, Tim calculates the time it is going to take the dog's master to climb out of his hiding place and cross the entire room to get to safety. Timmy, the dog's master, and the dog all seem to come to the same realization at once. The dog is blocking his master's escape path, and he will be on him before the man can make it out. The dog lunges for the man, who is bleeding from numerous bite marks about the head, face, arms, and shoulders. The desperate man plunges back inside his fragile shelter, frantically pulling piles of rags back on top of himself. The dog commences to shred the cardboard, something it obviously had been doing before the fire department arrived. As the dog pulls pieces of box and clothing apart, the man inside struggles to plug the opening with more rags from the inside. Tim can see the situation can't go on like this, so he smacks the dog's rump with the pole from the noose to distract him. It did just that all right! The dog turns, charging toward Tim, who quickly slams the door in his face.

Lt. Connelly now orders everyone to regroup. The still smoldering rope on the end of the dog noose is cut free and replaced with a new piece. The pole is cut shorter to make it more manageable to use in the crowded confines of the dingy hallway. The police officer is still unable to get an ETA for ESU. Connelly tells Kiernan to reload the Lyle gun (yes, with blanks only). "Let's try this again."

The dog responds immediately to the opening door, going right for the noose as the pole is inserted. Now he has the rope and the end of the pole in his crushing grip. The wooden handle starts to splinter as he gnaws on the device, all the while backing up, pulling Timmy and Lt. Connelly into the room. Suddenly, now able to see the people instead of just the antagonizing stick, he suddenly lunges for the door. Connelly is just able to pull it closed in time, catching the dog's snout between the door and the jamb. Tim tries to maneuver the noose around the neck but this dog has been tied up like that for nearly all of its life, and again he deftly dodges the noose. The voice in the box pleads again, "Just open the door," and again Timmy, Lt. Connelly, and John Kiernan tell the man what part of their Celtic anatomies the man can concentrate his affections on. Lt. Connelly tells the man that they are going to shoot the dog again and this time the man should start out of the box exactly as they do so. "Ready, on the count of three! One, two, three!" Kiernan fires again. The man is already on his way out of the box.

Well, now this dog has been through this whole thing once before. Just like the first time, the blast sends him tumbling backwards across the floor. The remaining hair on the other side of his head is singed. But this time he knows what's coming! "Hell, the first time didn't hurt so bad, I guess I'll live through this one too." The dog is shaking off the effects of the stun round much sooner this time, and then he sees him—the brutal human who tortured and tormented him from the moment he acquired him. The dog's eyes lock on the man like the fire control radar of a surface-to-air missile. The dog is up and on his feet, even as the man takes his second step out of the cardboard refuge. Two more steps to go to the safety of the hall. The dog is up and moving like an ICBM launched from a submarine—he's up, then pauses as the booster ignites. Tim is screaming for the man to "move it! Here he comes!" Connelly is holding the doorknob, waiting to slam it closed on the approaching menace, while Kiernan feverishly breaks open the action of the single-shot Lyle gun and inserts another blank round in the chamber, just in case. The roar of the blanks has left everyone temporarily nearly deaf, which adds a distant, slow-motion effect to what happens next.

The race for the door is a tie. Ninety percent of the man is through the door, but the business end of the pit bull is neck and neck with the remaining 10%: the man's left leg. The dog cannot be allowed out into the crowded hallway where there is nowhere to run or hide. Connelly slams the door in the dog's face just as it locks on his master's heel. The man's scream pierces even the fog surrounding the traumatized eardrums. Timmy and John struggle to pull the man the rest of the way through the narrow doorway, while Lt. Connelly holds the door to keep the devil dog inside. This time he's not about to let go. The harder the people pull, the more it wedges him into the corner between the door and the wall. It's like locking a vise down. Finally, Timmy reaches his hand in through the crack of the doorway with a knife.

To free the bleeding man, he must reach in and slice open the shoelaces, in order to allow the man to pull his foot free of the powerful grasp. As soon as the last lace is sliced, the man's foot pops out. Tim pulls his hand back just as the Lieutenant slams the door closed.

The man has lost so much blood that he has to be carried to the street and the waiting ambulance. He is bleeding from several wounds that appear to have narrowly missed major arteries, including the carotid and the femoral. An inch either way, and "Cujo" would have gotten his revenge. The man's sorry state does not garner him any sympathy from anyone on the scene: firefighters, cops, or neighbors. Even his girlfriend says the same thing, "That bastard got what he deserved, treating an animal so cruelly." Training those dogs to fight to the death is a crime in New York State and it's too bad more of these two-legged mutts aren't bagged for it.

Guess who shows up, just as John Kiernan walks out the front door with the Lyle gun? You got it, ESU. "Holy shit! Now the damned firemen are carrying guns!" one was heard to exclaim. The boys just walked back to the rig, satisfied with a job well done. One life definitely saved, and a whole new story to tell around the circular table in the kitchen on Bergen Street. The cops can deal with the pissed-off pooch at their leisure.

CHAPTER 29

Popes, Presidents, and Other Assorted Dignitaries

A unique responsibility that comes with working in Rescue Company 1 is providing crash rescue protection for the helicopter landing zones of the President of the United States, the Pope, and the heads of state of some of the most powerful nations on earth whenever they land in New York City. Whenever the Commander-in-Chief lands at one of Manhattan's three heliports, or at a non-protected site such as the middle of Central Park, Rescue 1 is there with Hurst tools and other extrication gear ready to go in case of a crash. This duty can, at times, be burdensome. One day we spent 14 straight hours, from 7:00 a.m. to 9:00 p.m., safeguarding a series of helicopter landings as Pope John Paul II flew to and from a series of masses around the city and in New Jersey. Generally, though, the members of Rescue 1 accept this task without complaint and, in fact, revel in the knowledge that it is they to whom all eyes will look should the lives of the most powerful leaders on earth be placed in mortal danger (fig. 29-1).

These presidential standbys have resulted in a number of interesting events: some quite funny, others more sobering. During the Pope's 1995 visit to New York City, there apparently was a very serious threat made by some Islamic terrorist group that an attack on the pontiff would be made using chemical weapons. The Secret Service took the threat very seriously, and the next thing you know, Rescue 1 is being trained in chemical weapons counter-terrorism. Suddenly, we were issued chemical protective

Figure 29-1. Rescue 1 is among the units standing by as Pope John Paul II arrives in NYC. Preparations include the presence of the firefighters in aluminized chemical protective clothing. From left: Tom Baker, Lloyd Infanzon, Paul Hashagen, Joe Angelini, and Battalion Chief Tom Fox keep a vigilant eye as the pontiff alights from a helicopter.

clothing (CPC), nerve gas detection paper, military CAMs (chemical agent monitors), chemical detection equipment, and all kinds of gadgets we could never get our hands on before. Then, as part of our standby routine, two members of our crew were detailed to be suited up inside their chemical protective suits and carrying a mask for the Pope whenever he alighted from the chopper until he was safely in the motorcade enroute to his secure destination. The CPC is a sealed, fully encapsulated plastic garment, containing its own atmosphere. It is aluminized on the outside, to provide some flash protection when dealing with flammable materials. As a result, the members inside the suits look like the Pillsbury Dough Boy wrapped in aluminum foil.

On the day of our marathon 14-hour standby, Lloyd Infanzon and Cliff Stabner are the two guys assigned to the suit duty (fig. 29-2). It's a very warm, sunny day, and the time spent inside the suits is very uncomfortable.

Figure 29-2. Papal visit preparations include these two "baked potatoes" in aluminized chemical protective clothing: Cliff Stabner (left) and Lloyd Infanzon.

The rest of the guys are on Lloyd's and Cliff's case too, saying they look like two tinfoil-wrapped baked potatoes. Each time we unzip them from their sealed cocoons, both of these guys are soaked with their own sweat. I ask them if they want to take a break so I can rotate other members into the suit, but each time they refuse. Both of them agree that they want to be the ones in the suits because, "If something actually does happen, TV cameras all over the world are going to get *our* pictures: two 'baked potatoes' carrying one 'Very Important Polish guy' to safety, then dancing and slapping high five's while still dressed in aluminum foil." The excellent work of the Secret Service, FBI, and all of the other counter-terrorist agencies paid off and nothing happened; the world missed the potato dance, but we were ready.

Security for these dignitaries is, of course, a primary concern. The only vehicles permitted in proximity to the frozen zone are Rescue 1's rig, a crash truck, and specific cars from the Secret Service motorcade. When we bring the apparatus out onto the pier, a bomb-sniffing dog goes through it to make sure we aren't a Trojan horse. But I used to wonder how the Secret Service could be sure that some of the 20-odd firefighters hadn't been killed and replaced by terrorists who then took our place on the flight line. The answer to my question came almost nonchalantly one day as we waited for the next incoming flight. I asked this question of one of the "men with the wires in their ear" and his reply came quickly. "Look around yourself, guy, and tell me what you see." There in the corner of the pier stood several groups of firefighters in their smoky, soot-stained bunker gear, all laughing and telling jokes or war stories, as though they hadn't a care in the world. A little to their left, at rigid attention, stood the Marine Corps security detachment of the Presidential helicopter squadron, HMX 1, in their sharpest dress blues. In front of them was a detail of Port Authority Police officers, also in their best uniforms, including white dress gloves, also at an unsmiling, rigid parade rest, and everywhere the ubiquitous men with the wires in their ears, each with their prescribed lapel pin and not a hair out of place. I had no idea what I was looking for, but the agent certainly did. He said, "We know you're all firefighters. Look at yourselves. Nobody on this pier behaves as calmly and casually as you guys. The minute we see a bunch of firefighters line up at attention and not playing grab ass and telling jokes, is the minute the hardware comes out and we start hosing them down, cause we know they can't be real firefighters." I know he meant it because I've seen these guys in action.

President Bill Clinton came to New York to celebrate Memorial Day and its "Fleet Week" observances, where a number of American as well as foreign naval vessels dock at the piers along the west side of Manhattan. The presidential helicopter was going to land on the deck of the USS *Intrepid*, a World War II aircraft carrier that now serves as a sea–air–space museum. As usual, Rescue 1 was assigned to the detail. Since this was not a normal landing zone and access was extremely limited, some modifications would have to be made to our usual routine. Obviously, we couldn't take our rig up onto the flight deck, so all of our needed equipment would have to be carried up and secured against prop wash (the wind caused by the whirling rotors of a helicopter as it ascends or descends) on the exposed catwalks that skirt the edge of the flight deck. We would have to share this space with the men with the wires, and in this instance, with a counter-sniper team. Normally, we never see the counter-snipers out on the pier heliports, but in this case, because the landing zone was surrounded by water on three sides, they were placed at the edge of the flight deck with the rescue team.

Everyone in the Secret Service obviously takes their job very seriously. Dignitary protection might require an agent to make life and death decisions with worldwide implications, and they might have to make the supreme sacrifice when protecting their "principal." Most of them are very down-to-earth people, good judges of character and, once you get to know them, relatively friendly people. During one long, hot day out on the sun-soaked sponson, we struck up conversations with several of these fine men, shared a few cold Coca-Colas that we had hauled up from the fireboat of Marine Co. 1 that was tied up below us to provide firefighting water, and shared a few stories. Suddenly, there came a shouted order from one of the snipers, "Everybody get down!" Remington Model 700s were shouldered and beads were drawn on a U.S. Navy helicopter that was making its way up the Hudson en route to the active aircraft carrier USS *America*, which was berthed right next to the *Intrepid*. I scrambled behind a quad 40 mm gun mount, the same kind that my father, "Butch," had crewed as a gunner's mate aboard the USS *Block Island* first in the North Atlantic, then later in the Pacific during WWII. I figured if it was good enough for him with U-boats and kamikazes shooting at him, it would be good enough for me.

When no shots rang out, I peeked out from behind the gun mount at the guys who only moments earlier had seemed as relaxed as any of the rest of us. One agent was chattering into a mike in front of his lip while he and the other shooter tracked the chopper in their scopes. Finally, the Sea Knight was clear of the designated security zone and the butt plates came away from the shoulders.

"Get that dumb son of a bitch on the deck and square his ass away!" was one of the selections that were playing through the agents' earpieces, audible even three feet away. When things had calmed down, I asked if the agents were really serious, drawing down on a marked US Navy helicopter. Their answer was unequivocal: "He was some place he wasn't supposed to be." I pressed them a little more. "Would you really have fired at somebody flying along seemingly innocently?" and "What effect would those .30 caliber rounds have on a presumably armored helicopter?" Again, their voice and demeanor made it clear they were not just posturing, "We weren't firing without some indication of intent. We had the pilots in our scopes and one of us had the door. If anybody poked anything out of the sides or door, we would do what we had to."

I was still curious. "What if the chopper itself appeared to be the threat?"

"There are others prepared to deal with that," was their curt reply. I decided not to press my luck with any more questions. The Secret Service didn't need a busybody in the landing zone.

We finally returned to quarters after a long day out in the sun. Our faces were sunburned, but the rest of our bodies were only flushed, and the bunker gear and hoods did a good job of protecting us from the radiant energy of the sun. Unfortunately, they keep in a lot of the body's heat. It's kind of like putting on a child's snow suit, complete with mittens, galoshes, scarf, and wool hat, and then climbing into a sauna instead of a snow bank. I asked everyone if they were all right. Of course, I knew that unless they were dying, no one was going to admit that they were anything less than in perfect shape. Joey was working, and to admit to being hurt or tired was to invite enough teasing that it just wasn't worth it.

KEEPING UP WITH JOEY

Joseph "Joey" Angelini Sr., was Rescue 1's senior man. In 1997, he had over 37 years on the job, including 19 with Rescue 1. Joe, age 60, was a white haired gentleman, diminutive of stature but huge of heart. I was always concerned about Joey. Firefighting is a young man's job, but I knew he would never let on if anything was bothering him. He always had to do more than his share, whether it's pulling ceilings, forcing doors, or tunneling through debris. When I first got to Rescue 1, I used to think it was an act of "showing off," just to prove he could still keep up with the young bucks nearly half his age and double his weight. Eventually, I realized this was no act. This was just the way Joey *was*. Paul Hashagen put everything into perspective when he told one of the guys, "Not that Joey's old or anything, but when he went to his first presidential standby, the president came into town on a stagecoach." Yeah, but he was fun to have around, especially when you needed a guy who wanted to get dirty.

One afternoon we respond to a fire in a fancy restaurant on the ground floor of a six-story apartment house. The fire began in the kitchen as they so often do and extended into the ventilation hood over the stove. This is a common occurrence, and for years, these hoods have been required to be equipped with an automatic dry chemical fire extinguishing system that discharges a fine powder throughout the duct work and down onto the stove and cooking surfaces. At this fire, the system had activated, but because the fire had started away from the stove, and had gotten a good head start, the extinguishing system did not stand a chance of stopping the fire.

As soon as Engine 23 has knocked the fire down with their 2½" line, Ladder 4 and Rescue 1 are inside, opening ceilings and walls looking for extension, and among the first places we look are at the ducts that lead from the hood. Sure enough, there is heavy fire present in the duct work, feeding on the layers of cooking grease that coat the insides. That's a serious problem in these situations, because while the restaurant is on the ground floor, the galvanized steel ductwork that exhausts the cooking fumes has to travel all the way up through the building and terminate above the roof line.

If this were a high-rise, the floors would be made of reinforced concrete, but in a six-story building, it is what we call Class 3 or ordinary construction, meaning the exterior walls are made of brick while the floors, walls, and most interior construction is made of wood. The grease-coated steel ductwork draws the fire all the way up through the building, and the red-hot steel conducts heat to all the wood it is in contact with, igniting fires on every floor and in the cockloft nearly simultaneously. The result is often a large multiple-alarm fire, unless we can immediately get water going up that shaft and get truck companies to the location on every floor where the ductwork passes through.

As soon as I see the fire in the hood, I tell 23 to get their line going into the duct, while calling Battalion 9 to let them know of the situation and the need for a second alarm. We are going to need a lot of people on the upper floors. I also know it is important for somebody who has actually seen the location of the vertical ductwork to go up above and show the units on these upper floors exactly where it is located so they don't waste time trying to find it themselves, since it is nearly always concealed from view. I take David

Weiss and Billy Henry with me to the upper floors to begin opening up. The second alarm companies arrive, hoselines are stretched to every floor, and the fire is contained fairly quickly: a great job by all. Now I return to the first floor for one more look around.

When I get to the kitchen, there is still water pouring down from the ductwork, and I assume there is a hoseline operating down the shaft from the roof: a very good tactic in these situations. Then I notice that there is a length of 1¾" hose now attached to 23's line: and the line itself goes up over the stoves and disappears into the exhaust hood. I look around for my crew; one guy is missing. "Where's Joey?" I ask. Nobody will meet my eyes, nor say a word. "Where's Joey?" I repeat, with more of an edge to my voice. Finally, one of the firefighters from 23 points up into the hood. "Holy shit!" I stand up on the stove and look into the ductwork, which travels across the ceiling horizontally about 10 feet. I can just make out the soles of a firefighter's boots as the water pours out around them. "Joey!!!" "Yeah, Cap," comes the reply. "Joe, come down out of there right now!" "But Cap, I think I should give this just a little bit more of a bath." "No, Joe, come down! Now!" When he finally wriggles himself back to the top of the stove, he is filthy: soaking wet, and covered in grease and extinguishing powder from head to toe. I have never seen a sorrier example of a human being in my life: and also the greatest example of a firefighter imaginable. His size makes him the only guy that can fit in there, and his heart makes him the only guy that would *want* to stay. He is just the kind of guy a dignitary in trouble would want coming to drag them out of a bad situation—but a captain's nightmare.

CHAPTER 30

V.I.B.s–Very Important Buildings

Sometimes the V.I.P. treatment is extended to more than living things. What you might call the V.I.B. treatment, for Very Important Buildings, is extended to some of the most famous landmarks in the city: Grand Central Station, Penn Station/Madison Square Garden, The United Nations Building, The Empire State Building, and St. Patrick's Cathedral. Each of these is among the buildings that are included in a yearly familiarization schedule. All of the first- and second-alarm companies either walk through the facility, looking at all the fire protection features, or else participate in a simulated fire at the selected location. I didn't know it when I first walked in the door at Rescue 1, but one of these buildings would have a tremendous impact on my family's life.

One of the "target hazards" in Rescue 1's territory is the United Nations General Assembly building. With all the politicians gathered there, it makes a tempting target for a few loonies bent on wreaking havoc and gaining world-wide notoriety. It has been the scene of such events in the past, including an attack with a mortar, and it probably will see similar events in the future, despite the best efforts of our intelligence agencies. I am most concerned with the present though, since it's my ass and those of my men that is left hanging out in the wind if something happens.

One such event was almost comical in how it unfolded. It seems a young woman from California became upset at the way (your choice): (a) whales,

(b) baby seals, (c) the lesser spotted loon, or (d) the three-peckered billy goat were being treated, and decided that the General Assembly of the United Nations was the only proper forum for hearing her complaint. So, she jumped in her 10-year-old van, reminiscent of something hippies would have been proud to own in 1968, and drives cross-country to air her complaints.

This woman is obviously not totally unbalanced, since she knows enough not to bring a new car into the city to do battle with the dreaded yellow peril—New York City taxis. Anyway, somehow she manages to pull right up to the guarded entrance to the main plaza directly in front of the main building in this huge complex, and then simply drives unimpeded right past the barriers and guards, right up to the front of the main entrance, and parks: the only vehicle in this huge plaza. Eventually, the crack security team at the UN (Mo & Joe's Cut-Rate Guard and Lawn Mower Service?) realizes this van with California plates and 3,000 miles of road dirt just looks a *bit* out of place. A guard wanders over to tell her to move and discovers the woman busily opening one-gallon cans of camp stove fuel (white gas). Now he realizes something is not right and decides to call out the "Uh-Oh" squad, who proceed to the scene, conduct a full size-up, and then collectively gasp, "UH-OH!"

Now the ball is rolling. The NYPD Emergency Services Unit is called and deployed in a screen around the van armed with shotguns, high-powered rifles, and submachine guns, all aimed to kill this woman who is threatening to kill herself if her demands are not met. Fortunately for her, the NYPD has a long history of dealing with crazies, both homegrown and the out-of-state variety. The Hostage Negotiating Unit is brought in and discovers that the woman is threatening to immolate herself (a bad thing) and her van (probably a blessing for humanity and other drivers in particular) if her demands are not met. She claims the van is packed with propane cylinders and the aforementioned white gas. The United Nations compound has its own fire brigade that deals with most of the routine incidents—alarm activations, smoke investigations, minor fires, etc.—but this is definitely not a routine incident so the FDNY is called.

A full first-alarm assignment: three engines, two ladders, a battalion chief, a deputy chief, and a rescue company are assigned. Rescue 1 is busy at another alarm, so Rescue 2, where I am working that day, is assigned. The PD briefs our bosses on the situation, who in turn brief us and ask for comments. The engine companies have stretched a pair of handlines anticipating a possible explosion. The cops are standing up alongside the van, talking with the woman, and can see the van has about 10 one-gallon cans of white gas and a similar number of 20-pound propane cylinders in it. The woman has only rolled the window down a crack, but the cops don't smell any propane. I ask the deputy for permission to approach the van for a better look and to bring an explosive gas detector up to see what the danger inside the van is.

Tim Higgins and I approach the van from the rear, where the two back-door windows are painted over. Timmy gently slides the probe of the explosimeter through the rubber seal at the base of the doors. He pumps the squeeze bulb the required 10 times and gives me the thumbs down signal. No explosive gas. Looking down the passenger's side of the van, I can see the driver in the mirror. She is still going about her business, watching the cop at her window, but still just about oblivious to everything else. We can also see that the propane tanks are spread around the rear of the van, but otherwise it looks like she has settled down for a long siege.

I make my way back to the deputy with a proposal. Since we can walk right up to the van, why don't we give her the bum's rush? I propose to take myself and three members up to the back doors again, this time with two Halligan tools and two large dry chemical extinguishers. The two 2½-inch handlines should be fitted with fog nozzles and foam eductors and told to stand by at their present locations, about 100 feet away on each side of the van. A team from Police Emergency Services will accompany us to our location. When the cops give the "go" sign, the guys with the Halligans will take out the rear and side windows, immediately following which, the two dry chemical units will open up into the van's interior, filling it with a blizzard of fire-killing Purple K powder and a shrieking noise from the high velocity discharge. The effect will be to prevent ignition of the gasoline while simultaneously disorienting and momentarily blinding the woman before she has any chance to react. With this accomplished, the cops can reach through the driver's side window and yank this nutjob out of the van. "Where the hell did you come up with that idea?" you ask. The answer is simple; it happened to me, so I know what it does to a person.

As a teenager, I worked for Tri-County Fire Equipment in Inwood, recharging fire extinguishers, installing range hood dry chemical systems and sprinkler piping, recharging air bottles, etc. More than once, a faulty valve on a newly recharged extinguisher, especially the old style Safe-T-Firsts with their multiple lever arrangements tripped, discharging their 20 pounds of powder inside the 4-foot × 6-foot recharging room. But the incident that really got my attention involved one of the large, 300-pound wheeled units that we had picked up in our service van. It seems a fire had occurred at a truck-loading rack at an oil storage depot in Greenpoint, Brooklyn. My boss, Neal Metz, gave Robby Frank and me the task of picking it up and bringing it back to the shop for refilling. The unit had been operated for only a few seconds during the fire and was still mostly full. Robby and I humped this monster up into the back of Tri-County's step van, secured it with ropes, and started back on the hour drive over the bumpy roads of Brooklyn and Queens for the shop back out in Inwood. These units are termed semi-portable, in that the whole thing weighs about 550 pounds, and has two four-foot diameter steel wheels to allow you to get the thing relatively close to the fire. And then it has 50 feet of one-inch (inside) diameter hose, similar to a booster line hose, and a nozzle with a normal horseshoe handle shut-off on it to allow the operator the mobility needed to work around the fire.

Well, unbeknownst to us, as they say, as we were driving back over every pothole in two boroughs, the nozzle had fallen out of its bracket. The nozzle was probably opened when it fell, but nothing happened. Since the fire had occurred Friday night, and we weren't called to pick it up until Monday, the powder sat packed in the hose all weekend. It apparently was caked in the hose quite tightly. Finally, as we crossed the Queens-Nassau border on Central Avenue, one final bump (or maybe it was just the sudden lack of bumps) loosened the powder enough to where it began to flow out of the nozzle. Thankfully, we were just preparing to stop for a red light at Doughty Boulevard. When this whirlwind of powder suddenly blasted into the van, neither Robby nor I said so much as a word—we both dove right out the open side doors of the still-moving step van. We dazedly watched the van as it slowly coasted toward the curb across the intersection, trailing a huge cloud of purple haze. I thought something had blown up. When the van finally stopped, Robby and I cautiously approached, coughing and sneezing the fine purple powder from our lungs, mouths, and noses. "What the hell was that?" we both asked. When the dust cloud finally settled, there was the open nozzle, with a trail of powder pointing right to it. Well, it certainly scared the shit out of us. Now I was counting on it having the same effect on the woman in the van, like a flash-bank grenade, but with absolutely no chance of a fire.

The deputy listened thoughtfully to the plan as I laid it out. Then he asked, "Where the hell did you come up with that idea?" Since I didn't have the time, and he didn't have the interest, for me to go through the whole explanation, it was just about dismissed out of hand. Actually, I think he did propose it to the cops, but they gave him the same look that he had given me. Anyway, nothing was going to happen as long as the hostage negotiator could still keep the woman talking. So we sat back and waited, and waited, and waited. Eventually 6:00 p.m. and the end of the tour came, and we were relieved by the night tour of Rescue 1. I was going off-duty and home. When I came in to work three days later, the company had gone back over to relieve other companies twice on four-hour stints. Eventually the cops wore the woman down, probably by continuously babbling at her. She surrendered without incident after nearly 48 hours in the hot, smelly van. Thankfully, it was another fine job by the NYPD Hostage Negotiation Unit.

The UN building was the scene of another interesting event for Rescue 1 one evening in 1997. I had gotten a "tickle" from a friend in the federal government one day, asking what I knew about "dirty bombs," and inquiring what preparations the fire department had made to deal with one. A dirty bomb is really a nasty device, as its name implies. Basically it is a conventional explosive that is wrapped with radioactive materials. It doesn't produce a nuclear blast. In fact, the visible damage is no greater than the blast from the conventional explosive. The insidious part though is the blasting of the nuclear material into the surroundings. This has two effects. It contaminates the atmosphere, so that anyone who breathes in any of the smoke or dust also inhales radioactive materials. Not a good thing for maximizing your retirement pension. Second, the radioactive matter is blasted in fragments into the nearby structures. Some estimate that if a terrorist were to "dirty bomb" a building using a tiny amount of plutonium, that building would be uninhabitable for 1,000 years! It would also cause a wretchedly horrible death in about one week for anyone who happened to inhale any of the contaminated dust or smoke: New York's bravest. You see, while we wear masks that would protect against such inhalation during structural firefighting, using them during the search of a collapsed structure the size of the Oklahoma City Federal Building bombing site would be impractical, due to the short working duration, great bulk when crawling through tight voids in a debris pile, and the debilitating effect of wearing them while performing long periods of heavy work, like collapse shoring or tunneling.

Just the fact that my friend was asking about such a possibility was disturbing enough, but his wanting to know what we would do about one seemed like more than just idle curiosity. I told him we would do our normal collapse search and rescue effort, with extra emphasis on wearing the masks, decontaminating everyone and everything that entered the blast area, and limiting the exposure of the people we send into the site: standard precautions for dealing with radioactive incidents. In addition, we would use the old standbys of time, distance, and shielding whenever possible, but that this all would depend on our recognizing that such an event had occurred. His reply was brief and to the point, "Well, you'd better recognize it, because somebody's trying to do it to the UN." I knew enough not to press for any details except to ask if there was any date or time. "Anytime from yesterday on," was the reply before he hung up. "Oh, that's just great," I remember thinking. While hazmat and a number of other units have radiation detection equipment, Rescue 1 and most of the other first-alarm units that respond to the UN don't. I immediately set out to get us some.

The department has a radiological officer: a position that goes back to the Cold War era. He is in charge of coordinating the operation and maintenance of a number of "monitoring stations" set up around the city in various engine companies, designed to report the levels of fallout after the "Russkies" atom-bombed New York. I try to get Rescue 1 issued a full monitoring kit through this officer, but he is not very cooperative. He wants to know why Rescue 1 needs a kit now, when they haven't been a monitoring station in 40 years. I can't just come out and say, "Because a little birdie whispered in my ear that a bunch of no-good, filthy terrorists is about to blow up a New York City landmark using nuclear materials." Even if I could, he'd probably tell me that Engine 76 on the upper west side is the nearest monitoring site to Rescue 1 and I should get fallout readings from them. Next I turned to Chief Jack Fanning of the Hazmat Operations group. Jack's people were very sympathetic but they didn't have any spare equipment to loan us. Running low on FDNY resources, I turned to Chief Mark Rolon of the Inwood Fire Department for help. When I was the chief in Inwood, I had gotten us two monitoring kits from New York State Civil Defense, just because they had offered them to us. While these were 1950s era technology, I always figured that something was better than nothing and said we'd take them. Now, applying that same logic I asked Mark if I could borrow one kit to take to work until I could make other arrangements. Mark said sure, just don't forget that they already have a home.

The next night I brought the whole kit into Rescue 1, found a good case to package it all, replaced all the batteries with fresh ones, and proceeded down to the kitchen for drill. I started the training by explaining about this potential terrorist attack on the UN and how it might involve nuclear materials. One of the guys popped in that, "It's OK, Cap, the UN has their own fire brigade. They probably wouldn't call us even if they blew the place over into Brooklyn." Without trying to sound like a wise guy or an alarmist, I told him that I didn't care what the UN guys do, we in Rescue 1 are going to be ready for whatever happens. I had to admit though, he was right. I had been at Rescue 1 for about three years and the only time I had ever been to the UN complex was that time six years earlier when I was in Rescue 2 with the psycho in the van. I proceeded to show everyone how to use the radiation detectors, how to differentiate between alpha and beta particles and gamma rays by opening the window on the detection probe, and how to charge and read their personal dosimeter. Because this was the first tour these devices were in place, and the other groups were not trained in them yet, I had Billy Henry carry them back upstairs to the company office so that I could work on writing up a lesson plan to give to the other officers to train their people. I had just dished myself out a bowl of Chip Tate's famous cherry vanilla ice cream when the teleprinter bell in the kitchen announced that we had a run. Paul Hashagen ("Hash") was on watch and you could hear the edge in his voice as he read the ticket over the intercom, "Phone alarm, 42nd Street and FDR Drive, an explosion at the UN building!"

"Holy Shit! This is it!" someone exclaims. Before I get into my bunker gear, I run up to the office and retrieve our new case full of nuke goodies. Back at the rig, I pass the box up into the back to the boys, but not until I take out a dosimeter for myself and take a quick look through it to get its "zero reading." I yell for everybody to take a dosimeter, and get its zero, and then call them up to me as we head out the door. As we head east on 42nd Street from 11th Avenue, I'm scrambling to record each man's reading next to his name on the BF-4 on the dashboard. Not all of the dosimeters are set at zero, some have "creeped up" due to their being in storage, unused for such a long time, and it will take a few days of continually charging and zeroing them to get them set. The important thing now though is to get their starting points and make sure none of the men absorbs more than 25 rems of radiation or, if they do, to be able to document it so their survivors get a line-of-duty death pension for the radiation sickness that killed their husbands and fathers.

As we cross Times Square, I am alternating between listening to the dispatcher on the department radio and yelling back at the men to "make sure everybody has all their exposed skin covered," and "Decon yourself before you take your mask face-piece off," as well as other pointers on "Nuclear Explosion Behaviors for First Responders 101." When we cross Sixth Avenue (Avenue of the Americas for the tourists), Engine 21's 10-84 shows on our MDT screen (mobile data terminal—a computer screen that sends and receives messages to and from the dispatcher). "Well, at least they're not screaming for help," I remark to Hash. As we reach the Public Library at Fifth Avenue, 21 calls a 10-7: a request for the dispatcher to call back the complainant to verify the address or get more information. The UN complex is a very large site, and while it's possible there is something happening where 21 can't see it, I tell Hash to slow down. This doesn't sound right.

With the UN's own security on scene, they are supposed to guide us right to the proper location.

Battalion 8 arrives and the dispatcher reports that the original call did not come from UN security and that the callback number they gave is busy. Battalion 8 replies that all units are investigating, but UN security has no incident on their property, although they do report hearing a loud explosion on the East River side of the complex. Eventually, Ladder 2 locates a manhole with its cover blown and smoke seeping from the opening. Chief Jensen transmits the 10-18 signal and Hash swings the rig south onto Second Avenue, heading back to quarters. A simple manhole fire, a very common occurrence in New York, especially after a snow storm, has added about a hundred grey hairs to the heads of each member of Rescue 1, who were sure we were on our way to an event that would rank right up there with the 1993 World Trade Center and 1995 Oklahoma City bombings. Fortunately, we have not experienced a dirty bomb attack, and the extensive preparations for dealing with terrorist attacks of all kinds in the wake of September 11th have left the FDNY much better prepared to deal with radioactive incidents. Today, every unit is equipped with radiation detection equipment. Hopefully, they will only have to respond to events that also turn out to be manhole fires.

Another of midtown's most recognizable landmarks is St. Patrick's Cathedral, the immense home of the Catholic faithful in New York located on Fifth Avenue. St. Patrick's is a firefighter's nightmare, and another terrorist target. I could not guess the role it would play in my life when I first walked through the huge vaulted doors in 1994.

It has been said that my middle son, Patrick, is my clone. Growing up, he has my looks, my build, my mannerisms, and even my personality. He is a lithe, wiry, athletic youth who plays high school football and lacrosse, wrestles, and lifts weights. Not that I did any of those things, mind you. One morning when he was a sophomore, he woke up feeling very listless. Since our oldest son, John, was complaining of a cold also, my wife, Jeanne, told Patrick to stay home from school and she would take them both to see the doctor.

While waiting to be examined, Patrick's condition deteriorated rapidly, and he began to complain of chest pains, specifically his heart. The doctor examined him briefly and attributed the pain to a pulled chest muscle, since he'd had a wrestling match the afternoon before and had then lifted weights. Pat insisted though that this was his heart. My wife, a registered nurse, recognized by his pallor and totally weakened state that this was much more than just a pulled muscle. Patrick was breathing heavy and his lips were taking on a dusky hue. She asked for and applied oxygen to Pat while the doctor ordered an ambulance. My wife called me at home.

I arrived at the office just as the crew from the Merrick Fire Department loaded Pat aboard. I told my wife I'd go with Pat, and that she and John should meet us at the hospital. In the Pediatric Emergency Room at the Nassau County Medical Center, the attending physician, a middle-aged woman, again initially "pooh-poohed" the notion that anything was seriously wrong with Pat. I had watched his condition worsen though, and he was begging for help. After about 45 minutes of no real activity, I'd had enough. I cornered the doctor in the room and told her that if she "did not have a heart monitor on my son in two minutes, that she and I were going to have a major problem." "Calm down, Mr. Norman," she said. "Fourteen-year-olds don't have heart attacks." That did it. "Don't tell me that crap, lady!" I had just attended the wake and funeral of a beautiful 13-year old girl that dropped dead of a heart problem in her family's kitchen while making Christmas cookies. She was the granddaughter of dear friends of ours from Rescue 1. "You better hook him up to a monitor right now or else!" She could tell from my eyes that "or else" would not be fun.

Hospital security arrived right after the heart monitor, which had to be wheeled in from the adult ER. The guards conferred with the doctor and kept an eye on me, but the EKG technician called the doctor over as soon as he got his first strip. All of a sudden it was assholes and elbows: the ER was like a totally different place. You could see the look on the doctor's face. What she saw on the monitor had obviously scared her more than I had, because now she was asking me all sorts of questions about Pat's past medical history, the onset of this problem and whether we had a pediatric cardiologist. Again, I felt like choking the shit out of her, but the presence of two security guards and my wife dissuaded me, as much as the need to have Pat taken care of. I settled for a sarcastic reply, "Oh yeah, lady, I always keep one of them in my pants pocket. Hold on I'll bring him out. Of course not, that's why we brought him to a hospital." My first instinct when I'd climbed in the back of the ambulance had been to take him to St. Francis, which is renowned as "*The* Heart Hospital." But that is located all the way across the county from Merrick. I had settled for NCMC only because my first and second choices, St. Francis and South Nassau, had been ruled out by the ambulance crew as being too far away or "too full." South Nassau's ER was "on diversion," so we were stuck at NCMC this fine Wednesday morning, with my son having a life-threatening cardiac arrhythmia that no one could explain.

After several hours of treatment with IV medications, Pat is stable enough to be moved to the Pediatric Intensive Care Unit. There we meet the pediatric cardiologist that I didn't have in my pants pocket. Finally, things had taken a turn for the better. This fine gentleman was also treating a teenage girl with similar symptoms. He said that the tests on her were inconclusive, but he felt she probably had a type of viral infection that was attacking the heart. It would be at least a day before Patrick's blood work would be completely analyzed, but that was his tentative diagnosis. The bad news is that viruses can't be treated with antibiotics or drugs. Just like the common cold or the influenza virus, the patient's body would have to fight this off by itself. Treatment was limited to supportive therapy, letting the body heal itself, by making its other functions easier; high flow/high concentration oxygen, intravenous feeding, and bed rest was about all they could do. So we waited overnight, then all day Wednesday, then all day and night Thursday. My wife took off from her job and didn't leave Pat's bedside unless I was able to give her a break. By Thursday night, Pat's condition remained unchanged for 36 hours. On Friday night, conditions were still described as critical, with no clear diagnosis as to the cause of the arrhythmia. When I left at around midnight that night, the doctor said that if conditions remained stable, maybe on Monday Pat could be transferred out of the ICU to a normal ward, which would at least allow visits by his brothers and the rest of our family: something that is not permitted in an ICU.

I was due back at Rescue 1 on Saturday morning, and though I would have been granted emergency leave, I felt that getting into the firehouse with the brothers would be beneficial to my mental wellbeing. When I checked the day book and found that we were scheduled for a familiarization visit to St. Patrick's Cathedral, I was upset, because it meant I would be away from the telephone in case my wife tried to call me with news on our son. At 10:00 a.m., the whole first-alarm assignment gathered in the vestibule where we met our "tour guide," an older Irish priest who was one of the caretakers of the magnificent edifice. As we stood there waiting for the entire group to assemble, I made a comment to this wonderful man. "Listen, Father, I have a son named Patrick who's in pretty tough shape right now. Do you think you could put a good word in for him with The Boss upstairs?" In a split second, a twinkle came into the eyes and a smile swept over the priest's red face as he replied with more than a hint of a brogue, "Son, we have a direct pipeline to The Boss from here. I'm sure he'll be taking good care of things. I'll be saying a prayer for your son myself now." "Thanks, Father, he can use all the help he can get right now," I tearfully murmured as the tour began.

We started in the Fifth Avenue vestibule, as the priest unlocked a small door that opens into a very narrow, winding staircase. We all traipsed up the stair to the choir loft where we saw the huge glistening pipes of the organ, and our guide showed us the overall layout of the historic church. Next, we proceeded back up the staircase to the south bell tower, where several members felt compelled to use their fingers to add their names to those of hundreds of others who've made this trek over the years, writing in the coating of dust and grime that covers the inside of the stained glass windows. I can't bring myself to be interested in this; I have more pressing things on my mind. From the bell tower we make our way out onto the south triforium, a narrow balcony high overhead above the rows of seats below. I can't help but recall Nelson DeMille's book *Cathedral* in which a group of Irish terrorists take over this holy place on St. Patrick's Day, taking the cardinal, a priest, and some dignitaries hostage. I think of how the terrorists had purportedly posted snipers in these high vantage points that look down on the altar of God, and I whisper a little prayer that "I hope nothing like that ever happens here."

From the triforium, we ascend yet another flight of narrow, circular stairs before we finally reach our main destination, the attic over the main hall area. This huge, undivided wooden space runs the full length of the cathedral, nearly a city block long, and it is laid out in the shape of a cross, with two short side projections, the transepts, about ¾ of the length away from our entrance. There are two narrow wooden catwalks that run the length of the attic. Along the outsides of the catwalks, the wooden lathe and plaster of the arched ceiling of the main floor plunges downward to join the sidewalls. Overhead, the heavy timber truss roof pitches sharply up from the same sidewalls creating a huge area that, from my perspective, looks like an indoor lumberyard—and a borough call waiting to happen. I marvel at the fact that the church has steadfastly refused to install an automatic sprinkler system in this area. I know that if a fire begins or extends into this area, that there is just no way short of divine intervention that we are going to be able to do anything about it. This place will go up like a Roman candle (no pun intended)—author DeMille's plan of using helicopters to deliver high expansion foam hoses through the roof scuttles notwithstanding.

We make our way to the north bell tower stairs and descend to the north triforium, then cross back over the choir loft to reach the main level. As we make our way forward to the altar area, and we pass row after row of magnificent artwork and statues, as well as hundreds of candles. Even though it is 11:00 a.m. on a Saturday morning, the church is fairly crowded with worshippers, sightseers, and an occasional homeless person sleeping on a bench. This last fact disturbs me greatly, not because I think it's irreverent to sleep in church (I have probably been tempted to nod off during mass now and then myself), but because I know that a large percentage of church fires are the result of arson, and that emotionally disturbed people are often responsible. As we conclude our tour at the rear altar, our tour guide wishes us all his blessing for doing our jobs safely and gives me a wink and a nod from across the hall. I want to thank him again but the handie-talkie blares out, "Rescue, you've got a run!"

We all sprint down 50th Street to the parked apparatus, and in a moment we are off to a 10-75 on the upper west side. As we pull into the fire block, Battalion Commander Walter Brett of the 11th Battalion tells me to bring the thermal imaging camera and meet him on the third floor of a brownstone. Entering the parlor floor, we smell wood smoke, but don't see smoke. At the third floor, the Chief shows us where they had found a small fire burning in the wall. Engine 76 and Ladder 25 had extinguished it, but the smoke keeps reappearing throughout the building. They have checked the floor below for fire, but it is negative. Behind the wood lathe and plaster is the brick wall of the building. There is no obvious source of ignition for this fire, no wiring to have shorted or lamps placed too closely or anything of that nature. One of the new men, Jay Leach, has the thermal camera and after a few seconds he says the whole wall has a series of red hot spots on it. Going up and down, we open the wall where Jay points and find more charred wood, although no hint that it was actively burning. Jay takes the camera upstairs and finds similar spots in the walls on both the fourth and fifth floors. I have an idea of what it might be and request Chief Brett to have someone check the building next door, to see if they have a fireplace going. Sure enough, Ladder 25's chauffeur reports that on the third floor of exposure 2 the tenant had had a fire in an old fireplace that they had just restored. Well, that was it. The old flue pipe had probably not been used in 30 years and now a new tenant moves in and starts a rip-roaring fire for the "effect" of an open fire. The flue pipe must have been cracked on the side facing our building, and when a fire shot up the flue, it ignited a pile of built up creosote that found a crack in the adjoining wall. Apparently, this had probably happened before, since the char on several bays did not appear as active burning. It's just that this time the smoke got out of the wall and into the apartment. It is a good thing this happened at 12:30 p.m. and not 12:30 in the morning!

Chief Brett thanks us for our service and dismisses us. On our way back to quarters we stop at Strako's Deli on Amsterdam Avenue to pick up sandwiches for lunch. Usually we would drive over to Central Park to eat on a day like today, but I want to get back to quarters in case anything happens with Patrick. As we get ready to back into quarters, Brian Foy, who had gone in to open the overhead door, calls out to me: "Hey, Cap, department phone!" My heart instantly leaps into my throat. I fear the worst. I fairly fly through the door and snatch the phone from Brian's outstretched hands. "Captain Norman," I speak into the mouthpiece. "Dad, it's me," is the reply from the speaker. "Who, Patrick?" I ask disbelievingly. "Yeah, it's me, Pat. Mom said to call you," he says. "How are you feeling?" I inquire. When I saw him the night before he was still completely unconscious. This must be some amazing recovery! "Fine," he says, "the doctor sent me home." "Sent you home? What do you mean, 'sent you home'?" "I mean he sent me home. I feel fine. I woke up when the doctor came in at around 10:30 this morning; he looked me over and said I could go home." Almost crying with relief, I'm nearly too stunned to think, but I manage to ask, "What time did you say the doctor came in?" About 10:30," he says. I smile wryly and say a little prayer of thanks as I think of our smiling Irish "tour guide's" comment, "We have a direct pipeline from here to the 'The Boss.'" I'll say! "Pat, I'm glad you're OK. I love you. I'll see you later. Could you put Mom on, please?" Before my wife can get to the phone though, the *tooo dooo* of the voice alarm sounds in my ear. We've got a run: a report of a man hit by a subway train at 34th Street and Broadway. Out the door we go, before anyone gets a chance to finish lunch. Given the state of the guy we're about to work on though, that's probably another blessing, only this time in disguise.

CHAPTER 31

Harlem II

Each promotion in my career has been met with some sadness at leaving the unit that I had come to love, some trepidation about what to expect next, and excitement at the challenges of the future. When I was studying for lieutenant, I really wanted the job and studied very hard for the test. I succeeded in writing the second highest written score out of several thousand test takers. I was similarly motivated when I took the captain's test, thinking naively that, since a captain works exactly the same shifts in a firehouse as do the three lieutenants, but gets paid substantially more, that a captain was just an overpaid lieutenant! Boy, was I naive!

Once I was actually assigned as the company commander of Rescue 1, as opposed to being a covering captain working in Rescue 1, I came to understand what a huge difference in responsibilities the official captain of a unit has. Not only is the captain responsible for all the fire duty and paperwork that is generated on their shift, but also for reviewing the activities of all three other shifts; ensuring the proper requisitions of supplies and equipment to make the unit function; maintaining the apparatus and quarters, personnel selection, and discipline; setting company policy and goals; and a myriad of other issues that a lieutenant simply does not have to deal with. I liked being the captain. Unfortunately, before I was assigned to Rescue 1, the department gave the examination for battalion chief.

I really was not looking to move up to BC yet. I barely had two years in rank as a captain, but given the uncertainties of life, I had to take the test, if only out of self-defense. You see, the department can make you work in a company that you really don't want to work in. They just assign you there, and for the most part, you have to wait until you get enough seniority to request a transfer, and once you are assigned to a unit, they don't really like officers going from one assigned unit to another. Now I had my heart set on assignment to a Rescue Company, but there is no guarantee that I would get one. If it came to be that I was assigned to a slow company in a place that I did not like, I could be stuck there for quite a while. But there are three sure-fire, guaranteed ways to get out of such assignments: retire, die, or get promoted. Since neither of the first two options appealed to me, I decided to take the BC's exam.

When the chief's list came out, I was nowhere near the top, and I was very happy with the results, because in the interim I had been assigned to Rescue 1, making me wish I had not even taken the test. I did not give promotion much thought until the list started to get closer to my number. I thought about turning down the promotion, I was thrilled to be doing what I was in Rescue 1. I mentioned my plan to another great mentor, Deputy Chief Vincent Dunn, commander of the Third Division in midtown Manhattan. Vinny is an absolute Hall of Fame fire officer: he has written many books on firefighting and building construction, including *The Collapse of Burning Buildings*, which I tell everyone who aspires to be a fire officer they must make a point of studying until they know it inside and out. Chief Dunn and I have known each other for at least seven years at this point, having taught seminars together for our great friend Jimmy Curran, who was the head of the New York Firefighters Burn Center Foundation, raising money for the Burn Center.

For the past four years, Vinny had been the commanding officer at many of the fires and emergencies that I had responded to with Rescue 1. When Vinny heard that I was thinking of turning down the promotion, he was upset. He came to Rescue 1's quarters one evening after finishing his shift and took me up to the company office to have a little "Dutch Uncle" talk with me. "John, look at the career you have had, working in some of the

greatest fire companies in the city. Don't you think you owe the job something for that privilege?" Of course I did, but I thought I had been paying
it back along the way. Anytime anybody asked, I stepped up: I taught the
department's Foam Coordinators course from the time that I was a firefighter in Rescue 3; I taught the hazmat class at the lieutenant's First Line
Supervisors school; I taught the Collapse and Confined Space Modules in
the Rescue Technician course; I had written several bulletins on each of
these topics. In short, I thought I was paying my dues. Chief Dunn wasn't
sold. "John, the job needs good bosses—guys with real fire experience like
you. Who do you want to be reporting in to when you get to a job, a guy
like yourself, or a guy who has only worked in slow places and passed the
tests?" I still wasn't sold. But then he threw in the kicker: "You know what
is happening next year, don't you?" I had no idea what he was talking about.
"Next year it will be '*get on line in '99.*' All the old timers like myself will be
retiring at a pace where there is going to be a line out the door of the pension office. The job is going to have a mass exodus of the guys who went
through the 'war years,' and it will need a bunch of you young guys to step
up." Due to a contract settlement, there were going to be extra incentives
for senior guys like Vinny to retire. He did in 1999, and I got promoted to
BC that March.

Being in Rescue 1, we got to work with the great chiefs in the First and
Third Divisions: guys like Joe Grosso, Al Turi, Jim Wendling, Walter Brett,
and Bob Schildhorn. When I filled out my "wish list" for assignment as a
new BC, I listed the 3rd Division at the top of the list. I saw what these
chiefs did at serious fires and said to myself: "That's where you can make
a difference." Once promoted, every officer covers for a time. In the case
of battalion chiefs, that often means all over the city, which is not that different from working in SOC. I like seeing different things. I was initially
detailed to the 12th Battalion for a few months, working in lower Harlem
with great companies like Engine 35/Ladder 14 (whose quarters they share),
Engine 58/Ladder 26, and 91 Engine. Then I went to the Bronx for a few
vacations, working three weeks at a time filling in at places like the 14th
and 26th Battalions; then I was surplus a few times, going everywhere from
the 22nd Battalion in Staten Island to the 40th in Brooklyn, to the 51st

in Queens. Finally, I was detailed to cover a vacation in Battalion 9, on Manhattan's west side, near Rescue 1. I worked there for three weeks. I did not smell smoke at all but was run ragged due to the automatic alarms in the high rises, averaging 15 responses every night tour. I was disappointed. On Father's Day, I was surplus again, available to be sent anywhere they needed a BC. I was sent to Battalion 16 in Harlem for the night. We did three runs all night. Every one of them was a working fire, where we were first- or second-due. (Being second-due, the "all-hands chief" is often more fun than being first-due, since the all-hands chief becomes the interior opera-tions commander.)

Then one day I received a call from Assistant Chief Donald Burns, asking what I had planned for the fall, and did I have any travel plans that couldn't be changed? "No, sir." "Good, you're going to be detailed to Headquarters for four months, working on the Y2K Task Force." "What is the Y2K Task Force?" "You'll find out when you get here. You start next Monday in the Chief of Departments conference room at 8:00 a.m."

Being taken off-line to work on projects is nothing new to me; as I said, I teach a lot, and that is nearly always off-line. That means that you are not working the normal shifts in the firehouse. You work straight eight hour days, five days a week: which, for someone who has gotten used to working 24 hours on and 48 to 72 hours off, is somewhat inconvenient. But most of my prior details had me reporting to the Division of Training ("the Rock") on Randall's Island. This job is to be at FDNY Headquarters, located in downtown Brooklyn, sandwiched between the approaches to the Manhattan and Brooklyn bridges that lead into lower Manhattan. Making it even more of a commuter nightmare, we will be commuting in at the same time as the rest of the rat race. Oh, woe is me!

The Y2K project was the city's attempt to ensure that when the clock struck midnight on New Year's Eve 1999, that we would be prepared if all the computers that run so much of modern life suddenly shut down or went haywire, as some computer programmers thought might happen. To prepare, the department created a task force of about 18 personnel, headed by Deputy Chiefs Sal Cassano (later chief of department and fire commissioner) and Al Turi (later assistant chief of safety); Battalion Chiefs John LaFemina and

Orio Palmer; several lieutenants and captains; and me. We were assigned to conduct an inventory of all apparatus, equipment, and quarters, and develop a plan for how the department would continue operations in the face of a total lack not just of computers (which are critical in our dispatch operations) but also of telephone and radio services—oh, and do it on one of our busiest nights ever: New Year's Eve at the start of a new millennium! Our group worked very hard, war-gaming every conceivable scenario, and then grading each one as likely, possible, unlikely, and catastrophic. We next set about creating a list of options for the fire commissioner and mayor to consider what they wanted to invest in for preparedness.

As hard as it is to believe, in 1999 not every firehouse or police station in NYC had emergency power generators built in. One of the first concerns if the computers went haywire was loss of electrical power. Select sites were chosen for installation of generators, which would recharge batteries for portable radios and other critical devices. Loss of power could also have an impact citywide, affecting street and traffic lights, private fire alarms and fire protection systems—especially pumps that supply water to the upper floors of high-rises and to life support systems of individuals in hospitals, nursing homes, and private residences. All of these events could create thousands of additional responses, as auto accidents resulting from traffic lights being out, elevators stopping in mid-shaft, and life support equipment suddenly failing. The task force developed a 106-page operational bulletin that would ensure that additional resources were on hand to cover virtually any problem that arose.

By the beginning of December, the task force had been whittled down to John LaFemina and me; everyone else had gone back to their assignments. We were still working 10 hours a day, sometimes six days a week, ensuring that all the moving parts were coming together and that critical personnel were trained on what was expected. In the end, John and I were reporting directly to Chief of Operations Dan Nigro, and Chief of Department Ganci, since Sal Cassano and Al Turi had both been promoted to staff and were taken off the task force. As the New Year's Eve celebration was beginning around 9:00 p.m., Chief Nigro and Chief Ganci met us in the war room. Both were very familiar with the plans we had laid and said, "Okay, guys, you've done enough. Go home. It's out of your hands now. Whatever happens, happens. We'll take it from here." I was disappointed. I wanted to be around to watch if things started to spiral out of control. They were adamant: go home, you've worked hard and deserve a break. They thanked us for our efforts, wished us a happy New Year, and said, "If you think of a better way that we can say thanks, let us know." Ever since I worked in Battalion 9 and then Battalion 16, I was torn: do I want to work in the prestigious 9th, and run like mad but not go to a real fire, or do I want to go to a lot of fires and not run like mad? That was how I came to be assigned to Battalion 16 in the first week of January 2000. I'm not crazy.

The units of the 16th Battalion were truly great fire companies, Engine 59/Ladder 30, Engine 69/Ladder 28 (where the battalion is quartered, "the Harlem Hilton"), and Engine 80/Ladder 23. These units were led by excellent officers, which is almost universally true of all great units. I have found that the firefighters will do whatever the leaders ask, and if a unit is sub-par, look at the leadership and you will find the cause. One of the officers that I truly admired, but had a friendly, ribbing kind of rivalry with, was the Captain of Ladder 28, Bobby Morris. Now he and I would see a lot of each other.

Captain Morris and I had known each other for a number of years, again through our friend Jimmy Curran. Bob also taught at the same Burn Center fundraisers. He is a nationally known expert on ladder company operations in general, and forcible entry specifically, teaching for more than 20 years at venues such as the annual Fire Department Instructors Conference (FDIC) in Indianapolis and the Firehouse Expo in Baltimore. Whenever we would get together, Bob would tease me, saying that my rescue background meant that I didn't know anything about first-due truck work, and how that is the heart of the job. Bob had worked for a brief time in Rescue 3, when they were on 181st Street with Engine 93 and Ladder 45, which truly did not do any first-due truck work. He found that he missed the excitement of being first due and so he returned to his prior unit, believing that a rescue company did not do first-due work. I tried over and over to explain that it was only Rescue 3 that did not do first-due, because they were in the quarters with a ladder company, and that Rescue 1 and 2, as well as 4 and 5, did plenty, but he was not having it. I could see why Bob loved 28 Truck: they were doing a lot of work, the firehouse was in good shape, and they had a great reputation, built on years of excellent firefighting.

For a chief, the 16th was a great spot. The companies knew their jobs; the officers ran tight ships, and the aides made the chief's job simple. Even the fill-in aides knew their job, handling the staffing and hiring for the units very smoothly. Jimmy Carney was a step-up guy, a senior man in Engine 69, who would fill in if a regular aide was off and ran the office like a pro. I was really enjoying life.

One of the things that impressed me was the lack of vacant building fires, after spending so much time in Brooklyn's vacant buildings. Not that central Harlem did not have vacants; it's just that they did not have fires in them. In 2000, the battalion had over 1,200 vacant buildings in our administrative district. We had 12 vacant building fires that year! The culture in Harlem was different from that of Brooklyn. The people understood the value of a building to the neighborhood, as opposed to a vacant lot.

One afternoon, we had a serious collapse of a large portion of a crumbling six-story vacant building, where parts of the exterior walls fell four floors and smashed through the roof of a two-story building, narrowly missing several occupants. After searching the smaller building and ensuring that no one was trapped, our next priority was to remove the hazard. There were several remaining unstable sections of wall that were a threat to surrounding buildings. They were beyond our capability to shore up. The only feasible answer would be to remove the threatening walls, which would require heavy demolition equipment. The deputy chief on duty responded and agreed with my assessment. We placed an emergency call to the Department of Buildings, which sent an inspector who also agreed that this was an emergency and ordered an immediate demolition of the remaining structure. The Building Department has the authority to hire a contractor under these circumstances, and to order the demolition. With the area taped off with warning tape, and the threat to life stabilized, I left the scene in the hands of Buildings, with the understanding that the contractor was on the way, and that demolition should be done overnight.

When I returned for my next set of tours, I drove back over to the scene to check on the status. I was stunned. The building was in exactly the same condition as when I had left, except for some small sections that had collapsed further. I called the building department and reached the inspector who had been at the scene. He explained that he had been overruled by the borough superintendent, who had been contacted by the community board, who expressed their dismay that the building was to be demolished. Their viewpoint was that they did not want Harlem looking like the huge swaths of the Bronx and Brooklyn where vacant lots outnumbered occupied buildings. It was all about aesthetics. The old buildings looked like they belonged there; a new building would look out of place. As long as you aren't killed by a collapsing building, I guess it makes sense. It made it a lot easier for a chief to keep our firefighters out of burning vacant buildings if they never caught fire. Life in the 16th Battalion had settled in to a very comfortable pattern: go to work, go to fires, go home, repeat. I was already planning how I would spend the rest of my career as a battalion chief on 143rd St, trying to line up a permanent aide, since my current aide Bobby Laird had become gravely ill. I went on vacation on September 10, 2001, planning to spend three weeks working around the house and relaxing before returning to the Harlem Hilton for a very enjoyable 13½ years until retirement. Father Michael Judge once had some great advice about planning the future: "If you ever want to make God laugh, tell Him what you're doing tomorrow."

CHAPTER 32

The World Trade Center

The morning of September 11th, 2001 came way too early for me, but thankfully, also too late. You see, that morning was supposed to be my first day of a three-week vacation, and with my wife Jeanne going off to work, the older boys, John and Patrick, away at college, and Conor off to high school, it would be my first opportunity in a long time to sleep late. I planned to make the most of it.

The night before was a regularly scheduled meeting night for Hose Company #1 of the Inwood Fire Department. While I had stopped being an active firefighter with the IFD nearly eight years prior when I moved 12 miles further east of town, I still stopped by on occasion to see some of the great friends that I had fought fires with for nearly 25 years to do some training, to swap lies and tell war stories, and share a few laughs over a few beers. That particular night there was a great cast of characters in the house: my brother David (Dave as we call him), Mark Rolon, Bert Addonna, David Vacchio, Butch Borfitz Jr., as well as a lot of newer members whom I knew only in passing. As the night went on, things just got funnier and funnier, but since nearly everyone else had to get up early the next morning for work, inevitably the festivities started to wind down as one after another of the guys said, "See you later." My brother Dave was one of the first to go. He should have stayed with me.

Having moved 12 miles away meant I couldn't stay too late myself, since I didn't want to risk driving under the influence. Around 11:30 p.m., I headed home in a great mood, having spent some great time with guys that I really love—having crawled down the long, hot hallways together, as well as spending afternoons at each other's homes celebrating holidays, birthdays, and christenings. I was still pumped up when I passed a local "watering hole" that is within walking distance of home and spotted the cars of several of my new friends, including a couple of guys from "the job." I decided that since I was off the next day, I would park the car at home and walk back to stop in and say "Hi." When I got home several hours later, I turned off the ringer on the telephone and turned the volume on the answering machine down as well. I was going to sleep in. Good plan, poor timing.

At 9:55 a.m. I awoke to a very low voice almost whispering in my ear. It took a couple of seconds to register that it wasn't my wife, but the words "recall" and "World Trade Center" finally caught my attention. It was Tom Roby, the chief on duty in the 16th Battalion. I reached for the phone, but it was too late. Tom had already hung up. He was responsible for notifying all the off-duty chiefs, battalion aides, and covering lieutenants assigned to the 16th to report immediately to the firehouse. He didn't have time to waste on giving lots of detail into an answering machine. I glanced at the clock as I reached for the replay button on the machine—9:55 a.m. Turning up the volume, the sense of urgency was clear in Tom's voice as he barked, "John, I know you're on vacation, but there's a total recall because of what happened at the World Trade Center. Get in here!"

I listened in an uncomprehending fog, thinking, "total recall, what the hell is that about?" We've never had a total recall, even after the 1993 bombing at the Trade Center, which was a big deal. "What the hell happened at the World Trade Center?" I asked aloud. I rolled across the bed, searching for the TV remote on my wife's side. I hit the power button and on came Fox 5 news, usually the last thing we watch before bed. The images that appeared are too terrible to be real. Both towers are ablaze. I jump out of bed, now wide awake, and start looking for my uniform clothes, but since I am not due back in the firehouse for three weeks, they are all down in the cellar in the dirty laundry. Damn! My eyes fall on my FEMA "go bag" which, as part of New York City's federally funded Urban Search and Rescue Task Force I have kept stowed in my closet for years, in preparation for running out the front door at a moment's notice, headed for an earthquake or some other catastrophe somewhere else in the United States, or for that matter the rest of the world. Now I would need it right here in New York. The bag is packed with three days' worth of work clothes, clean socks, underwear, T-shirts, and all the other necessities to operate self-sufficiently in a disaster environment. It does not have any FDNY uniforms in it, since it is a joint FDNY-NYPD venture, and we want to encourage joint teamwork rather than individual organizational loyalties, but I figure clean clothes are better than rumpled dirty ones pulled out of a hamper any day, and some of the other things tucked inside might be useful, like the full facepiece respirator.

I am still in the bedroom, pulling on the navy-blue battle dress uniform (BDU) trousers with the button fly, when an anguished cry comes over the television speaker as the talking heads call out in horror: "The South Tower of the World Trade Center has just collapsed!" I stare at the TV, momentarily in shock. "What the hell was that?" I shout at the screen. Having been asleep for the first hour and 10 minutes of the attack, I have no idea how long it has been going on, or what had transpired to that point. I had heard the anchors talking about airplanes striking the towers but wrongly assumed it was an accident.

I think back to 1960 when two airliners collided in mid-air over New York City, one falling onto Staten Island, and a DC-8 crashing into Park Slope, Brooklyn. A picture of the tail section of that plane, burning in front of the "Pillar of Fire" Church hangs conspicuously in many of the firehouses that fought that disaster, including Rescue 2's kitchen. "That must be what happened," I say to myself. Then the talking heads describe eyewitnesses saying two separate planes had clearly targeted the buildings: a terrorist attack! I scream out, "You murdering bastards!" as the clear intent of the attack becomes plain. I grab the FEMA bag and run down the stairs two at a time, heading to the latest battle of the war on terrorism.

I throw the bag in the passenger's seat, then go to the trunk and pull out my spare set of bunker gear and helmet, throwing them into the front seat as well. The television people have been reporting on various aspects of the attack, including the recall of all off-duty cops and firefighters, EMS technicians, and other critical workers. They also report that the city has been put on total lock-down, with all of the bridges and tunnels sealed to all but emergency vehicles. I figure the Uniformed Fire Officers Association (UFOA) placard on the dashboard and the white chief's helmet may be needed to get me past any of the security checkpoints. I regret not having a clean white FDNY chief's uniform shirt on and hope the relatively unknown FEMA USAR patch on the blue BDUs will help and not confuse things. I start to head for my assigned unit, the 16th Battalion Headquarters on West 143rd Street in Harlem.

As soon as I get on the Southern State Parkway, heading west toward the city, I can tell this will be a trip like no other. The entire western sky is a huge pall of smoke. Traffic is at an absolute standstill, with thousands of drivers sitting in their cars in shock, their eyes on the sky in front of them and their ears glued to their car radios or cell phones. It is about 25 miles from my house to the firehouse on 143rd St in Manhattan, and on a good day it is a 45-minute trip. This is not a good day. It looks like it will take a helicopter to get me there today, and I am fresh out of helicopters. Just then, a Nassau County Police Department Emergency Services truck pulls alongside of me, driving like a madman on the grass alongside the highway. He is followed by several more NCPD vehicles, an ambulance from a Suffolk County Volunteer Fire Department, and a string of private cars all flashing their headlights and honking their horns and displaying a placard similar to the one on my dashboard or issued by a law enforcement agency identifying them as emergency responders. When the first break in that line appears, I make my way out and join the long line of emergency responders heading toward that evil-looking gray cloud on the horizon. We snake along the shoulder, being forced back out onto the gridlocked highway about every half-mile when the overpasses of local roads block the "grass highway" we are creating. Several times I have to lean heavily on the horn and basically shove the nose of my car into the traffic lane, as the motorists stranded in that lane don't want to let us in. They want to get out of there themselves. After waving my chief's helmet at them and letting go with a string of appropriate expletives about their intelligence and threats to ram them, backed up by a string of other firefighters and cops, we get around the impeding overpass and back onto the grass highway.

Finally, in the vicinity of Valley Stream on the Nassau County/New York City border, the reason for the gridlock becomes visible. New York State Troopers have the entire highway blocked and are detouring all traffic except emergency responders off onto local streets. Our whole line of emergency responders pulls up to the checkpoint, one at a time. As the trooper comes toward me, I wave my FDNY identification card and chief's helmet, saying "I'm a battalion chief in Harlem on my way in." With hardly a look he briskly waves me to an open lane. "Go with God," were his only words.

Once clear of the checkpoint, I have a decision to make: Do I head north up the Cross Island Parkway toward the bridges connecting Long Island to upper Manhattan and Harlem, which are also locked down to all but emergency vehicles, or should I continue along the South Shore toward the Trade Center? The radio says that the North Tower has collapsed just after I clear the checkpoint. The firefight is pretty much over, I think to myself at that point. Now it is a building collapse rescue operation, something that I have been doing and training others to do for the past 17 years. I decide to head directly for the site.

After the Southern State checkpoint, the road is wide open in front of me, a couple of other firefighters and cops, an occasional out-of-town ambulance or police car, and that was about it. The Belt Parkway past JFK airport in Queens and into Brooklyn looks like it is 3:00 a.m. on Monday morning, and the speed limit is forgotten as we race to get close enough to help before it is too late. I decide to get off at North Conduit Avenue which heads due west toward Manhattan, rather than continuing all the way "around the horn" of south Brooklyn. I figure if I hit traffic, I can work my way through detours easily enough. I grab the digital camera that I keep in the car to take pictures for training purposes. "This is like Pearl Harbor," I think aloud. "Fifty years from now, somebody is going to be looking for first-hand information on what happened today and I want to do my part to show them." I take a picture through the windshield showing the smoke cloud on the horizon over the East New York housing projects and the deserted roadway (fig. 32–1).

Figure 32–1. Looking through my windshield on a deserted North Conduit Avenue, the smoke cloud of the World Trade Center blows south. The lack of westbound traffic lasts another two miles before encountering total gridlock. (*Photo courtesy of the author*)

Traveling along the local streets now, I hit heavy traffic again. The cops have the limited access highways pretty well under control, but there is no way to clear the local streets that quickly. I knew Atlantic Avenue, the most direct route, would be heavily congested, but after working right here in East New York with Engine 290 and Ladder 103, and later covering all of Brooklyn with Rescue 2 for years, I know the back routes through the neighborhoods like the back of my hand. I decide to take Pitkin Avenue, a wide thoroughfare that connects to Eastern Parkway, which is a main approach to Flatbush Avenue and the bridges into lower Manhattan. I get about five blocks on Pitkin before it too slows to a crawl. Honking and shouting does me no good with this crowd, only earning me "the finger" and at one point, a wave from a "9 mm." I turn north at the next corner, a southbound street I could see is clear up to the next intersection, so I head up. The block on the other side of the intersection is loaded with cars, so I turn west again, this time on Glenmore Avenue, an eastbound street. Traffic is relatively light and I make it to my next turn north to Liberty Avenue, again going the wrong way down a one-way street. Now there is no place to go. Liberty is jammed in both directions. I turn west into serious gridlock. After 15 minutes I have barely moved a block. I look up at the street sign: Hendrix Street. I know there is a firehouse about four blocks away on Bradford Street,

Engine 332 and Ladder 175. It is a 5-minute walk, or an hour drive at the rate I am going. I pull my car up onto the sidewalk and lock it up. I hope it will stay there for the next few days. I put on my bunker gear and grab my FEMA bag and walk to the firehouse.

At Bradford Street, I identify myself and check in with the officers on duty. I then sign myself in to their company journal as "RFD," Reporting for Duty. By this time, it is clear that New York City and the United States is under attack: the Towers are down, the Pentagon has been slammed, there are reports of numerous missing aircraft, and rumors are flying wildly. I want some official record that I am on duty in case I am never to return home. Since Jeanne and Conor were both gone when I left, no one in the family knows where I went or what time I left.

The fire department has rapidly instituted a plan to use and control the recalled personnel, who normally would have reported to their own firehouses. Now many were like me, just showing up at the nearest firehouse they could get to. Each division headquarters was designated as a mustering site, where all the personnel would be recorded and broken up into groups of four or five firefighters and one officer, then assigned five units at a time to a battalion chief.

The officers at 332 and 175 had already started splitting their people up and stripping the firehouse and apparatus of any kind of tools or equipment that could conceivably be of use at the disaster site: extra air cylinders, old fashioned claw tools from the cellar, bandages and other first aid gear, anything that wasn't essential to the on-duty crews and their apparatus, which might themselves be called to the scene at any second. They also sent guys out looking for a way to get us there.

A few of the 332/175 troops went north to Atlantic Avenue, while others went south to Liberty, looking for a truck or bus to carry us and all this gear they had gathered. By now, about 40 off-duty guys had shown up. The group on Liberty Avenue had the first success. A city bus was inching its way to the corner, and the guys flagged it down and basically hijacked it, saying they were commandeering it for emergency use. The driver, a young guy about 28 years old, was all for it. Most of his passengers had long since left the bus, realizing they could walk to their destinations faster than this bus could get them there. He ordered the remaining few people off at the corner. They looked like they didn't mind at all.

I write a quick description of my car and where it was virtually abandoned on a piece of cardboard and poke my keys through it, then hand it to one of the on-duty officers who was remaining at the firehouse with the apparatus, asking him, "Once things calm down a bit, can you have someone go get it and bring it up closer to the firehouse?" I knew from my years in 290/103 that a car without any occupants left on the streets (or sidewalks) of East New York was like a flashing neon sign to the bad guys saying, "Come steal me!"

We jump on the bus heading for the 15th Division mustering site at its headquarters on Livonia Avenue and East 98th Street in Brownsville. Liberty Avenue is still packed with traffic, so several firefighters get out, wearing their bunker gear, and start walking ahead of the bus, directing people to pull over or up onto the sidewalk to make way for our newly designated "emergency vehicle." With the help of our foot soldiers, we get to the Division in pretty decent time, taking back streets the whole way. As we pull up to the Division, a number of fire trucks, city and school buses, and a civilian flatbed truck are already there and are being split up into various task forces, some going to a forward staging area at the Manhattan Bridge, with others being held in reserve to man spare apparatus that is being readied at the FDNY repair shop in Long Island City as well as reserve apparatus stored at firehouses around the city. I know I want to get down to the World Trade Center as soon as possible so I go looking for the chief in charge. Fortunately for me, it is Deputy Chief Seamus McNeala, whom I have known for several years, each of us having taught various parts of our First Line Supervisors Training Program and our Chief Officers Development Program. Seamus knows that I teach building collapse rescue. When I say I want to get down to the towers right away, he points to a bus that is just pulling away and says, "There's your ride, catch it. It's going right downtown." Then he calls into his portable radio, "Hold up a second, I've got another passenger for you." I thank him and ask him to make sure to put me down on his roster of people dispatched into "the hell downtown."

I grab my FEMA bag and run up East 98th Street. Climbing aboard, I look around for familiar faces, hoping to find some Special Operations people from Rescue 2 or Squads 1 or 252. I see quite a few faces I recognize from around Brooklyn—guys I knew from 290/103 and from going to fires with Rescue 2—but the only guy I recognize with Special Operations training is John Fox, another battalion chief. That is good and bad. John is one tough son of a bitch: a former Vietnam combat-decorated Marine, and a genuine hero. At the first World Trade Center bombing in 1993, John (a lieutenant in Squad 1 at the time) was lowered by rope into the blazing four-story-deep blast crater that remained of the sub-cellar levels of the complex in order to reach the seriously injured Firefighter Kevin Shea. Shea had fallen four stories into that same crater and was in desperate need of assistance. John was instrumental in his rescue, and would receive the FDNY's highest honor: the James Gordon Bennett Medal. It will be good to be with John today. I just wish the bus was full of 50 more of him. The guys on it are all great firefighters, but they don't have the collapse and confined space rescue training they will need for what lies ahead. They will, however, get it through on-the-job training over the next weeks and months to come.

The bus works its way down Eastern Parkway and Flatbush Avenue, many times with the aid of the walking firefighters directing traffic. At the Brooklyn side of the Manhattan Bridge, right in the shadows of Fire Department headquarters, we report in to a staging area run by Deputy Chief Dave Corcoran, who knows lower Manhattan and the Trade Center better than the back of his hand. He orders us to wait while he gets in touch with the command post by radio for directions. After a frustrating 20 minutes, he finally gets through the chaotic radio traffic to reach Deputy Chief Tom Haring, who is running a forward staging area at the corner of Broadway and Vesey Street, right on the eastern edge of the collapse zone. Haring tells Corcoran to send us over.

In the interim, I have been scouring the growing group for SOC firefighters. I run into Mike Pena, one of my lieutenants when I was the captain in Rescue 1, and Ray Grawin, a friend of mine from my Rescue 3 days. Ray is retired but knows his help will be needed. I thank God that Mike is alive, but worry about which of the Rescue 1 officers and firefighters are working;

with Mike here there are only three possibilities. Mike has spotted a few guys from Squad 1 and Squad 252. We decide that we will go over together and try to stay together until we can hook up with our units. We climb aboard the back step of Engine 264's rig, which has traveled from its firehouse in Far Rockaway, Queens—as far away from the World Trade Center as you can get and still be in New York City. They never in their wildest dreams thought they'd be responding to the World Trade Center, but this is no dream. It's a nightmare.

As we head across the Manhattan Bridge, I see another familiar face on 264's rig: Al Schwartz, who I know from Ladder 4 in midtown Manhattan (fig. 32–2).

Creeping across the bridge, I take my camera out of my BDU pocket and snap a shot of the guys atop the rig with the bridge in the background, trying to see the towers. Mike Pena and I continue to try to figure out which of the officers and firefighters would be working in Rescue 1 that morning. Mike had been relieved two days earlier by a covering lieutenant, who is filling in for the just retired John Kiernan, meaning it would either be my good pal Dennis Mojica, or my replacement as captain, Terry Hatton. Mike was the first to mention it. "You know this thing started at about a quarter to 9:00." That is right at the time when the officers change shifts and 15 minutes before firefighter shift change. "Terry likes to work his straight tours."

Figure 32–2. Firefighter Al Schwartz, who ran out of his house in the shorts he was wearing, joins a dozen other firefighters for a ride atop Engine 264's pumper as it crosses the Manhattan Bridge into Manhattan en route to the World Trade Center.

That meant that he did not do a lot of mutual exchanges, and therefore both officers would likely have been in the firehouse when the first plane hit. I know what that means: both of them were likely be on the rig as it bulled its way through traffic on West Street.

Up until that point, I actually did not know that the attack had started at 8:46 a.m. Being asleep, I thought it had started closer to 10, just before Tom Roby called. I did not see the situation developing on TV like most of the rest of the nation. I had no idea that so many firefighters and cops had been on the scene for so long, which meant that many, many of them would have climbed dozens of flights of stairs, closer to the danger over their heads, and farther away from any potential area of safety. Now it finally dawns on me that we are going to have hundreds and hundreds of dead firefighters and thousands of dead civilians. We ride the rest of the way down through Chinatown in stony silence, not wanting to think about what lies at the end of our journey.

The silent ride over the bridge prompts me to think about my brother David, who is an NYPD Emergency Services cop assigned to ESU Truck 1 in lower Manhattan. Shit! He said he was working today. I say a silent prayer that he is okay. Maybe he called in sick.

The rig drives down a deserted Park Row past City Hall into the war zone. The dust and burning debris in the street here are "only" a few inches deep, since this is about two blocks north of the WTC site and slightly upwind. There are cops with automatic weapons stationed around City Hall, but very few civilians. Chief Corcoran had told us to report in to Chief Haring at Vesey and Broadway, so we make our way south, stumbling through the clouds of smoke and dust. I am pretty familiar with the area, having worked there extensively as a fire protection engineer before getting on the job, and then from my time with Rescue 1, so I help to lead the way downtown (fig. 32–3).

Chief Haring is trying to organize the chaos that surrounds him. He is getting a mixture of personnel reporting in, intact fire companies from the far outer boroughs like Engine 264, regrouping survivors from units that had arrived prior to the collapse, off-duty guys from all over the city, as well as several volunteer fire units from New Jersey, upstate New York, and

Figure 32–3. Our convoy of fire trucks and city buses pulls up on Park Place, adjacent to City Hall. This area is several blocks away and largely upwind from the WTC.

Long Island, who had taken it on themselves to respond when they saw the devastation on television. Haring knows that command, control, and communications will be critical elements of organizing this mess and preventing further casualties.

I report in to him with some reservations. I have assembled a group of SOC firefighters, guys that are the most highly trained people that we have in the job for dealing with problems like collapses, high-angle rope work, and entry into confined space voids—all skills that I am sure we are going to need. They also make a point of being the best around at trapped firefighter rescue, something else I am pretty sure we will need. I do not want to see them wasted as just another group of hands. We will need to depend on Haring's command post for logistics and communications support. The off-duty firefighters have no tools, masks, or radios; basically, we have only a few small hand tools, the bunker gear on our backs, the knowledge under our helmets, and hearts, guts, and balls. We will need him for everything else.

I step up to the command board he is operating and catch his eye while he dispatches a group of companies to report to operations on the south edge of the sprawling disaster site. He waves me in. I start to tell him that we have organized an SOC group, but he could care less. All he is interested in is warm bodies. From the microphones of some of the on-duty firefighters' radios we can hear dribs and drabs of conversations. The radios are nearly unintelligible, with too many people trying to talk at once, cutting each other off or "stepping on one another." Several times we hear, "Mayday, May—" and the rest of the message will be cut off by somebody else transmitting another message, often equally important. It is sheer chaos.

Haring directs each of the on-duty units to start splitting up their assigned radios. Each engine company normally has three or four, while the ladder companies have five or six. Since the off-duty guys have none, it is essential to take some from those that do. This is a major shock, since the job makes a really, really big deal out of losing a $2,500 radio. Officers and firefighters who lose track of their assigned radios are formally disciplined, with penalties that include hundreds of dollars in fines. No one gives away a radio voluntarily. Now they were being ordered to do just that. Since these units will not be splitting up at an operation of this nature, but remaining together as a unit, if each group retains two radios, that will allow them to communicate effectively and still outfit several other groups. I end up with a radio from a ladder company from Queens and get another for Mike Pena.

On the way over, Mike, Ray, and I had discussed getting the NYC FEMA USAR Task Force gear down to the scene from its warehouse location out in Queens at the FDNY Technical Services Division. A USAR Task Force is a group of 72 specialists in collapse search and rescue. In 2001, FEMA sponsored 26 of them around the United States, and New York City is one of these. (Now there are 28 federal task forces and numerous state level task forces.) Being the scene of the attack means that the NYC task force will not be federalized. All the New York City firefighters, cops, and paramedics that make up the task force already have jobs to do here with our own agencies, but we sure could use the tools and equipment that is cached for that purpose.

Getting back to Haring, I find he *is* actually interested in what I am thinking. Maybe it's because I am wearing my FEMA BDU's instead of my FDNY shirt, but suddenly he wants to know where New York City's USAR task force is housed, and if our task force is on its way. "How do we get that stuff here?" Haring wants to know. I tell him it is stored at the FDNY shops in Long Island City and it will take a couple of hours to get, since it fills three tractor trailers that need to be loaded and driven here. He turns to an aide and tells him to contact the shops to make sure the gear will be brought to the scene. It is, in fact, already in the process of being loaded, but we don't know that. Next, he asks about the need for outside help. How much do we need? I have not seen the site yet and have not gotten a handle on the scope of the entire disaster—which will actually take a week to finally see all sides of the area—so I cannot make any logical estimate. After consulting with Mike Pena, though, we tell him that we should ask for at least four task forces immediately. We have heard about the attack on the Pentagon and other unconfirmed attacks, so we don't even know if we will be able to get that kind of assistance, but we figure it's better to ask and be told no than to let available resources go to waste. (Eventually, 11 FEMA task forces from all around the country would operate at the site.) Haring directs his aide to contact the dispatcher and make that request.

Right away we hear another May Day reporting trapped firefighters. I turn to Haring and say I want to take the SOC guys to that location, but he has other ideas. "I want you to stay here and give me a hand keeping track of incoming companies," he says. I start to protest, saying, "I've got a group of collapse rescue experts here that may be critical to reaching the trapped members—" He cuts me off, saying, "Send them. You stay here." I said, "Yes, Sir. Let me get them set up and on the way."

Mike and I had noticed Squad 288's rigs parked about half a block south of the staging area. We walk over there and strip it of anything that might possibly be useful: high-angle rope gear, dust masks, hooks, you name it. I decide to stash my FEMA bag in a compartment of the squad, figuring that I'd be tied up for a while, but first I give Mike the full facepiece respirator that it contains. "Here, you'll need this more where you're going than I will here," I say, passing it to him. It is not as good as a self-contained breathing apparatus (SCBA) regarding protection against smoke, oxygen deficiency, or toxic gases, but it does offer one big advantage: it offers what limited protection it does provide for eight hours. An SCBA only lasts 30 or 40 minutes. We all know we'll be here for months if not years.

Mayday messages keep pouring in, and I am chomping at the bit to get out of the staging area and over to them. Finally, after about 15–20 minutes, the operations post requests additional companies and chief officers be sent to West Street and Vesey Street, on the far side of the pile. Haring tells me to take a group and head over.

I am very familiar with the World Trade Center and vicinity. My first job out of college as a fire protection engineer was to run a portion of an 8-inch-high pressure standpipe line through the sub-cellar levels of one of the smaller plaza buildings that surrounded the twin towers: Number 5 World Trade Center. Number 5 WTC was basically a 50-foot-deep hole in the ground when I started there in January, 1973. The Towers themselves had been "topped out" (their structural steel framework completed), but with no windows or interior walls in place on the upper floors. For the next 6½ years, I was in and out of the growing complex on nearly a daily basis, delivering pipe and supplies into the sub-cellar loading docks, measuring area after area for the sprinkler systems we would be installing, learning all the hidden hallways, service elevators, and backrooms that the public would never see.

One of my jobs was designing the sprinkler system for portions of 5 WTC, as well as for the observation deck on the top floor of the South Tower. My company also did the same for the Windows on the World restaurant on the top floor of the North Tower. I remember getting off the express freight elevator, "the 50 car": one of the first installed and the only one in each tower that served all 110 stories. I was sent to take some measurements of the 110th floor of the South Tower, but there was no window glass in place yet, only the steel exterior walls and the gypsum walls around the elevator shafts and stairways. When I stepped out of the elevator, I thought the floor was on fire. It was like stepping into a heavy smoke condition. You couldn't see five feet in front of you. We were in the middle of a low hanging cloud at 1,300 feet above the ground. The cool mist permeated the entire floor. Having been a firefighter for a few years, already I knew how to move around in zero visibility—very carefully! I stepped gingerly onto the floor, sliding my front foot along the floor in front of me, and dragging one hand along the studs and sheetrock of the few existing core partitions, not wanting to step into an open shaft or out a window. Both were a long way to the next stop. Even with the floor plan in hand, I was unable to get my bearing as to which way was north, and I had no way to take any measurements because I had no idea which way to turn to get to my next destination. It was a totally disorienting experience, one that would be made worse by orders of magnitude nearly 30 years later on 9/11.

Even after getting on the fire department, I continued going back to the World Trade Center. Of course, it was a favorite destination for doing the tourist thing when I had good friends from out of town visiting. As a lieutenant in Rescue 2, we were the second rescue company to respond to any serious fire there, so we visited for familiarization several times as well as for a couple of real fires, including the 1993 bombing. Beginning in 1994, when I became the captain of Rescue 1, the situation became even more important to me personally, since Rescue 1 was the first rescue company assigned on any alarms at the complex.

For all this time, I was also an instructor at the Nassau County Fire Service Academy on Long Island, and one of the classes that we taught for many years was high-rise firefighting for the Port Authority Police Department's World Trade Center command. The PAPD acts as the in-house fire brigade for the complex. Since they had officers on vertical patrol throughout the complex, they would have officers on the scene of any fire for up to 10 minutes before the nearest FDNY unit could reach the upper floors. Don Hayde (later a BC in SOC), Sal Marchese (a Lieutenant in Ladder 113 in Brooklyn), and I, along with a few others, nearly all active FDNY officers, would spend several weeks each year teaching the PAPD guys how to use SCBA, how to attack live fires with a variety of fire extinguishers and standpipe hoselines, how the elevator recall worked, communications protocols with the FDNY, and other related topics. To make sure we instructors were current on the ever-changing conditions in the huge, still growing complex, we would all go down to the site beforehand for a walk-through with the complex's Fire Safety Director.

Later, in the fall of 1999, as a battalion chief working on the city's Y2K plans for the New Year's Eve celebration and possible computer failures, I would spend every Friday and a few other days each week in meetings at the Mayor's Office of Emergency Management (OEM), which was located on the 23rd floor of Number 7 World Trade Center. We would often take lunch out on the raised plaza in the center of the complex: a beautiful spot on a nice fall afternoon. In short, I knew the World Trade Center complex as well as any other member of the FDNY. None of that would prepare me for my next visit to the site.

When Deputy Chief Tom Haring sends me to report to operations on West Street, I have no clue what to expect. Our position at Broadway and Vesey Street is downwind of the complex, right in the path of the blowing smoke, dust, and debris. You can't see a hundred feet at most times, even though it is a bright, sunny, cloudless early afternoon. My best guess is that it is about 1:30 p.m. when Haring sends me over to West Street.

As we head west on Fulton Street, toward Church Street, a gust of wind momentarily clears our view, and I get my first glimpse of the actual destruction. There in front of me is 5 WTC: a nine-story building that I had helped to install sprinklers and standpipes in a lifetime ago. It is now a roiling mass of smoke, flame, melting aluminum, and falling glass and steel. "What the hell is that?" someone behind me asks. I don't know how to answer him. "That's not supposed to be like that," is all that comes to mind. I know it's not supposed to be like that. It had been my job to make sure it would never look like that, but now it does. My shocked comments inspire some strange looks from those who are close enough to hear (fig. 32–4).

Figure 32–4. Looking west along Fulton Street at about 1:30 p.m. on September 11th, 5 WTC burns freely. The white-shirted officers on the right are NY State Court officers, looking to help in any way possible.

My shock only deepens as we get closer. The Towers are major New York landmarks, clearly visible up to 20 miles away, and they help to orient New Yorkers to what direction they are facing. Now they are both gone. As the wind clears the obscuring dust momentarily, I keep looking for a glimpse of the Towers. Even though I had seen the collapse of the South Tower on television before I ran out the door, the brief view I had gotten only showed the top of the tower disappearing into the huge cloud of smoke and dust. I keep looking for the remaining portion, but I can't see anything, which is unsettling. "Where the hell had they gone? Surely two 110-story buildings did not simply disappear." There must be some very large sections remaining with their distinctive steel and aluminum facades full of 12-inch-wide windows. Only later, after getting around to the upwind side on West Street would the full, horrifying reality be visible. The twin gleaming towers had ceased to exist, crumbling more than 1,300 feet to a pile of blazing debris no more than 150 feet above the surrounding streets. It was beyond surreal.

We see there is no way through the remains of 5 WTC in front of us. The blowing smoke and dust obscure visibility to the south, but the wind is providing better visibility to the north, so that is the direction we head. Walking alongside the wrought iron fence enclosing the graveyard of St. Paul's Church on the east side of Church Street, I am startled at the sight of dozens of burned fire apparatus, police vehicles, and ambulances. It is not hard to imagine that we have stepped into Dante's Inferno. The street is ankle deep in powdered ash—the remains of pulverized concrete, plasterboard, and people. Burning papers still swirl through the air amid the dust, blown out of the collapsing towers at tornado-like velocities. It was these reams of blazing paper that got sucked into the air filters of the running diesel engines, which ignited the vehicles and several surrounding buildings. Explosions are all around as burning tires pop and ammunition in several of the police Emergency Service trucks cooks off. It seems like the natural order of life is being turned upside down. The graveyard is a sanctuary of calm amid the chaos. I again take out my camera to record the devastation (fig. 32–5).

Reaching Vesey Street, which I know is a wide through street that goes all the way to West Street, we turn west, but soon run into a group of distraught

Figure 32–5. Littered with debris and burned-out emergency vehicles, looking north on Church Street toward Vesey Street.

firefighters who tell us, "You can't go that way, it's blocked with a mountain of debris," and they add, "They don't want anybody getting too close to that big building on the right." I feel I have to see this for myself; it's too incredible to believe. I move forward about 50 feet, and as the wind swirls, I can see it was no exaggeration; the entire 70-foot-wide street is covered three or four stories high in smoldering debris. Off to the right I can make out a really big (47-story) office building pumping smoke, so I ask for their suggestion. Go up two or three blocks, and then head west again. We continue up Church Street, past the huge post office there, another block to Barclay Street, and turn left.

As we come to West Broadway, we finally have a bit of fresh air, since the wind is out of the west-northwest. We can see south down West Broadway for a bit. I recognize 7 World Trade Center, where I had spent so many Fridays two years earlier. Now it's clear that it is on fire as well, but it is hard to see much besides heavy smoke issuing from many, many floors. The street is still full of smoldering vehicles as well: mail trucks, police cars, and more burning fire trucks. We decide to continue west. More burning debris blocks the most direct path at Greenwich Street, so we detour north another block to Murray before getting into the "clear" air at West Street (fig. 32–6).

We make our way over to West Street and turn left, heading south to the intersection of Vesey Street where a group of senior chiefs has started to re-establish command. I report in with my mixed group of on-duty

Figure 32–6. Looking south down West Broadway, 5 WTC burns in the background. Building 7 WTC is at the right.

and off-duty personnel. I report in to Assistant Chief Frank Fellini, who instructs me to assume command inside the 30-story Verizon Telephone switching center at 120 West Street, which is the main telephone switching center for lower Manhattan. Chief Fellini tells me there are already two engine companies and two ladder companies operating inside, trying to stop fire from extending into the building from the burning debris pile that used to be the North Tower, which is now filling Greenwich Street several stories high. I envision a nightmare, thinking back to another telephone switching center fire in lower Manhattan 23 years earlier. It was a devastating fire that took days to contain and resulted in hundreds of firefighter injuries. The burning polyvinyl chloride wire insulation caused cancer in more than half of the hundreds of firefighters that fought it, including my good friend, Danny Noonan. "Please God, not another phone company fire!" Little did I know what horrors awaited us (fig. 32–7).

I have several units with me, including engine and ladder personnel with basic tools, rolled up lengths of hose, Halligans, and axes. But we have no pumping engine to supply water, and very few of the tools that we will really need for a serious high-rise fire, like enough radios, air masks, spare cylinders, thermal imaging cameras, and rabbit tools to force large numbers of locked doors. We will be in severe trouble if we find a fire.

There are several pumpers already in the immediate vicinity of course, but I do not see a single hoseline supplying a Siamese connection for the

Figure 32–7. Looking south along West Street, the Verizon Building is on the left foreground, with 6 WTC and the north footbridge in the rear. Notice that few of the firefighters have SCBAs.

standpipe or sprinkler systems, and there are no engine chauffeurs at the pump control panels (they have been driven out of the area by the collapsing towers, and most are in no shape to return, having sucked in lungs full of the pulverized pumice). I start to ask if there are any pump operators in the group with me, but I am quickly told to forget it. The collapsing towers have sheared several major water mains in the area and the hydrant system is virtually useless. Somebody mentions that hoselines are being stretched over to the Hudson River, where several fireboats are tied up, but I am told not to wait for it. When that water does arrive, it is going to be used on the debris pile in an attempt to keep any survivors in the rubble from burning to death. We will have to make due with whatever water is available from the buildings' gravity tanks: water specifically stored in the building for the purpose of supplying standpipe and sprinkler systems. Having been a fire protection engineer, I know that this will only be about 5,000 gallons—barely enough to supply a single 2½" hoseline for 20 minutes. Without another source of supply, we are going to be in deep shit if we meet a serious fire.

We go in through the west side, into a towering, ornate, art-deco style lobby. The building is a massive pre-World War II heavyweight building, designed in an era when everything was overbuilt with lots of redundancy in its design and systems. We quickly spread out, some firefighters heading across this huge open lobby for the east side, facing Washington Street, where we had seen burning debris piled high up against the building. Others head for the south side facing Vesey Street, which we haven't seen, and still others head for the stair towers that will take them to upper floors. Since radios are in such short supply, I make sure everyone knows to stay within voice contact range of somebody with a radio. We were told that the upper floors had been struck by the collapsing North Tower, so it is a high priority to get a look at those impact areas, to see if there are people trapped or if there is fire extending to those floors.

We know it is a telephone building, but I have no idea what to expect on the upper floors. I had been in a telephone center fire several years earlier in Brooklyn, and I am dreading what we might face: giant racks of plastic insulated wires laid out like the inside of a toaster, with very tight spaces between opposing racks of highly combustible, noxious wire insulation, and holes between floors connecting the racks. I am pretty sure we do not have a serious civilian life hazard, since the upper floors are most likely equipment spaces. Or so I think. The first floor makes me start to doubt that because there are a lot of offices, some very heavily decorated with wood paneling. Is this actually an office building? We might have a higher life hazard.

We organize several teams and get a pair of hoselines operating into the first and second floor offices where the blazing debris from the collapsing North Tower has blown in through the windows and started several fires. I call one of the units in the stairwells for an update. I had told them to walk all the way to the top floor, over 300 feet up, sticking their heads in quickly on each floor to make sure they weren't bypassing heavy fire or trapped people, and then make their way down more slowly, doing a more thorough search. One of the groups says they are on the 15th floor and everything is empty: no fire and no one trapped. They want to come down. "No, take your time and do it right. Make sure we don't miss anyone. Check the roof and any setbacks too," I order. "Chief, there is nothing up here at all; every floor is empty," was their reply.

Fortunately for us, the digital revolution that has occurred in the intervening 23 years since the last telephone company fire has removed the threat of another such blaze. The rows upon rows of cable trays that held all that wire have been replaced by miniature computerized switches that take up a tiny fraction of the space the old system required. The floors are mostly empty shells, with bare concrete walls, floors, and ceilings. Well, at least we caught one break today! I try to report this good news to command, but the radio traffic makes it impossible. Instead, I send a runner (a firefighter who will make a face-to-face report).

When the runner returns, he tells me that command wants me to take a look at 7 World Trade, to see if I think it is in danger of collapse. I make my way to the east-facing wall on the second floor and find the windows blown in and a group of firefighters playing a standpipe line out onto the burning debris in the street. It is a hopeless cause because of the massive volume of debris and the limited reach from the hose stream. I look up at 7 WTC as far as I can see across the narrow street, and I see smoke issuing from several floors of a modern office building. It doesn't look too bad from that vantage point, certainly not in imminent danger of collapse, but as the wind shifts a little bit, I can see that a large portion of the southwest corner of the building has been sheared away by the collapsing North Tower (fig. 32–8). I can see about 20 feet of building is missing at the front corner, starting about ten floors up from the street and continuing up for maybe 20 floors. That puts a whole new light on things. It is now apparent there is

Figure 32–8. Looking up at the southwest corner of 7 WTC from inside the Verizon Building: heavy smoke and structural damage is visible at right edge. (*Courtesy of the author*)

heavy fire on numerous floors of that building as well. The groups on the upper floors have now made it up to the 23rd floor and report conditions unchanged except for several pieces of red-hot steel that have punched several holes right through multiple floors. I order the upper floor teams to come down and the first-floor firefighters to tie the hoselines in place and evacuate everybody back to the lobby on the west side of the building, away from 7 WTC. Then I go out to report to command (fig. 32–9).

When Chief Fellini hears about conditions in the Verizon building and conditions visible at 7 WTC, he makes up his mind. We will withdraw everyone from the collapse zone around 7 WTC because it is too dangerous. He and Chief Frank Cruthers have already decided that there is no way they can put out the fire in 7 WTC since the collapsing towers have sheared several large water mains surrounding the area, eliminating water supply. These chiefs have decided that since we have just had two modern steel-framed high rises collapse, that this third one is just as likely to do the same, given the amount of fire and visible impact damage present. They have received several reports from other chiefs and fire officers who were inside 7 WTC and observed serious structural damage, with large sections of numerous floors and walls missing, loud groaning and clanking sounds indicating that the building was moving, and no water left in the buildings' fire protection systems.

Figure 32–9. Firefighters start to tie off a hoseline from the second floor of the Verizon building as they prepare to evacuate that building before 7 WTC collapses. (*Courtesy of the author*)

Several chiefs had tried to organize attacks on what, at the time, were relatively small fires, only to find there was no water in their hoselines. Searchers had rescued several civilians trapped on the eighth floor, including the city's head attorney (the Corporation Counsel), a Housing Authority staffer, and a retired NYPD detective who headed security for one of the building's tenants. The building was believed to be fully evacuated and searched up to about the 23rd floor without finding any civilians other than the three mentioned. In addition, the resources available were a mixture of on-duty and off-duty personnel, with few of the tools like SCBA and radios needed to conduct a safe, successful operation.

Standing at the corner of Vesey and West Streets, as the wind periodically shifts, we can see huge amounts of fire venting from the south façade of 7 WTC. The command staff decides to abandon the building, although at least one is vigorously opposed. "What the hell are we supposed to do, let all of Manhattan burn down just because there is fire in it?" was one comment I overhear while awaiting orders. This is an absolutely unimaginable situation: the FDNY is just going to walk away from a building and let it burn! But it is the right thing to do. There is simply no way we can put that out at this point. Now the job is to minimize further casualties.

Several years later, while working on an investigation into the collapse of 7 WTC, I interviewed Assistant Chief Al Hay, among dozens of other firefighters and officers who had been inside 7 WTC after it was struck by the North Tower. On 9/11, Al was a newly promoted deputy chief. It was his evaluation and recommendation, along with several other deputy chiefs, that we not try to fight the fires in 7 WTC, which undoubtedly saved many more lives. Al was put in charge of monitoring conditions and maintaining the collapse zone around 7 WTC. He recalls this waiting time period as a very stressful one for him personally. "I kept saying, 'I hope that thing falls down, I hope that thing falls down,' since I didn't want to be called onto the chief of department's carpet a few weeks later to explain why I decided not to fight a fire in a building that did not fall down," he said. Thankfully, he was right.

Chief Fellini points to the remnants of the north pedestrian bridge: a huge structure that has been smashed down to the ground across West Street. It is blocking the command post's view of all the major parts of the site—both Towers (1 and 2 WTC) and the Vista Hotel (3 WTC). He is getting radio reports that there are numerous people trapped there, and assistance is needed. He sends me to get him a description of what is going on, what needs to be done, and what kind of resources will be needed; and most importantly, to help clear the collapse zone around 7 WTC.

I head south, to start clearing the collapse zone. There are dazed firefighters, cops, and a few civilians spread out over the hundred yards from Vesey Street to the bridge. We are telling everyone to forget what they are doing and get back, either north of Vesey Street behind the Verizon building, or west to the river, behind the World Financial Center buildings. As I reach the remains of the footbridge, I see Rescue 1's apparatus: its cab crushed flat to the ground by the collapsing bridge. I shudder as I look at it. How many times had I ridden in that very cab? What is the fate of the guys that had ridden it here today?

The north footbridge is a major impediment that totally blocks access from one of only two access points into the site. I ask several passing cops and firefighters how to get around to the south side. None has any clue. I realize that for all my knowledge of the World Trade Center, I am now totally clueless as well. I cannot get my bearings. Few of my familiar landmarks are visible, and those that are visible are blocked by solid debris, on fire, or both. Someone suggests that I circumvent it by going through the lobbies of several buildings in the World Financial Center complex between West Street and the Hudson River.

Emerging onto West Street, south of the footbridge, I get my first view of the remnants of the towers. I am stunned. I find myself at the southwest corner of the 16-acre WTC complex. Across the street stand two skeletal structures about 150 feet high that form two L shaped walls, all that is left of the massive towers. A four-story pile of burning debris is all that remains of the 22-story Vista Hotel. The roadway of West Street is wall to wall steel—the exterior walls of the towers laid out flat for the most part, covering every surface with multiple layers. Now I have to make my way back north about 1,000 feet to get to the footbridge and start evacuating the collapse zone, which is easier said than done. In several places, the three-fingered wall sections have buried themselves vertically, right through the concrete sidewalks.

We see numerous fire apparatus buried under this debris and expect to find firefighters in the rubble. Crossing the smoking, jagged, dust-covered pieces of twisted steel that cover the entire area poses some challenges, and trips and falls are a constant occurrence, but then I meet the real challenge: the hundreds of firefighters and a few dozen cops that were heavily engaged in rescue efforts and do not want to leave. The FDNY, like the USMC, has a tradition of never leaving a brother or sister behind. Our tradition also demands that once a body has been recovered, the deceased member's unit is given the honor of removing their fallen comrades from the scene. Both those traditions are about to be sorely tested (fig. 32–10).

Dust-covered firefighters, some survivors of the two prior collapses, are dropping down through the smashed-out windows of the wall sections to get to the apparatus, looking and listening carefully for any sign of life. Several times I approach groups of rescuers working feverishly, digging into the smoldering debris with bare hands, trying to get to a trapped firefighter. "Guys, we have to move back, we're establishing a collapse zone here.""Chief, we can't go; we got a brother trapped here!" is the reply several times. "Let me get a look," I say. "We almost got him," is a common response. In every case, what they had was a lifeless corpse, buried inextricably under hundreds of tons of steel. Bunker pants and boots sticking out from under the crushed remains of a pumper that was itself smashed flat, adding its 30,000 pounds to

Figure 32–10. In the aftermath of the collapses of Towers 1 and 2, looking north on West Street, stunned rescuers search the debris field for survivors. Fewer than 20 would be found inside the rubble field and the remains of the towers.

the tons of steel above it. "Guys, he's dead," I say as gently as possible. (All 343 FDNY members lost this day are males.) "There's no way you can cut him out of there without some heavy steel cutting equipment and big cranes. Let's go." Some of them know in their heads that I am right but cannot find it in their hearts to leave. A couple of times the rescuers don't want to leave, even after they realize that the person is dead. "Look, guys, we already lost a bunch of people today, we don't want to lose you too, not to retrieve a body that is already dead." It works most times. Only once do I need to seek reinforcements.

I meet Deputy Chief Nick Visconti, who is also leading the evacuation of the collapse zone but has been diverted from that mission by the discovery of the body of Chief of Department Peter Ganci. Nick had worked the vast majority of his career in very busy units in the Bronx and Harlem (another former battalion chief in Battalion 16 where I am assigned). He is tremendously popular and universally respected. Currently he is the chiefs' representative on the Uniformed Fire Officers Association union board. Guys knew he would never leave anyone behind unnecessarily. They listened to Nick.

Chief Ganci's body is now reverently being removed from the debris field. The discovery of his remains and those of First Deputy Fire Commissioner Bill Feehan had drawn numerous firefighters and police officers back into the debris field, seeking survivors as well as trying to retrieve the remains of their compatriots. But the process is being complicated by the blazing hulk of 7 WTC on the north end, a 47-story giant that is itself in danger of collapse, and 90 West Street, a 25-story fiery torch on the south end.

As I make my way through the debris field trying to clear the collapse zone around 7 WTC, I encounter numerous groups of rescuers who are heavily engaged in void exploration, dropping down through the mass of twisted, smoking steel looking for other survivors. I tell them we have to retreat back out of the next collapse zone. Many of them stare at me uncomprehendingly. They look in the direction I am pointing but all they see is the blazing shell of 6 WTC, a nine-story building in the foreground. "Chief, we're a good 150 feet from that. We're safe over here, aren't we?" they plaintively ask. The smoke and dust swirling over the entire site hides the real threat looming 600 feet over them on the other side of 6 WTC.

When planning collapse zones, the recommendation is to clear an area at least 1½ times the height of the threat. We're still too close. Trying to move hundreds of stunned rescuers who are looking for friends and comrades (and finding them, although deceased) is a task that is nearly impossible. At times, we have to order guys to leave a body they have been working to extricate, telling them we are not trading live firefighters for dead bodies. I pray at times that we don't find a live victim that is seriously trapped, for I know I will not be able to get some of these guys to retreat if we do. Fortunately, at several critical moments, the wind shifts, allowing a view of the real threat posed by the looming 7 WTC. With that ability to recognize the danger now, slowly we are able to clear the remaining collapse zone. Now, we wait.

Unfortunately, there were plenty of bodies to be recovered inside the collapse zone, including the body of Chief Ganci. Chief Ganci and the rest of the command post staff had survived the collapse of the South Tower by running down the ramp into a parking garage under the World Financial Center. They emerged from that refuge to the numerous mayday messages for trapped firefighters in the wreckage of the 22-story Vista Hotel, which took the brunt of the west wall of the collapsing South Tower. The hotel was being used as an access point into the South Tower by fire units responding to that building, since the connected lobbies offered an entry that was sheltered from the falling debris and bodies of the jumpers. Now, the collapsing tower had cleaved a gash deep through the hotel, trapping firefighters under tons of debris. Some of them were still alive and urgently calling for assistance.

Chief Ganci and others, such as Chief of Special Operations Ray Downey, and Safety Battalion Chiefs Larry Stack and Brian O'Flaherty, made their way through the jumble of tangled steel and blazing debris and quickly sized up the situation: there were numerous trapped firefighters in the lobby, but the nature of the debris precluded a rescue with the resources that were available, and the North Tower, which had been struck by the first plane and had been burning for 15 minutes longer than the South Tower when it collapsed, still loomed ominously over the entire scene. Ganci knew we were whipped. As groups of dust-choked survivors of the first collapse struggled through the debris in an effort to reach their trapped comrades, Ganci gave the order instead to retreat. He knew that the two towers were built in identical fashion, and that if one could collapse, so too would the other. The only question was when.

If there was anything good about this tragic day, it is the fact that the second collapse did not occur for another 29 minutes, giving the people still streaming from the North Tower and the surrounding structures time to complete their escape. Due to Chief Ganci's order to withdraw, enforced by Ray Downey and Stack and others—men who chose to risk and ultimately lose their lives so that others would live—the firefighter death toll, as terrible as it was at 343, was not made even worse by dragging additional rescuers into the kill zone that the second collapse created (fig. 32–11).

The only group that was not pulled immediately from the collapse zone was one spearheaded by BC Mark Ferran and Lt. Glen Rohan of Ladder 43 from Spanish Harlem. They and several others had somehow scaled the blazing remnants of the North Tower in response to the repeated mayday calls from my friend Capt. Jay Jonas and his crew from Ladder 6. Jay and his company, as well as Lt. Mickey Cross and several men of Engine 16 and Lt. Jim McGlinn from Engine 39, and Port Authority Police Officer David Lim had been in the B staircase helping a lady that was having severe difficulty

Figure 32–11. This shattered four-story hull is all that remains of the 22-story Vista Hotel. With nightfall, the first heavy equipment begins picking away trying to clear a path through the debris. Huge quantities of massive equipment toil on site for nine months. It was here that Chief Ganci gave the order to evacuate, saving the lives of countless rescuers after the first collapse.

descending the stairs. The woman, Port Authority bookkeeper Josephine Harris, had started walking down from her office on the 73rd floor, but her health had finally brought her to a halt just as Jay's unit reache her on about the 15th floor as they themselves descended.

Mrs. Harris couldn't walk as fast as everyone else, and periodically had to rest. Ladder 6 had stayed with her, carrying and cajoling Mrs. Harris down to the 5th floor before her legs gave out entirely, causing the entire group to halt. At that point the entire North Tower collapsed around this small band of stragglers, entombing L-6, E-16, Mrs. Harris, BC's Richard Prunty and Rich Picciotto, and Officer Lim. Miraculously, the staircase they were in, along with several adjacent elevator shafts, had remained reasonably intact, providing a sheltered void allowing the 14 survivors to see tomorrow. Now no one was leaving until these survivors were all safe. Some rescuers guided the first survivors down from their precarious position, out across the devastated moonscape of twisted steel, burning debris, and wind-blown dust and ash to the safety of West Street. Lt. Rohan and others descended into the dust-choked remains of stairwell B to widen the escape route for FF's Jeff Coniglio and James Efthimiades, and brought in a Stokes stretcher to place Mrs. Harris in before carrying and lowering her to follow the others to safety. Chief Prunty would unfortunately succumb to injuries he received before rescuers could free him. The last survivors of stairwell B had now all been evacuated and the work of clearing the collapse zone continued.

As we are waiting for 7 WTC to fall, everyone is asking for information on friends or their units. "Did you see my brother?" or "my cousin?" is a constant refrain. "Does anyone know where Engine 54 is?" "How about Ladder 16?" The air is thick with questions, but not enough answers. Nobody knows what is really happening. "Has anyone seen Chief so-and-so?" "Oh, yeah, I saw him about 15 minutes ago over on the other side of the bridge there." As many times as not, the answer is wrong. Guys reported as dead turn up fine later. Others reported as walking away uninjured will be found months later, as the debris is excavated. Some are never found.

It does not take that long, maybe 30 minutes, before the agonizing sound of shrieking steel and imploding concrete and glass reverberates over lower Manhattan for the third time in just over seven hours. In a way, this acts as a release of pent up frustration and expectations of sorts. Command tries to quickly ensure that everyone is safe and that no further personnel are injured or trapped by this latest collapse, and after some of the smoke and dust from this pile settles out or blows farther away, they give the okay to begin moving back in to the areas we had just recently evacuated. After 7 WTC fell at 5:20 p.m., the troops pour back into the evacuated area.

Now we are going back to get the remains that we had left earlier, and hopefully to locate more survivors. The rescue of Captain Jay Jonas and his group from the stairwell inspires hope that other similar voids may be found in the other stairwells of both towers (each had three main stairs). Crews are already scrambling back over the blazing mountains of debris, seeking any opening that will allow them to penetrate deeper into the piles. Bucket brigades are organized, where one or two rescuers digging with hand tools and even bare hands, claw their way into promising locations looking for survivors. But using only hand tools and portable power equipment like saws and even the Hurst hydraulic rescue system means that we are seriously overwhelmed in terms of what we can achieve. We are simply not prepared for the size and nature of the debris we have to deal with. Even with the arrival of the FEMA cache of tools from the FD shops, we have nothing that can penetrate and move this kind of debris. Many of the steel columns are three feet wide, five feet deep, three inches thick and 40 or more feet long, weighing upwards of 25 tons. We are no match for the steel. Still, we have to do something!

I make my way back through the World Financial Center buildings to the location of the command post. I report back to the assembled chiefs describing conditions on the south side as I observed them. We need to get really heavy equipment on the scene to begin clearing access to the remains of the two towers.

As more and more responders arrive, a plan starts to take shape. We will use our standard collapse rescue plan consisting of five steps: first, scene size-up, followed by surface victim removal, then void exploration to identify and reach trapped victims, selected debris removal to free those victims, then general debris removal to ensure we do not miss any victims. It is a great plan, and one which we have used successfully to rescue numerous trapped persons from many collapses for over 50 years. The problem is this collapse is unlike anything we have ever experienced in terms of the nature of the debris. Still, that is what we do instinctively. Searchers have already begun penetrating as many voids as can be found. They start passing debris back out of their way, to create access points in to other potential voids. This activity attracts others who are standing around in shock, looking for some guidance on what to do. Soon we have a series of lines stretching hundreds of feet, passing individual pieces of debris by hand back out of the immediate void area to dump sites at the perimeter.

The story of Ladder 6 makes its way around the site. It helps to make our priorities clearer. We can't cut our way through the solid steel that blocks so many pathways. We know that each of the towers have three main staircases. Perhaps each of the other five that haven't yet been searched will contain similar pockets of survivors. We have to access the top of the stubs of the buildings that remain and cut our way into their staircases.

By this time, numerous pieces of heavy equipment are beginning to arrive: city-owned bulldozers and front-end loaders from the sanitation and highway departments, as well as some smaller mobile cranes. Many independent equipment owners and contractors had watched the collapses on television and knew immediately that their equipment would be essential to the rescue efforts. Many spontaneously got in their rigs and headed downtown. Now they are staged up and down West Street and on all the side streets in lower Manhattan. The first order of business is to clear an access lane through the mass of building steel, abandoned and burned-out vehicles, shells of fire trucks and police cars and ambulances that litter the disaster site, many of whose former passengers will never return. Those that aren't too badly damaged and can be started are driven to empty parking lots several blocks away and parked with little semblance of order. Their owners, if they ever return, will spend hours scouring the borough for their cars. Heavy-tracked bulldozers from Sanitation throw chains around the burned-out wrecks and tow them unceremoniously to clear spots where they are then bulldozed up into piles of so much scrap, with no resemblance to the once proud chariots they had been, transporting their crews to their doom.

Since it is obvious that clearing this access path is going to take many hours, Command decides to look for other ways into the structures.

Many of us who are familiar with the complex know that there are multiple below-grade levels (six) that are served by several subway lines, as well as train lines from New Jersey. These seem to offer sheltered voids where there might be survivors. I am ordered to pick a crew from the hundreds of firefighters awaiting orders in the staging area along West Street and see if we can get into the concourse area under the buildings by entering from one of these tunnels. I walk toward the milling throng, looking for some SOC guys I might know, but they are all working at other tasks already. I see a group of about 15 people from Engine 35 and Ladder 14 in Harlem. I had spent about four months covering in the 12th Battalion, which shares their quarters, and know them to be a good bunch of steady, reliable people. I gesture to their leader, Lt. Mike Haddock, to bring them over for a briefing, telling them I know the complex pretty well and have several ideas for approaches that might be successful. I find out that they are looking for two chiefs from the 12th who responded together that morning and haven't been heard from since: BC Joe Marchbanks, who I went to probie school with 22 years earlier; and Freddy Scheffold, the Battalion Commander to whom I owed a tour for working when I needed to be off. I assure the folks from 35/14 that we will do everything we can to find them.

We make our way up to Church Street and Park Place, where there is a subway entrance, and proceed underground into total darkness. Everyone has a flashlight, but I have no idea how long we might be in there, so I tell them all to hook up with a partner and share their lights, turning one of each pair of lights off, so that we will have fresh batteries in half of our lights to find our way back out. We come into the concourse at the northeast corner, under 5 WTC, one level below the street. The lights here are out also, but a surprising amount of daylight is visible through the roof, which is the plaza area between buildings, and has many holes smashed through it. The area that we enter through is relatively intact except for about an eight-inch deep layer of dust and some scattered debris, light fixtures, ceiling tiles, and the like that had blown down. I knew that off to the right is the main bank of escalators that led down several more levels to the PATH trains to New Jersey. We head that way.

The ceiling over the escalators is hinged downward from the far side toward the top, forming sort of a sloping wall. Several members set to work cutting holes through the plaster and wire mesh with Halligans and axes. On the far side, daylight shines through numerous holes, and I can make out part of the massive TV antenna that used to sit atop the North Tower sticking up through the plaza area. At the bottom of the ceiling we hit layers of solid steel. The holes through the roof were caused by massive wall sections from both collapsing towers. We look for other ways in. We find several intact staircases, good sheltered voids, and work our way down until we get about four levels below grade. These lead to a variety of storage and mechanical spaces, but none of them lead upward. It is obvious that we are the first responders into these areas, as the dust is absolutely undisturbed, like gray snow, but just as obviously, anyone in these areas when the planes had hit had gotten out of the area on their own. We head back up to the street, exiting past the Port Authority Police booth in 5 WTC.

We then go over to another subway entrance on West Broadway and make our way south, still hopeful of finding a route into the base of the towers. We make it about ¾ of the length of the subway platform before we run into a solid plug of steel and concrete that fills the whole tunnel, from floor to ceiling. We again retreat to the street. I decide I had better report our findings to command. The members of 35/14 want to keep looking for Joe and Freddy, but I don't have any useful information to give them to narrow their search area, other than that they most likely responded to the South Tower, which was on the far side from where we are now. I wish them good luck and ask them to be careful as we go our separate ways (fig. 32–12).

When I make my way back to the West Street command post with my observations, Chiefs Fellini and Cruthers put me in charge of removing the north foot bridge that blocks access to the site. I have about 30 firefighters assigned to me, as well as a contingent of ironworkers from Local 40, several small cranes, and a pair of backhoes. We start clearing a path from Vesey Street heading south. The ironworkers are a tremendous asset. They too are unsolicited volunteers, but they come with a "command staff" of their own: several shop stewards from the union who knew their people and came to

Figure 32–12. This is the wall of debris that blocked access into the subway tunnels below the WTC complex. It would take months to clear and years to restore service to the tracks.

the fire department command post for orders. They have their own tools (taken off whatever jobs they were working on that morning) and by late afternoon are getting set up for what they recognize is going to be a monumental task. The backhoes use their buckets to rip the sheet metal cladding off the bridge, exposing the heavy steel structure, and allowing us to send several search teams inside to look for victims. None are found.

My next concern is to try to get access to any voids that might be present under the massive structure. I feel that if I were out on West Street when the collapses began, I would have looked at this bridge as a potential refuge and sought shelter beneath it. We spend several hours cutting holes through the steel floor and searching in a few voids there, with no results. (Several weeks later, we are shown the videotape shot by French videographer Jules Naudet, who was riding with BC Joe Pfeiffer when the planes struck. Pfeiffer, Naudet and several others are seen traversing through this bridge after being driven out of the lobby of the North Tower by debris from the collapsing South Tower. They were the last to use it before the collapsing North Tower smashed it flat. It was also evident from the tape that after the collapse of the South Tower, there was no one lingering in the vicinity. I wish we had had access to that tape in the first 24 hours. It would have saved some time and effort.)

With the steel framework at least partially exposed, oxy-acetylene torches are deployed to start cutting through the massive columns and girders.

These first torches prove to be too few and too small for the task. I send a messenger back to ask for bigger tips and more torches. It is just as well that we have to wait. We really want to get a complete look at the entire structure before we start making cuts that might precipitate a secondary collapse. Unfortunately, the limited reach of the backhoes prevents us from reaching the upper half of the structure, and much of the ceiling inside is still intact.

Around this time, another larger crew of iron workers from Local 40 has arrived: about 50 guys with more equipment, including some really big torches. All will be invaluable. The foreman of this group is a large Polish-American fellow named Walter. I truly regret that to this day I do not know his last name. I had written it and numerous other details that I knew I would want to remember in a steno notebook that I kept in my back pocket. I lost that book somewhere on the pile in mid-October, and would give anything to have it back. Alas, Walter is the best I can do, but he and everyone who worked that job will know who he is. Thanks, is all I can say.

The iron workers start scrambling over the debris, scaling the steel columns, cutting away more of the sheet metal cladding to expose more of the structural steel skeleton beneath. We begin to get a clearer picture of the scope of the problem. The bridge is approximately 150 feet long, 30 feet wide, and 30 feet high, and was supported by a rectangular box structure consisting of absolutely huge steel girders that span the width of the highway below, and smaller, though still very large beams spanning the shorter directions. Cutting torches will need a lot of time to get through that mass of steel.

About this time a contractor shows up with what he claims will be the answer to my prayers: a large, articulating boom with a set of hydraulically powered shears at the upper end. The damned thing looks extremely strange and slightly menacing in the gathering dusk. It kind of reminds me of a mechanical monster from a 1950s vintage Japanese horror film. I immediately dub it "wreckzilla." It belongs to a large demolition contractor, Mazzochi Wrecking, and the owner assures me that if we could just get him access and a place to set up his machine, he will make quick work of the bridge. He says his shears will be able to simply "snip" the steel beams and drop them to where they can be carted away. Having just come from a close-up inspection of the structure, and having seen the size of the steel that made up the girders, I am somewhat skeptical to say the least, but I decide we have to try anything that might speed access to those trapped in the remains of the smoldering towers. I order the dozer operators to work with the wrecking guy to clear access and an operating area for his machine.

About that time, I am summoned to the command post. FDNY Fire Commissioner Tom Von Essen wants an estimate of how long it will be before we get the bridge out of the way and can begin setting up the really big cranes to access the towers. I had no idea of what resources would be available or the exact make-up of the structure, but from what I had seen so far, I knew it was a very substantial steel box frame. I tell him that if we get the necessary resources we need in terms of oxygen and acetylene supplies, as well as sufficiently large cranes and dump trucks to haul the steel out of the way, I think we can get it done in 18 to 24 hours. My estimate turned out to be wildly off.

When I return to the bridge, everything is in total chaos. The hydraulic shear, instead of being set up and cutting to speed the removal process, has brought the entire operation to a grinding halt. The iron workers, seeing the huge shear approaching, have climbed down from the bridge and gathered around the machine in angry knots. Making my way through the crowd to the loudest point, I find the wrecking boss and a gray-haired iron worker screaming at each other menacingly. "What the hell is going on here?" I ask. "These guys won't let me set up," replies the wrecking boss. "Goddamn right we won't!" yells a voice behind me. The gray-haired guy turns to scream at me "If you put that thing to work here, I'm pulling every iron worker off this entire site. We ain't workin' with them scabs and we ain't touching that steel after that thing chops into it." The odor of whiskey on his breath in such close proximity is overpowering. I am dumbfounded. This guy wants to turn this catastrophe into a union squabble! I tell him to get out of my face and turn away as I scan the sea of hard-hatted faces for the foreman, Walter. Thankfully, he is just making his way up behind me. "What is this about?" I ask. "Well, we don't really work together," he replies.

"There are two ways to take down a building," he explains, "our way, where we go in and rig the steel up, then cut it loose so that it stays in controlled locations, and their way, where they go in and grab it and shake it until something pulls loose, not always where they planned. It's a different way of doing things, and we don't feel safe around it, and they use non-union workers to boot." "That's not true!" the wrecking boss yells. "All our guys are union." The gray-haired whiskey breath starts screaming and threatening again, stirring the pot. "Who is that guy and what's his problem?" I ask Walter. "Oh, he's a pain in the ass, but he's one of the senior guys in the local, so a lot of guys listen to him." "Is he a boss or something?" I ask. "Nah, just another iron worker." "Well, tell him to shut up or I'll have his ass locked up and hauled out of here in handcuffs if he keeps stirring up shit." "You can't have me locked up!" the loudmouth screams. "I'll shut this job down." Now I've had enough. "Get him the hell out of here and sober him up, and if anybody wants to walk off with him, go right ahead, but if you think for one second that you will shut this rescue operation down, you'd better think again. There are hundreds of union members trapped in that pile of shit right there: cops and firefighters. If you walk out on them now, I am going right up town to every newspaper office and television station in town and I am going to tell them how the ironworkers are trying to stop this rescue." "Hold on, chief," Walter says. "No one is shutting this job down." He gives a nod to several others, signaling them to haul the troublemaker away from the scene. With him gone we can work on finding a workable solution for all.

"What can you tell us?" I ask the wrecking boss. "Like I said, all my guys are union guys, that's all I'll use here. We have some non-union guys that we use at other jobs, but everyone here is union." Walter said he can live with that if someone checks their cards. "What about the shear?" I ask. "Can we use that on some areas without affecting the iron workers' safety?" "Absolutely. I won't cut anything without them giving me the okay." Great, this just might work.

Just then a major monkey wrench shows up: a Port Authority employee who tells me, "You can't set that thing up there—that's right over the ramp to the parking garage, you'll cave that ramp in and end up in the cellar if you try." Wonderful! Eventually, with a lot of cooperation between all parties—FDNY, Port Authority, iron workers, wrecking crew, and numerous volunteers—the shears are set up in a safe location and begin snipping away, clearing the rest of the sheet metal cladding and cutting easily through the smaller beams that make up the bridge structure. The heavy box beams are another matter. The thickness of the steel is just too great. Wreckzilla had met its match. The iron workers, with some newly arrived exothermic burning bars, will be essential to clearing the wreckage. By the time everything is set up and into operation, the night somehow eases into dawn. I have no idea where the time went. Progress was much slower than I had hoped, but at least we were making progress.

A little after the sun is fully up, I am leaning against Wreckzilla when a young fellow in clean and pressed blue BDU's and an orange safety vest walks up, asking who is in command. "Well, I'm in charge of this sector right in front of us here," I reply. "The overall incident command post is back that way," I point. "North of Vesey Street about a hundred yards. What can I do for you?" "I am Lieutenant so-and-so of the United States Coast Guard, and the Captain of the Port sent me here, so I am now the safety officer for this site." "Great! Welcome aboard. We can use all the help we can get," I say. "Where should I start?" "Anywhere you like. Pick a spot," I reply with sweeping waves of my arms in large arcs. The young man slowly does a full circle, looking at the mayhem that surrounds him, pauses briefly, a look of dismay dawning on him as he gets his first look at some of what was going on around him, does another 360, then abruptly announces that he thinks he has to go make some calls. I never see him again.

Looking around me now, with the sun pretty high overhead, my eyes catch on a large section of steel protruding from an upper floor on the southwest corner of 3 World Financial Center. I hadn't seen it through the smoke and the dust in the fading daylight hours yesterday, and the darkness overnight was near total outside the glare of the few portable floodlight trailers we had brought in. Now however, with daylight and a westerly breeze, it is clearly a potential "widow-maker" (a piece of debris precariously balanced that could fall at any time). I tell everyone to start moving east along the bridge, out from under the danger, then go back up to the command post to report what I see and what I think we need to do. There, one of the staff chiefs looks at my mismatched dirty outfit and dust-narrowed eyes and asked, "How long have you been here?" "I guess since about one o'clock," I replied. "This morning?" he asks. "No, yesterday afternoon." "Do you know what time it is?" "I guess about nine o'clock." "It's almost noon. You've been here almost 24 hours. I'll get somebody to relieve you in a little bit. Get ready to go home" (fig. 32–13).

I go back over to the bridge to await relief. I start thinking about what has happened here so far, and what I want to tell my relief. By about 1:00 p.m. on Wednesday, we have peeled much of the sheet metal skin off, and taken down about half of the upper beams of the long box-like structure. I am starting to feel the effects of going non-stop for over 24 hours; the initial adrenaline rush is wearing off. (Many responders worked well over 24 hours without relief, until the FDNY Command organized a system of rotating personnel to ensure there were rested people making difficult decisions.) About an hour goes by before another battalion chief shows up. I brief him on what we have done to that point and point out the widow-maker, and tell him that I would keep everyone over on the east side until somebody gets a closer look at it to see if it is in danger of coming loose, and if it is, to secure it. I introduce him to the new foreman of the ironworkers (Walter had left earlier that morning) and the wrecking crew and give him a short update on what has transpired between those groups. Then I wish him well and walk back up to the command post to check out of their accountability log.

Figure 32–13. "Wreckzilla" works on cutting away the north footbridge, just above the remains of Rescue 1's crushed apparatus.

At the command post, I meet Chief Cruthers, who asks when I am due back in my assigned unit, Battalion 16. When I tell him that I am on vacation and don't have any scheduled shift coming up for three weeks, he says, "Great, consider yourself detailed here for the duration. Now go get some rest and come back prepared to work." I head north on West Street, with no particular destination in mind at first. Since I have no car or means of transportation (the subway system and pretty much all of Manhattan is shut down now), I just keep walking up the wide six-lane boulevard, looking to get away from the choking dust, noise, and devastation. I decide to make my way to Rescue 1's quarters on 43rd Street. I get to Canal Street when a police car pulls up alongside and asks if I wanted a ride. "Yeah, I'd really appreciate it," is my weary reply. I climb in the backseat next to another firefighter and a cop. "Where to?" The driver asks. "Anywhere on the West Side near the *Intrepid*," I respond. We drive north along West Street in silence, passing throngs of stunned civilians who are crowding the avenue peering south, looking for a glimpse of hell. No one without authorization can get past checkpoints set up at Canal, at least in theory. Many of these people want desperately to help, to do anything they can to assist in the unfolding tragedy. Some have brought coolers of bottled water, soda, or other beverages. Some have brought boxes of sandwiches they made themselves or bought at a local deli or bodega and are passing them out to rescuers leaving the site. Others wave handmade signs thanking all the responders for their efforts. Still others just stand there in shock and gaze at the smoking hole in the southern skyline of Manhattan. I am too tired to take it all in, my eyes red and gritty from the previous 24 hours. I close them and don't reopen them until the cop at the wheel says "Okay, boss, we're at 12th and 42nd, is this close enough?" I glance up across 12th Avenue to see the looming hulk of the USS *Intrepid*—a World War II aircraft carrier, itself a victim of kamikaze aircraft piloted by madmen bent on destroying the United States. Now it sits there like a toothless tiger. "Thanks for the lift," I say as I climb out. "You guys be careful now, you hear me?" I add as I shut the patrol car door. "You too, boss," is the nearly simultaneous reply from all three occupants as they drive farther north, away from the unimaginable scene of death and destruction a few short miles south. I whisper a brief prayer: "Please God, watch out for those guys in the coming days." Then I think, "We're all gonna need you to watch out for us."

I glance up at the *Intrepid* again as I walk up to 43rd Street. I think, there it sits, unable to do a damn thing to protect its permanent home port, just as it had been unable to fend off the swarms of Japanese kamikazes. But then it gives me some comfort to see it there: that not only had it survived that onslaught but it had come through even stronger, modernized, and served a distinguished career for many years afterward. I hope we will do the same.

As I come east along 43rd street, the scene at Rescue 1 is almost as bad as what I had left, minus most of the dust. There is a small gathering out in front of the open doors of the firehouse—my old unit that I had left a little over two years earlier—grim-faced firefighters from the company, covered in the detritus of what soon would come to be called "the pile"; cleaner firefighters that I do not recognize; and knots of wives and family members clutching each other, weeping in despair. The tears on their faces tell of the non-stop terror they have been living with for going on 30 hours now. The remaining members have gotten in touch with everybody they could reach, and now the toll is becoming clear: the company has 11 members missing out of 29 assigned. There are other civilians just standing awkwardly off on the fringes: people from the neighborhood who have come by to offer support but don't know how to do so in the face of such grief.

There is a priest just inside the open apparatus door, Father McGrath, a true gentleman and longtime friend of the company. He is a parish priest in Massachusetts, and a veteran fire buff who comes to New York City several times a year. He would hold mass in Rescue 1's quarters on Christmas Eve and on the morning of the department's annual Memorial Service in October, but mostly he was just there to follow us to fires and talk to us in the kitchen afterward. Now his presence takes on a new dimension. He would become a source of consolation and strength in the coming months, and many of us would seek absolution from him before we returned to dig at the pile—one of many precautions we would come to take "just in case something goes wrong." Some guys took to writing their names on their extremities in case that was all that was found of them.

I get a look inside the apparatus bay, and I see Squad Company 61's rig backed in where Rescue 1's rig should be. It is one of only two Special Operations Command units left in the city, out of the original 13 (5 Rescues, 7 Squads, and Hazmat 1). Squad 61 has been relocated here from their quarters in the northeast Bronx since the collapses, when the dispatchers had been unable to contact any of the other units. Captain Mike Banker tells me that they and Squad 270 in eastern Queens are all that is left; all the other SOC units are gone. "What do you mean, gone?" I ask. I had seen Rescue 1's rig largely crushed under the north footbridge that we had worked on all night, so I know it is out of commission, but I had also seen Squad 288's two rigs largely intact over on Broadway. Surely, they're not all gone. "Well, chief, I don't know about rigs, but SOC has been calling every company about once an hour, trying to find out who was working and who has survived, and it looks like every SOC company that responded was wiped out to a man, no survivors so far." I am dumbstruck. I want to say that's impossible; we can't have lost everyone. There have to be survivors, maybe in hospitals that no one knows about yet. But the words won't come. The devastation that I have seen in my limited forays around the edges of the disaster tells me that nothing is impossible.

As I make my way back to the firehouse kitchen behind the squad's apparatus, I glance up at the "riding list"—a white board that shows the members working on the shift, with their assignments for the tour (fig. 32–14). It is normally updated twice a day, at the start of each shift, 9:00 a.m. and 6:00 p.m. I see that it has not been updated. The six names glare down at me as I pause before it.

Officer	Lt. Mojica
Chauffeur	O'Keefe
Can	Nevins
Irons	Henry
Hook	Montesi
Roof	Weiss

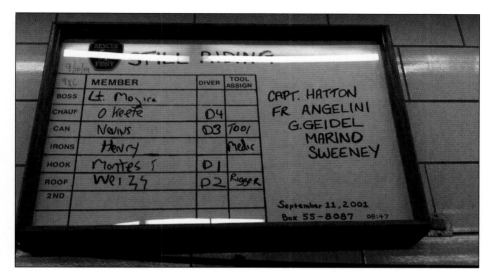

Figure 32–14. The riding list is repeated in dozens of firehouses in the city.

Shit! Not Dennis! Then I catch myself, not any of them! I had selected Jerry Nevins, Billy Henry, Mike Montesi, and David Weiss to come to the unit. Patty O'Keefe was a senior man and he had been there before I got to the company in 1994, but he too was one of my favorites. Then I see another list on the whiteboard alongside, it just gets worse—five more names unaccounted for: Captain Terry Hatton; Joe Angelini, the senior man in the entire department with over 41 years on the job; Gary Geidel, another solid senior member who exuded calm and confidence at any job; Brian Sweeney, a new man in the unit who lives near me on Long Island and whose father is a friend of mine from my days in Ladder 111; as well as another new man, Kenny Marino.

In the kitchen, the mood is grim, strained. Several family members have come back there seeking information on their loved ones from the firefighters who have been working down at the disaster site. Unable to bear the strain alone, they seek solace in numbers and in the comforting words of their husband's, father's, or child's coworkers. "Don't worry, we'll get him," I hear repeated over and over. Normally these words from such highly trained experts in collapse rescue would be a sign of complete faith and confidence, but having been there and seen the scope and nature of what will be involved firsthand, I can sense the first seeds of doubt in my own mind. I want to say, "We shouldn't build their hopes up too high," but I see the looks in the eyes of both firefighters and loved ones, cautioning any outsider from intruding. Each needs to hear the words. It is what will allow us to face the coming months. "Don't worry, we'll get them."

I make my way up the stairs to the company office as I had done countless times before—my former office—now as a visitor. It is empty. Half of its rightful occupants, Terry Hatton and Dennis Mojica, had left about 30 hours earlier heading for the World Trade Center. Of the other half, Mike Pena is down there now looking for them, while Lt. John Kiernan had retired just weeks ago, leaving a vacant spot waiting to be filled. I go over to the desk to look at the day book to see who was supposed to be working the day shift. That was part of the reason for the extra missing personnel: the attack had occurred at 8:46 a.m., just before the 9:00 a.m. shift change, extra personnel riding to their deaths.

I strip off the muddy, dust-coated uniform and climb into the shower, spending an extraordinary amount of time under the steaming water, trying to wash the grit, cement dust, and smell out of my hair, eyes, ears, mouth, nose, and every other nook and cranny it had invaded. Even after 20 minutes of intensive scrubbing I still feel grimy when I step out of the shower. I manage to scrounge a pair of clean underwear from Mike Pena's locker, then make my way into the officer's bunk. I need sleep.

Climbing between the sheets, I cover my head with a pillow to block out the light and the sounds of the distraught families from the apparatus floor below. In spite of my over-tired condition, sleep does not want to visit me. My mind is racing, replaying endlessly the horrific images of the past 30 hours in vivid slow motion, the horrifying details jumping from one scene to the next—the TV images of the two towers burning; the first collapse; the dead firefighters' legs sticking out from under the crushed engine; Rescue 1's rig smashed on West Street with its cab, the one I had ridden in thousands of times, crushed nearly flat to the ground; a scene worthy of Dante's Inferno stretching as far as you can see in every direction; myself riding the "50 car" elevator up to the observation deck on the 110th floor of the South Tower. Eventually I must have drifted off, but not for long.

I wake up to hear a familiar voice in the distance. I struggle to pull myself back into consciousness. I drift back and forth, into and out of my nightmare, before recognition of the far-off voice finally pulls me all the way back into reality. It is the voice of Maria Baretto, Dennis Mojica's fiancée. She is calling for him now, "Dennis, Dennis, where are you? Please don't leave me now."

Only a few months earlier, my wife Jeanne and I had met Dennis and Maria at their favorite restaurant in Greenwich Village, Sevilla, where they finally answered my wife's perennial question: "Dennis, when are you going to marry this girl?" They had been going out and living together for nearly 10 years, and now they were ready to announce that they were finally going to take the plunge and tie the knot. "November 10th. Would you please be there?" We shared a few more hours of joyous celebration, planning the long-awaited event. I promised them that I would try to have either Rescue 1 or Rescue 2's rig at the ceremony. I had first met Dennis 12 years earlier, when he was my driver in Rescue 2, before his promotion to Lieutenant. Now I will have to make another promise, one that I am not sure I can keep. I dress and make my way downstairs, each step taking me closer to the mournful cries. When Maria sees me, she immediately comes over and embraces me in a clinging, tearful hug, in spite of the cloud of dust that comes off my clothes at the slightest touch. She asks what I know. Where is he? Why hasn't he called her yet? What will happen next? How were we going to rescue them? All the same questions for which there are no good answers that are being repeated over and over again in hundreds of firehouses, police stations, and private homes and apartments across the New York and Washington, D.C. regions. Like everyone else in those circumstances, I don't know what to say. I find myself mouthing the same lame answer being given too many other times this day: "Don't worry, we'll get him."

Fortunately for me, before she can ask too many more probing questions, and look into my eyes as I try to find answers, piercing the illusion of hope, the voice alarm interrupts everything, with the blaring announcement: "Attention Squad 61, acting Rescue 1, respond to One Liberty Plaza for a report of a building collapsing." "Oh shit, not again!" is the silent (or not so silent) plea of everyone in the building. I can't picture One Liberty Plaza in my mind, but I know it's another large high-rise downtown, near the Trade Center. I grab my gear and jump on the backstep of Squad 61's pumper for the ride downtown.

We take Broadway downtown, toward the still billowing cloud of smoke, dust and human souls that continue to soar heavenward. We stop just north of Vesey Street, almost in the same spot where I had reported in to Chief Haring the previous day. One Liberty was still two blocks south, but after the prior day's collapses, no one wans to risk having another high-rise building land on them. The Squad officer brings out a pair of binoculars as we head south to meet up with Deputy Chief Pete Hayden, a survivor of the attack, who came over the one block from the WTC incident.

> "John, I think this is nonsense. It seems like a bunch of people who were standing near here thought they saw the building moving and started screaming that it's collapsing, but we don't believe it. I think it was just a bunch of dust blowing off the window sills. The city and the building's owners have had their structural engineers go over the whole building and they say it's sound. Take a look and let me know what you think."

Squad 61 takes a walk through the building and confers with the building owner's representatives.

It turns out that it is a false alarm (one of several in the coming days) due to people's overactive imaginations, coupled with an optical illusion caused by looking up at a steep angle along the side of a towering high-rise. Add in dust blowing off roof tops and window ledges, and white clouds scudding along against the bright blue sky, and the impression that the building was moving is complete. I for one am thankful for the false alarm, but for a different reason than most of the other responders that dread another collapse: this run had extracted me from a very painful situation that I did not think I could handle without breaking down. With the determination that One Liberty Plaza is in no danger, Squad 61 is directed to return to Rescue 1's quarters. I decline their offer of a ride back, preferring the tragedy in front of me to the one back in the firehouse on 43rd Street. I decide to return to the north footbridge to see what kind of progress has been made in the five hours since I left.

One Liberty Plaza is located diagonally opposite the north footbridge, across the huge disaster site. Since I have already seen the north and northeast perimeter, I decide to take a different route back, along the south and west side. Here too, the destruction is indescribable: in some ways even worse than the other areas. I start across Liberty Street, which I know will run through to West Street, but which turns out to be impassable once you reach Greenwich Street right in front of the firehouse of Engine 10 and Ladder 10. Huge chunks of steel create a mountain of burning debris several stories high. I detour further south, behind 130 Liberty Street, the Deutsche Bank building. This area, downwind of most of the collapses, is almost waist deep in ash and debris, making it impassable to anything but heavy equipment and intrepid pedestrians like myself who trudge through the area much the same way you would through waist-deep snow, slowly and carefully. I then make my way west across Albany Street, once again in dismay at what I am seeing.

Out on West Street I come across a group of fire officers who are just exiting a charred high-rise on the corner of West and Liberty streets known as 90 West Street. I recall seeing it blazing away through much of the previous afternoon while I was trying to evacuate the collapse zone around 7 WTC. At that time, I had been directing people to head south, past that south footbridge until someone pointed out the burning 24-story building, which was sheathed in pipe and plank scaffolding. "Chief, they're setting up a collapse zone around it too," someone had pointed out.

Now, just over 24 hours later, I see that this high-rise has not collapsed, despite obvious signs of heavy fire throughout, smoke and steam still emanating from many windows, heavy soot stains, and signs of heavy flame impingement over numerous windows all over the façade. One of the chiefs notes in passing, "It pretty much burned itself out; we didn't mount an attack on it until this afternoon." I later found out that this was another building that Command had decided was unsaveable. It had been struck by tons of flaming debris from the South Tower collapse, and units had been ordered to stay away due to the fear that the steel and wood scaffolding would likely collapse. Units had operated several hose streams from a hotel across the very narrow Carlisle Street to stop the flames from spreading south into the Wall Street area. Now, however, it is just one more smoking hulk that surrounds the perimeter of the main problem area.

I take a look at the south footbridge for clues that might help me understand the makeup of its northern counterpart, noting the massive concrete piers that served to elevate it above the highway below. These were not visible at the other bridge, having been punched up through the steel of the floor by the massive impact of hundreds of tons of steel falling hundreds of feet that smashed it to the ground. I also notice that the "widow-maker" still projects out precariously from the side of 3 World Financial Center. From this angle, with the evening sun shining on it from the west, it looks even more perilous than it did in the morning looking directly up from underneath it (fig. 32–15).

As more and more responders show up, additional voids are entered and more lines are formed. Soon the whole site is crawling with people, passing debris by hand. Eventually, pallets full of 5-gallon buckets are delivered to the site, allowing loose debris to be moved more easily. By Thursday, these bucket brigades are employing thousands of responders. From an elevated vantage point, like the lunchroom on the 4th floor of 3 World Financial

Figure 32–15. The "widow-maker" projecting from the upper floors of 3 World Financial Center created delays in clearing a path through the debris field below until it was secured in place by a FEMA USAR Task Force from Ohio.

Center, where I proposed to set up our command post, the scene looks like ants crawling over a discarded sandwich. The bucket brigades are a total waste of time and effort. At the rate we are moving debris, I figure we might be taking ½-inch per day off a debris pile that is hundreds of feet deep. More importantly, they prevent the use of the heavy machinery that will truly be needed to move the massive steel that is covering access to the areas we want to get into, the stairwells. It is a feel-good exercise, however. People desperately feel the need to help, to do something, anything. It is going to take us days and weeks to get the kind of heavy equipment in place to make the needed lifts, so the bucket brigades continue, and command comes to realize they are a useful tool for giving people an assignment so that we can control their whereabouts, keep them focused, and prevent them from wandering into trouble.

This issue was a critical matter of safety. The fact of the matter is that the last survivors were rescued from the debris pile Wednesday morning, the 12th, approximately 26 hours after the first plane struck. By Thursday morning, the 13th, virtually every accessible void had been explored, often multiple times, and no further survivors had been located. Of course, we did not know definitively that no other survivors would be located, but we did know that there were numerous extremely hazardous areas around the perimeter that had many severe hazards, like secondary collapse dangers and toxic environments. We did not want anyone else entering a space that had already been searched. The risk was too great to justify. But what was happening was that as new people arrived on site, they had no idea of all the dangers and how the areas had been searched, so they would see a void opening and decide to go exploring for themselves. Hundreds of unsolicited responders showed up, not part of any requested group, and just wandered in to operate on their own. We had no way of knowing who they were, where they were going, or how to contact them. It created numerous problems for command. By organizing the bucket brigades, we had a way to control some of this activity.

What we really needed was site security. As it was, in the early days following the attack, anybody could drive up and get very close to the scene. Take, for example, one such case out of hundreds: we had one group drive a "gumbo trailer" up from the deep south and set up shop serving food to any responders that walked by. While I appreciated their desire to help, think about that for a moment. They were able to drive a truck and trailer right past any checkpoints and set up in close proximity. One of the concerns that we had initially was the potential for a secondary attack using explosive-laden vehicles. Even assuming somebody did check their truck and trailer for explosives before passing it through the checkpoints (doubtful), there was no plan for where they were positioned. They were in the way of operations and we had to move them in order to stage heavy equipment like cranes and dump trucks. Then there was the issue of sanitation; they were set up in an area that was ankle deep in swirling dust and debris. The firefighters and other responders were coming directly from digging through toxic debris full of pulverized bodies, and stopping there for a plate of food. In one previous disaster, an entire FEMA USAR task force was put out of commission by food poisoning from improperly handled food. The issue of unsolicited responders and volunteers was just out of control.

I make my way back along the Hudson River waterfront to the command post at West and Vesey Streets. There I report in to a deputy chief that I do not recognize. He tells me to stand by while they check for an opening for me. I make my way a little bit south, closer to where I had spent the previous night. I can see there is very little activity there. The bridge is still largely as it was six hours earlier. I am told the entire operation has been halted due to the fear that the "widow-maker" might break loose and crash down on the swarms of rescuers operating below. The Ohio Urban Search and Rescue Task Force has been assigned to attempt to secure it with wire rope to prevent that from happening: a very ticklish task 30 stories up. At about this time, word reaches the command post from the USAR folks that they have it tied back with cable to several interior columns and that it should be considered stable. I volunteer to go back into that sector, explaining how I had spent 14 hours there the previous night and was familiar with many of the issues.

Once the groups filter back into their tasks, I am happily reunited with Walter, who promises that his men are glad to be helping and further promises that there will be no repeats of the issues of the previous night. All hands are eager to get back to work clearing this major obstacle, hopefully making up for some of the five hours lost due to the dangling "widow-maker." Mother Nature has other plans, however.

At about 11:30 p.m. that night, the radio crackles with an emergency message from command. The National Weather Service is tracking a line of severe thunderstorms that is crossing New Jersey, bearing right down on the disaster site. Lightning, high winds, hail, and heavy rains are predicted to strike the area in about 30 minutes. Hail and rain are nuisances that we can deal with, but lightning and high winds are real dangers, especially with several cranes in operation. They will have to be made safe, lowered as far as possible, and their crews evacuated until the winds die down. The winds also pose another serious hazard: all around the perimeter of the site stand dozens of seriously damaged high rises, their tens of thousands of windows largely broken, consisting of millions of razor-sharp shards of glass, lying on ledges or held to the window frames only by thin strips of glazing compound. Pieces have been sporadically falling for two days, and personnel have been cautioned to wear their helmets or hard hats and avoid lingering around the perimeter of damaged buildings. Scores of workers, many volunteer firefighters from Long Island, upstate New York, and New Jersey have been working to clear or secure the worst hazards all day, but there are tens of thousands of broken windows. Now the looming storm promises to turn the area into a maelstrom of flying glass shards. All personnel are ordered to cease operations and seek immediate shelter inside substantial structures, away from any damaged windows. The warning comes none too soon. Many workers needed time to secure their gear and climb down off the debris pile, and the wind is picking up considerably by the time the last people in my sector pass me enroute to the interior of 3 World Financial Center.

I want to make sure no one was left behind in my sector, so I make a quick loop of the entire perimeter, then crawl up under a massive grappler, a kind of giant backhoe with a claw on the boom that can pick up steel beams, to wait out the storm between its steel tracks. I had taken the precaution of scouting the site earlier that day, in case an area of refuge was needed from any number of potential threats, and I know that the back side is shielded by the reinforced concrete wall of the parking garage ramp. I crawl in under the beast with my back to the wall, pull down my eye shields, and wait. In under a minute intense gusts strike, accompanied by pelting rain and hail. Glass and sheet metal fly in all directions.

Without the 30-minute warning that the Weather Service provided, we would likely have suffered hundreds if not thousands of additional casualties. As it was, not one person was injured by that storm. It also proved that the rigging job that the USAR Task Force had done on the "widow-maker" was up to the task, as it didn't move at all. An hour and a half after the first warning, everything is back to normal, or at least as close to normal as we would ever be.

The struggle to clear the north footbridge goes on the rest of the night of Wednesday, September 12 and on into the day of Thursday, September 13. At about 10 a.m. on Thursday I get another visit from the command staff, wanting to know what kind of estimate I have on clearing the remaining structure. Substantial progress has been made, in spite of all the disruptions, but the entire "floor" of the structure on the south side remains, impeding placement of the several large cranes that are now staged along West Street. I am having difficulty with estimating the pace of our progress; some tasks seem to just fly by while others progress at the pace of a disabled snail. I project another 24 hours at the current pace. Somebody asks me how long I have been there, but I can't answer. I have lost all track of time; the entire 48 previous hours just seem a blur. "Get ready for relief," I'm told. Three hours later it happens. Another battalion chief arrives, and we talk about what has happened overnight and thus far today, and what I have envisioned happening going forward. As I am walking away, he calls out, "You had better check in at the command post. They told me to tell you that they want you at headquarters at 6:00 a.m. tomorrow." What the heck is that about? I wonder. Does he mean at fire department headquarters in Brooklyn, or at the command post here? Since I am supposed to be on vacation, I don't think it's because I am being missed at the 16th Battalion, but I have no idea what else it could be. To tell the truth though, I am kind of glad for the change of scenery. That north footbridge has been consuming every waking moment, and many sleeping moments too.

I report in to Chief Cruthers at the command post, now set up inside a large tent on West Street and he tells me, "John, you've got Ray's Downey's job. We'll be meeting at headquarters in the morning to work out the command structure. Be there at six." I am somewhat unclear on exactly what that means, but I figure that I will find out more the next morning, so I had better head home and get some rest and a clean uniform for the meeting. On my way out I meet an old friend, Fred Endrikat, a Rescue 1 lieutenant from the Philadelphia Fire Department. Fred has responded as part of an Incident Support Team (IST) for FEMA, coordinating the response of the USAR Task Forces that are responding from around the country. Fred and I have known each other for nearly 30 years, having been student firefighters together at Oklahoma State University. I would come to rely heavily on Fred for the next nine months and beyond. He has been at the command post overnight and he also congratulates me. "For what?" I ask. He tells me that Chief Cruthers told him that I was now the new Chief of Special Operations in Ray Downey's absence. (At this time, Ray, like so many others, was considered missing in action, not dead, and I thought Chief Cruthers meant I would be Chief of SOC for the WTC site, not the department.) By the time I check out of the command post and hitch a ride out of the site, and then find a FDNY Suburban that is heading to Brooklyn that can drop me out at Engine 332's quarters, it is after 4:00 p.m. (fig. 32–16).

Figure 32–16. Two of my most important advisers during the WTC disaster are (middle) John O'Connell of Rescue 3 and (right) Lt. Fred Endrikat of Philadelphia's Rescue 1, in his role as a FEMA IST member.

Out at Bradford Street, I am amazed to find my car untouched as opposed to being wheel-less and up on cinder blocks, even though it is still parked on the street, several doors down from the firehouse. The attack downtown must have had a positive impact even on the bad guys in East New York, making them feel patriotic. I arrive home for the first time in almost 60 hours, just after dinner. I had not had a chance to call home after I left on Tuesday, since nearly all the phone lines and cell services in Manhattan were down. My wife had no idea if I was dead or alive and feared the worst, expecting an FDNY chief's car to pull up with a department chaplain to deliver the bad news. We had just gotten word that my youngest brother Dave, the police officer in ESU, was alive after surviving the attack. He had been on the 31st floor of the North Tower when the South Tower collapsed and had just barely escaped. Another brother, Warren, a retired NYPD Detective, had driven the 800-mile round trip up to Buffalo to bring home our two older sons, students at the State University of New York there. Everyone was holding an anxious vigil in our home. When I pull up in my green Chevy, everyone lets out a huge sigh of relief. Dinner may have been leftovers, but it never tasted so good or included such dearly missed companionship. Since leaving home three days earlier, all I have eaten is a half of a ham and cheese sandwich and about a dozen small packages of Oreo and Lorna Doone cookies provided by the American Red Cross and The Salvation Army volunteers. After dinner and a long, hot soaking bath, I fall asleep almost immediately in spite of the searing images flashing through my mind.

I report to the eight-story office building that is Fire Department Headquarters as ordered at 6:00 a.m. on Friday, September 14. I have served several previous stints there, working on several special projects like the High-Rise Fire Safety initiative in the wake of the fatal Vandalia Avenue and Macaulay Culkin fires, as well as the Y2K initiative, so I am familiar with the layout, and many of the senior staff. Now many of those senior Chief's offices are empty, including those of Chief of Department Ganci, and Assistant Chiefs Donald Burns and Gerry Barbara. Both are still unaccounted for, having last been seen in the lobby command post of the South Tower. A total of 22 chief officers are missing. The remaining senior staff is gathering in a conference room, among them Chief of Operations Dan Nigro, Assistant

Chiefs Frank Cruthers, Frank Fellini, and Joe Callan. A handful of deputy chiefs are also present, as well as one or two battalion chiefs like me. Hushed conversations fill the small room as clusters of men discuss the impact of the attack on the department in muted tones, discussing who is missing, who has been found in distant hospitals, and what units have been decimated.

Fire Commissioner Tom Von Essen calls the meeting to order promptly. The main purpose is to reorganize the department, shuffling personnel as needed to fill the gaps left by what is now looking to be the deaths of well over 300 members. In addition, plans will have to be made for handling the World Trade Center disaster site long-term. For nearly three days we have been operating on an ad hoc basis. Individual units and officers were being assigned to handle specific areas or problems with little coordination between units and little guidance from the command structure. That is about to change. Chief Nigro is appointed as the acting chief of department and Chief Cruthers is appointed as the World Trade Center incident commander, with Deputy Chief Peter Hayden as his executive officer (XO). A 2,000-person task force will be selected from within the department and dedicated solely to the WTC incident, while the remaining 9,500 members will continue to protect the rest of the city. I am designated as the search and rescue manager for the WTC operation.

Since the day of the attack and recall, the department has been operating on a continuous duty system, with all leaves cancelled. We are operating on a war-time footing, waiting for the next attack to occur. Firehouses that still have working apparatus are responding with double the usual complement of personnel in many cases. Additional surplus personnel take turns relieving these responding crews or are sending some people down to the Trade Center. Every reserve unit, a fleet of 20 fully equipped engines and 10 ladders, and every serviceable spare unit were put back into service immediately on Tuesday morning. Some had responded to the disaster, while others had relocated into the firehouses of units that had been lost to the collapses. Other firehouses were being staffed with a mix of recalled firefighters and mutual aid firefighters and apparatus from neighboring communities from Nassau and Suffolk counties on Long Island, Westchester and Rockland Counties in upstate New York, and communities in northern New Jersey.

This influx of outside resources was essential to maintaining fire protection in the city, since nearly 50% of the FDNY fleet was either destroyed or still committed to the Trade Center operation. It was not without problems, however: some relatively minor and some absolutely huge. To begin with, New York City had never been the recipient of such a large-scale mutual aid effort and had not planned for such an occurrence since shortly after the end of the second World War. As a result, there was no common communications network for the out-of-town apparatus to communicate with the FDNY dispatchers or with FDNY units on scene. This was partially resolved by stationing mutual aid units in quarters and having them respond as task forces with the few remaining FDNY apparatus, and assigning FDNY personnel to ride on the mutual aid apparatus as local street guides. Another major problem was the lack of common hose thread fittings, particularly for connecting to standpipe and sprinkler system Siamese connections. The FDNY uses a 3" connection on such appliances, while nearly everyone else uses 2½-inch threads. Work-arounds were developed in some cases for standpipe systems by supplying a first-floor interior hose valve and ignoring the Siamese. No such solution is possible for the sprinkler system, but by ensuring that at least one FDNY pumper was committed to the Siamese early in the incident, fires were successfully extinguished.

One serious issue that was indirectly related to the mutual aid was unsolicited responders showing up directly at the Trade Center site. This had been an issue at every major disaster for many years, but it had several serious side effects in this case: exposing many improperly protected people to danger unnecessarily and creating severe morale problems within the FDNY. Many fire, police, and EMS resources were requested to respond within existing mutual aid guidelines as described above. Some, however, took it upon themselves to respond directly to the scene without authorization and without checking in with New York City officials. We had no idea who they were, no way to communicate with them in case of an emergency like the thunderstorm, and no way to tell them what areas to avoid, except to send someone out to tell them face-to-face, needlessly exposing the messenger to unnecessary danger. Even months later I had to go out into the debris field, into an area that we had deemed off limits due to the danger of loose steel hanging over the area, to remove two firefighters dressed all in white turn-out gear. They clearly were not part of any authorized unit but had driven down from New England because they "wanted to do something to help." I thanked them for their efforts but told them they had to leave because they were adding to our problem, rather than helping. They did not like hearing that, but left.

The root of this problem was a lack of security for the site. We needed the area secured to prevent unauthorized access but never got it. There were rumors going around that Al-Qaeda had gotten hold of several fire apparatus and utility vehicles, packed them with explosives, and were planning to drive them down to the site and detonate them near the command post or other facilities, taking out a large concentration of expert responders.

The impact of these unsolicited responders went beyond safety con-cerns. The 2,000-member FDNY Task Force was working 12-hour on and 12-hour off shifts, giving us plenty of help once all the initial voids had been searched. All our people wear black bunker gear with yellow reflective stripes (and later, brown Carhartt overalls) that is clearly marked FDNY. The FEMA USAR Task Forces wear blue BDUs, also clearly marked. Many of these unsolicited responders from out of town wore bright yellow bunker gear that stood out like a sore thumb from everyone around them. We had over 9,000 other FDNY members basically being held hostage in their fire-houses, whom we had told, "Do not come down to the Trade Center, we don't want you here." They desperately wanted to be there and we told them not to come. We did that to facilitate our command and control over who we had on scene. Now, these 9,500 firefighters are sitting in firehouses watching the nonstop television coverage of the scene, and seeing firefighters in yellow turnout gear, who are clearly not FDNY people, doing a job that they have been clamoring to do: to be able to go downtown and help look for their friends, and in many cases, relatives.

The most enraging cases occurred when one of the out-of-town folks gave a TV interview, telling the reporters how heroically they had just spent eight hours digging New York City firefighters out of the debris. The FDNY guys did not take it well. In some cases, there were threats of rebellion: that guys in the firehouses would just leave, take their own cars downtown, and go to work. Others threatened to take their apparatus with them and the hell with letting somebody else collect our dead. It took a great deal of character on the part of many more reasonable firefighters, company, and chief officers to prevent such mutinies. We could not stop it entirely until we got site security, which we never did truly get, but having a perimeter fence erected during the second or third week helped.

With respect to the morale issue and unsolicited volunteers, even when we had the fence up and some decent folks at the checkpoint, it couldn't stop a guy from walking around the perimeter and giving an interview to the first TV camera that he saw, talking about all kinds of heroics that were absolutely made up. Several impostors were arrested. They used their firefighter's clothing to gain access to buildings that were unguarded, and basically looted the places. Effective security would have prevented so many problems.

The missing component was credentialing, with photo identification badges to authorize access. The first few attempts did not have photos, and did not have expiration dates or limits to access. All were part of the problem. Without a picture, anyone could hand off a pass to a buddy. Access to one part of the scene should not get you access to all areas, such as the command post, unless your job requires it. Access also needed expiration dates, easily determined by a checkpoint sentry trying to get as many needed people through the checkpoint as quickly as possible without letting any unauthorized people in. Eventually, we settled on color-coded badges that also required a separate color-coded wrist band that was valid only for certain work cycles.

Enforcement, however, was only as good as the personnel manning the checkpoints, which in some cases was not good at all. I drove past several unmanned checkpoints, because their assigned personnel wanted to see the site for themselves, or felt that they could be of more value adding another pair of hands to the bucket brigades. Part of the problem was that the security effort was divided among so many different agencies: NYPD, PAPD, National Guard, and police departments from all over the tri-state area. Everyone wanted to help, and all had good intentions, but if somebody walked up in some kind of uniform and gave a story about how they just drove for so many hours and really, really needed to get in there, or if they walked up with a box of donuts or sandwiches and coffee, sometimes the checkpoint personnel just looked the other way. It could have killed us in more ways than one.

After the headquarters meeting broke up, we returned to the WTC site and set about preparing for managing what obviously was going to be a major, long-term commitment. The command post tent was not going to be sufficient in terms of size or durability, as was made plain during the Wednesday night thunderstorm when it suffered substantial damage.

One of my next assignments was to locate a suitable long-term location to set up a CP for a large staff. The first place I found and liked was a large employee cafeteria on the third or fourth floor of the World Financial Center. It had a huge area overlooking the entire incident site. Unfortunately, nearly every window was blown in, the place was full of dust, and all the power was out. Not insurmountable obstacles, but command told me to look at other options. Next was the firehouse of Engine 10 and Ladder 10 at the corner of Liberty Street and Greenwich Street, just across from the South Tower. This also seemed to offer a few good features: two stories with a deep double-bay apparatus floor, offices on the second floor, and a roof that overlooked the scene from the south. We actually started setting up there and used it for the first few days, before the drawbacks became obvious. The street in front was totally impassable with debris and would remain so with ongoing debris removal operations for many months, and the close proximity meant that anyone and everyone who had a question or request simply made their way inside, interrupting whatever was going on. It was too close for command but would serve well as the Church Sector Operations Post.

On Sunday the 16th, while we were setting up inside 10/10, one of the most touching events of the tragedy occurred. A group of us were moving equipment and supplies into the building when I spotted a familiar silhouette slowly making his way across the dust-coated, twisted steel that covered the entire length of Liberty Street. I stopped a second to watch his approach, gingerly trying to pick his way through the treacherous tangle of sharp steel and icy slick sheet metal, while hobbling along on a pair of crutches, his left leg in a soft cast. I made my way over to assist my friend Denis Murphy into a chair on the apparatus floor. "John, I'm here to help. Just tell me what you need and I'll do whatever I can." Denis was the captain of Squad 288. He and I go back about 13 years, to when he was a firefighter in Ladder 111 and I was a covering Lieutenant there; then he transferred to Rescue 2, and I followed shortly as one of Ray Downey's lieutenants. We also taught together for many years at the Nassau County Fire Academy.

Denis had been severely injured three months earlier in an explosion at a hardware store fire that killed three firefighters, including our friend Brian Fahy, and seriously injured dozens of others: known in the FDNY as the Father's Day Fire. Now, responding to the recall, he was here to do his part, whatever is needed. The exertion of traversing the debris pile showed on his face. I asked him for news about his unit. I had seen their rig over on Broadway when I first arrived on Tuesday, so I knew that they were on scene before the collapse. Denis looked heartbroken as he replied, "They're all gone." At this early stage, we did not know that for certain, but it was the stalking fear in all our minds. While the surviving off-duty members of all the SOC units were climbing all over the debris pile looking for void spaces that might contain survivors, Denis could not possibly climb so much as a step ladder. I desperately wanted to find a job for Denis, to give him something to do to distract him from the mind-numbing reality that his entire company, as well as all but one member of Hazmat Company 1, which shares Squad 288's quarters, are missing. (The one survivor was a chauffeur who was sent back to the apparatus to move them farther away from the scene so they wouldn't be blocked in, in case they were needed elsewhere.) We did some arranging of equipment, setting up a communications network with telephones, portable radios and chargers, notepads, a whiteboard, etc., but the work took a toll on Denis. He was so close to the members of his unit, yet so very, very far away: it was eating him up. Finally, I said to him, "Denis, don't try to come back here tomorrow. It's too hard. Go to your quarters; your guys and the families need you there." He gave me a grimace. I knew what kind of pain the prospect of trying to console the inconsolable would be like. He was not looking forward to it, but realized this place was just too dangerous for him in his condition. I had just asked him to do what I did not want to do myself: face the families.

Denis was charged with the most heart-rending tasks of any command-ing officer: to try to make sense of a tragedy that has claimed the lives of someone's spouse, parent, or child to their grieving families. He has had to do it before, but this time it is in multiples, and the scale of the grief is unfathomable. We were all beginning what will be known as "The Year of the Four Fives."

The term *four fives* refers to the telegraph signal that was used in the past to notify FDNY members that a line-of-duty death had occurred. The tele-graph would tap out five bells, followed by a pause, then five more, another pause, five more bells, a final pause, then the last five bells. In the days before radio and voice alarm communications, it caused all activity in the fire-houses city-wide to stop as everyone wondered who had lost their life. The housewatch would lower the American flag to half-staff, and most would say a private prayer for the deceased. In the modern era, the practice continues with the four fives tapped out on a brass apparatus bell that is used at the graveside. The news today is delivered via radio, voice alarm, and teleprinter, which all deliver the same solemn message—i.e., "The signal 5-5-5-5 has been transmitted; all units stand by to copy. It is with deep regret that the department announces the death in the line of duty of Firefighter Ronald Gies, of Squad Company 288, which occurred on September 11, 2001, at Box 5-5 8087. Funeral arrangements will follow when completed." That same message was transmitted to every firehouse in the city every time a firefighter or officer was recovered. In many cases, it took months before a body was found. The family and the department then began the preparations for laying the dead to rest.

The wakes, funerals, and memorial services began within the first days after September 11th, with services for Chief of Department Ganci, First Deputy Fire Commissioner Bill Feehan, and Department Chaplain Father Michael Judge, who were all recovered within hours of the collapse. Their services were all held on Saturday, September 15th. Three true heroes were buried that day. I had told Chief Cruthers I would be taking a few hours off that day. I just had to go to Chief Ganci's funeral. I had worked for Chief Ganci many times: when he was a BC in the 57 Battalion while I was cover-ing in Ladder 111 (where he had served as a firefighter) and Rescue 2, and on several special projects working at headquarters when he was the chief of operations and later chief of department (the Bunker Gear Initiative, Y2K, and the High-Rise Task Force, among others). Those first funerals set the tone for the coming months: unendurable grief accompanied by a feeling of sadness that we needed to do more but just couldn't.

The wakes continued at a pace that would be beyond comprehension for more than three months. A line-of-duty funeral is a soul-rending affair, fixed in pageantry and protocol. We had just gone through it in June, when three members were killed in an explosion and collapse on Father's Day. Harry Ford and Brian Fahey of Rescue 4 were buried on the same day, a few miles apart. John Downing of Ladder 163 was buried the following day. More than 10,000 firefighters from all over the United States, Canada, and even Ireland attended the services.

Three months later we now have a funeral for the chief of department killed heroically in the line of duty, and only a few hundred firefighters showed up, and most of them were from volunteer departments on Long Island! It was not because we didn't want to go. We loved Pete Ganci. It was just because we had over 300 firefighters buried in "the pile" and everyone who wasn't working in the firehouses was digging for their trapped comrades. In those first few days, we thought there might be some void areas somewhere in the six sub-cellars that might hold live victims. The dead would have to wait while we tried to save the living. It was not to be. The last live victim was dug out of the rubble just 26 hours after the first collapse. Everyone else would be body retrievals (fig. 32–17).

The members didn't give up hope. After working their 24 hours in the firehouse and getting off at 7:00 a.m., they fell into a steady rhythm. If there was a funeral that day, they went from work to the funeral. That evening, they went to wakes of others being buried in the next few days. The following day they repeated the ritual. If there were no funerals for friends, they went to "the pile" on their days off. On the third day they went back to work. Even at that frenetic pace, it was impossible to do what you needed to do.

The worst days came in early October, after we had gotten a lot of the heavy equipment in place, which allowed us to start uncovering larger numbers of bodies. The peak was on October 5th: a Friday. I had vowed to try to

Figure 32–17. Lt. Vinny Ungaro of Squad 288 stands atop the North Tower wreckage to remove debris pinning two bodies from Rescue 1. Vinny was one of more than 200 FDNY members to perish due to cancer in the 18 years since the attack. (Courtesy of the author)

make as many of the funerals of personal friends as possible, but on that day, there were 27 firefighter funerals being held throughout the tri-state region, from eastern Long Island to Orange County in upstate New York. I scheduled a few hours to make those of two friends. The first was a morning service in my hometown parish for Ronnie Gies, a firefighter in Squad 288. I left Ronnie's service before the final procession and headed east along Sunrise Highway, heading for Ray Meisenheimer's service in West Babylon. Ray and I had been firefighters together in Hazmat Company 1 when it opened, and I really liked him. I had not gone a mile from Ronnie's service when traffic was stopped for the procession delivering Firefighter Adam Rand, also of Squad 288, to St. Barnabas Church in Bellmore. I had wanted to attend the funeral for another friend that day, but it was being held upstate, and the travel time would mean I could not make Ray's or Ronnie's, and would not be able to get back to the Trade Center that day. I passed two other services before reaching the church in West Babylon where Ray was being laid to rest. As I pulled in to the parking lot, I am stunned and distraught. I am too late: the crowd is at attention offering the final salute as the honor guard is carrying the casket, draped in the FDNY flag, from the church heading to the hearse. I make a comment to myself about being too late to say goodbye to Ray, but it is overheard by a firefighter in line next to me. He said, "No, don't worry. Ray is next on the schedule. That is a firefighter from a different company." That church buried three different firefighters that day.

The impact of the attack and the subsequent rescue and recovery efforts on the members of the Special Operations Command (SOC) is difficult to fathom: we lost 94 SOC members that terrible morning, out of approximately 450. The Trade Center attack wiped out more than 30% of the personnel in the rescues, squads, and the hazmat company. To cope with this loss, it was necessary to consolidate our remaining personnel. Normally, we have the equivalent of four shifts in our units, working 24 hours on and 72 hours off. (Not the official chart, but that's what most guys do.) With more than one whole shift wiped out, we were forced to put the members on a 24-on, 48-off system which is not a severe problem in itself, but the impact of the attack just kept compounding.

The situation escalated with the delivery of five real, weapons-grade anthrax-tainted letters to several media outlets in Manhattan at the end of September. The "other shoe" that we had been waiting to fall just did. Within hours, the flood of "suspicious powder" responses threatened to overwhelm our existing resources. In response, New York City created a number of joint police/fire hazardous materials response units that combined the needed hazard mitigation, evidence collection, and safeguarding and criminal investigation staff in a combined unit. The units handled over 300 runs per day to these incidents, the vast majority of which proved to be unfounded, but each of which had the potential for tragedy. We had to hire SOC firefighters on their days off to staff these additional hazmat teams, so now the remaining firefighters were spending even less time at home with their families. I began to get calls from wives, begging me to please let their husbands come home to see their kids. It got so bad that I had to order members not to go to the pile on their days off. By that point, it was obvious that there would be no survivors. We had to take care of the living.

But fate wasn't through with us yet. Our beloved Yankees won the American League Eastern Division, meaning playoff games at Yankee Stadium, a high-profile target for terrorists, meaning more personnel ordered to work on their nights off. The Yanks then won the American League Championship Series, even though many in the law enforcement and special operations fields were hoping they would lose in the early rounds of the playoffs; now, the World Series meant three games at Yankee Stadium. Then the General Assembly of the United Nations held its annual opening ceremony in New York, with lots of heads of state: more targets, more ordered overtime. November brought the New York City Marathon, a nightmare for planning, and the Macy's Thanksgiving Day parade, more SOC units deployed. Then another tragedy, American Airlines Flight 587 crashed on takeoff in the heavily populated Belle Harbor section of Queens, igniting a dozen homes, killing six people on the ground and all 260 aboard. Firefighters dug through the rubble for two days to locate all the bodies.

It was during this latest catastrophe that the impact of what these men were doing really struck me. I found myself standing on top of the foundation wall of a home that was completely demolished, supervising a group of firefighters and officers in digging through the debris. The plane had nose-dived straight down, with most of the impact centered in the small side yard between two homes. The rows of seats had just accordioned right into this space, one after the other. Bodies and body parts by the dozens were crammed in there, mixed in with the seat cushions, luggage, and plane parts. Everything was heavily charred and smoking, despite the numerous hose streams in operation. At one point, a young probationary firefighter was working just below me, the orange "pumpkin" on his helmet front piece still brand-new, with hardly a hint of dirt on it, probably only out in the field since August. This young man bent down and pulled on a pile of indistinguishable smoldering debris, when whatever was holding it suddenly let go. The opening he had just created allowed a new plume of acrid, burned flesh smoke to rise right up and engulf him, and a piece of the debris flew up and hit him square in the face. It was a woman's scalp and upper shoulders. He puked all over my boots.

I flashed back immediately over 25 years to my first plane crash on Rockaway Boulevard, and the insensitivity of the nitwit who told me to swallow hard. I tolf the probie, "Come on up out of there, son, we'll get somebody else to take a turn in that mess." "I'm all right, Chief, it was just the smoke that caught me. I'm used to this by now." "No, come up out of there now, and go over to the RAC unit and get something to drink. And don't come back unless I personally tell you to. I will tell your boss that is my order." To think that this young man has been asked to do the most grisly job you can imagine, and does so without complaint, is a testimony to the quality of the personnel we are blessed with. Here he is, a battle-hardened veteran of all of four months, and he'd seen and done more than anyone should have to.

I thought to myself that day, "We're going to be all right. If these are the kind of kids we're recruiting, we're going to make it." The senior men will move on and retire, but the new guys will step up and take their places. Words cannot express the love and pride I feel for these guys. They have been kicked hard, knocked down, and kicked again while they are down, but they never give up. It humbles me yet again to be a part of this magnificent group. We will be all right. I hope.

Emotionally and physically the ensuing months have shown how serious the toll is. Mike Pena is among more than 300 firefighters who are suffering severe respiratory difficulties: a young guy who probably had 15 more years of lifesaving ahead of him. He's going to be forced to retire. We have been told to expect a wave of emotional problems as the anniversary of the attack approaches. Suicides are a real concern: survivor grief. There is no explaining why some people survived while those in front and behind them in a line perished; and no explaining why two members swapped shifts the way they did, one member working for the other, going into the towers never to return. The absolute randomness of these incidents preys on the members' minds. After 18 years, it appears that the experts are wrong: suicides are not escalating, thankfully, although other problems do surface. I like to think that the experts' predictions were for the general public, not for as special a group as the members of the FDNY. We can do anything. Show us the door and we'll put out the fires of hell! We'll be all right. *We will be all right!*

CHAPTER 33

The Rebirth

The collapses of the World Trade Center towers had a devastating impact on the FDNY, the City of New York, the United States, and the world. But none of them ended that day. We all picked ourselves up, dusted ourselves off, and started putting one foot in front of the other, moving to where we needed to be. The journey was long and difficult, but with a lot of help from our friends, we kept making progress. In the case of the Special Operations Command (SOC), the needs were enormous: we lost several apparatus that responded, many crushed flat like an empty soft drink can. These were custom designed and built units that typically take at least a year to build after the specifications are finalized. Fortunately for us, Captain Terry Hatton at Rescue 1 had just completed spec'ing out a new apparatus for his company. We used that as the design for all five of the rescues, which helped tremendously. Terry never saw them. He was lost that tragic day, but they were part of his legacy, along with a child that he would also never see: his wife delivered a baby girl in May of 2002.

I stayed at the World Trade Center site for the first seven weeks, acting as the Search and Rescue Manager. I assumed full-time responsibility for the Special Operations Command at the end of October, turning over the role of Search and Rescue Manager to my good friend Donald Hayde, one of the FDNYs longest serving firefighters who retired in 2018 after 41 years. On the morning of September 10, 2001, SOC had seven chief officers assigned, and one acting battalion chief, all under the leadership of Raymond Downey. We also had 13 captains, 39 lieutenants, and approximately 335 firefighters working in the 13 land companies. By the afternoon of September 11th, Downey and four of his chiefs were dead, along with six captains, 12 lieutenants, and 71 firefighters. The sheer numbers don't tell half the story, however: the average service time of each of the 94 SOC personnel lost that day was almost 16½ years as a firefighter. To make matters worse, the department soon underwent a wave of retirements as firefighters looked at the memorial plaques on the wall of their firehouses commemorating the dead and realized that their friends were never coming back. The overtime that was generated in the fall of 2001 trying to satisfy all the demands, from staffing the units to special events, the hazmat teams responding to anthrax letters and other duties, drove many firefighters' earnings so high that if they did not retire very quickly, they would lose tremendous amounts of money.

After September 11th, the department immediately hired new firefighters and promoted officers from within the ranks to start filling the voids, but you cannot take a probie and place them in a SOC unit: it is too dangerous for all. SOC had to recruit from the vast pool of experienced firefighters, particularly those from busy ladder companies. The company commanders of those units were not happy having some of their best, most senior people being "poached" by SOC units. The companies that lost their own firefighters did not want anyone leaving during the grieving process. I centralized all requests for details into the SOC units under the control of BC Jack Spillane, and told our captains that we could not take more than three firefighters from any one unit, to avoid draining the experience from the engines and ladders. It still was not enough. I was working to recruit new officers to work in SOC. The first place I looked was to officers who had previously been firefighters in SOC and had since been promoted. Not all of them wanted to come back; the dangers exposed on 9/11, when nearly every on-duty SOC unit had been wiped out to a man, had shaken many families. One excellent lieutenant told me he wanted to come back, but his wife promised to divorce him if he did. We needed to incentivize the commitment.

One of the first officers who did agree to come to SOC was Captain Bob Morris of Ladder 28. Bob was extremely happy in his assignment at 28 and initially did not want to take on the role of company commander of Rescue 1. Bob always considered himself a first-due truck officer, and not a technical rescue expert, so it took some serious conversations to explain how much the job needed his leadership in that critical position. We had technical experts to do the task-level stuff. We needed leadership at the top to make it work, and I had an incentive for him that he could not resist. "Bob, there are doors in mid-town Manhattan that are the absolute state of the art in forcible entry challenges, doors that will not be seen in Harlem for another 20 years, they're just too expensive. You could be forcing them next week if you come down to Rescue 1." The hook was set. Thankfully, Bob stepped up and took the helm at a critical time, leading that unit on to even greater glories in his more than 10 years in command. For others, though, we needed a different approach.

For years, police officers assigned to the NYPD's Emergency Services Unit had been getting specialization pay in the form of promotions to detective, which is not a civil service test but simply an appointment. BC Bob Ingram and I went to Fire Commissioner Thomas Von Essen in the waning days of his tenure with a proposal to create specialization pay grades for SOC personnel: up to an additional 15% of each rank for certain positions. The commissioner was all for it, and while we were in his office, he called his budget director and told him to make it happen. The budget director replied that it was not possible, not because of money shortfalls, but because it would breach the union contracts—it was a mandatory subject of bargaining. The roadblock would be overcoming the objections of the many firefighters who would feel they deserved more money also but were not getting it. Commissioner Von Essen left office in January 2002 without being able to achieve that goal.

The next commissioner, Nicholas Scoppetta, had no firefighting background. He was a long-term municipal administrative figure (a lawyer) who had served in many capacities at various agencies over several decades. He was a great listener and evaluator of problems and solutions who had the confidence of the new mayor, Michael Bloomberg. Again, Bob Ingram and I laid out the plan to him, early in his tenure, and he too jumped at the plan to get it done, directing his labor relations people to make it a priority in upcoming contract negotiations. The unions soon saw the benefits of establishing specialization pay as a way to get extra money into the pockets of its members. There was already the precedent of chauffeurs pay for those who drove the apparatus, and CFR-D pay for those who performed first responder EMS duties. Soon, we had SOC specialization pay, which while we only received 12%, was a big morale boost and went a long way toward recruiting new personnel and even retaining some who were considering retirement. That was a win.

The next hurdle was training the massive numbers of personnel we were recruiting. SOC training is broken down into several disciplines: there is basic rescue technician, an 80-hour class; hazmat technician, another 80 hours; 40 hours of basic high angle rope work; and 40 hours of basic confined space rescue, which all new SOC members receive. Then there are the advanced classes that allow personnel to be able to do their jobs in field units: 80 hours of structural collapse rescue training, 40 hours of trench cave-in rescue, 40 hours of rigging, 120 hours of black water SCUBA rescue training for those assigned to Rescue 1, 2 and 5, the SCUBA units, training in WMD response (weapons of mass destruction—mostly chemical, biological and radiological attacks), shipboard firefighting, and numerous other specialties.

Not only did we not have the time to do all this training, we did not have the instructional staff in house any longer. Several of our instructors had perished or retired in the aftermath of the attack and recovery. We had to get creative to solve some of these issues. We received permission from the unions to hire retired members as instructors in several specialties, with the understanding it was only to last for one year until we could get additional instructors certified in those disciplines. We also received tremendous support from many outside agencies. The New York State Police SCUBA unit provided us with certified instructors in that field, thanks to NYSP Captain Steve Nevins, commander of their Special Operations Division. Steve's brother Gerry had been a firefighter in Rescue 1 and was killed that day. The International Association of Fire Fighters (IAFF), parent union of our own locals, provided us with hazmat instructors. The US Army, Navy, and Marine Corps provided us with specialized training in subjects such as terrorism and chemical weapons management, blending our students in with their personnel at highly advanced schools. This training was not all one-way; we reciprocated at every opportunity, putting personnel from these agencies through much of the training that the FDNY excels at, especially structural collapse rescue. We were on the road to recovery. It took just over two years to get the department as a whole back to the personnel strength that it had on September 10th. I like to think SOC made the level that we were at in less time than that. Of course, you cannot replace the experience of a 40-year member like Joe Angelini or Ray Downey overnight.

The units worked hard to make up for in training what the members lacked in seniority. The support of the administration was critical to making that possible. When I first met with the new Chief of Operations, Sal Cassano, he gave me simple instructions: "Don't come into my office and tell me what we can't do. Come in here with a solution. Tell me what we need to do to make it happen and I will get you the help you need." Who could ever ask for a better boss than that? This attitude applied to nearly every member of the FDNY family, both uniformed and civilian. Steve Rush, director of the Office of Budget and Finance, was a great example. Steve held the purse strings for the entire department, working with the Mayor's Office of Management and Budget, and the city was in fiscal trouble as a result of the attack and subsequent stock market decline (also due to Al-Qaeda's attack on its trading). Steve understood how critical not only rebuilding prior resources was, but also preparing for the newly emerging threats. I would go into Steve's office with a request for money to fund a new program: extra training in some discipline, or a new apparatus to expand our capabilities, such as the rebreather trucks that allow us to put firefighters into four-hour breathing apparatus for use in the many under-river tunnels. In the past, a request for a whole new type of apparatus would have had to go to the mayor's office, then through budgeting, then for review by other offices. It would literally take years.

When the Iraq war started in March 2003, I went into Steve's office with a request for several million dollars for two new vehicles and hundreds of new-style four-hour SCBA. Steve asked simply, "Do you really need this?" I said, "Yes." Steve said, "I'll find the money somehow." End of conversation—almost. "Steve, we need this yesterday."

"What do you mean, and what do you want to do?" he asked. I expected attacks on our subways to begin any minute. We could not wait two years for these rigs to be designed, spec'ed out, and built. The masks were an off-the-shelf item, we could get them in a few days, but they required specially shaped blocks of ice to be inserted into each one just prior to use, to cool the firefighters' breath as they inhaled the recycled air over long durations. I proposed renting some box trucks, buying some chest-type freezers, and putting them and the masks in the trucks. This would give us the basic capabilities within a week. Steve said he would find a way to pay for it. That is the footing the entire department operated under for the first five years after the attack.

None of this would have been possible without massive assistance from the federal government in the form of financial grants. When President George W. Bush stood atop the crushed remains of a fire truck just days after the attack, he vowed to bring the perpetrators to justice. That process took just under 10 years, but within weeks of the attack, the federal government mobilized a massive effort to rebuild the FDNY and expand the homeland security precautions of every city in the nation. At one point, SOC was administering over 70 million dollars in federal grants. The department had many specialists working behind the scenes to make that work: Robin Mundy Sutton, the Director of Technical Services, worked closely with firefighter Bill St. George at SOC; and Irene Sullivan, the department's initial grant coordinator, worked to navigate the labyrinth of federal purchasing and grant requirements. Soon, tools, apparatus, and equipment began flooding in, allowing the department not only to rebuild, but to expand our special operations capabilities dramatically.

Some of the plans that were developed then have been implemented with great success in the years since they were first introduced. One of the programs that we put in place designated 25 ladder companies across the city as SOC support ladders and an additional 26 companies as CPC (chemical protective clothing) units. These units were given an extra apparatus, about the size of an ambulance, stocked with equipment that would allow them greater capabilities at technical rescue incidents such as collapses, confined space rescues, and other emergencies. This idea was developed out of the plans that we had made for the Y2K event. It allows the department to increase capabilities during normal times, but also greatly expand them during disasters, by staffing these units with overtime personnel, allowing 50 additional units to respond to relatively minor incidents such as auto accidents, stuck elevators, and the like, which in turn frees up normal ladder companies for response to more serious incidents. These preparations played key roles in alleviating the burdens on the department and the residents of New York City at several large-scale events such as the northeast blackout in 2003, and hurricanes Irene and Sandy in 2011 and 2012. The recovery of SOC is complete, and the department has continued to expand its capabilities and achievements under the very capable leadership of those who have succeeded me as the Chief of Special Operations: Chiefs Bill Siegel, Bill Seelig, and John Esposito.

The City of New York has recovered also in many ways, turning the page on decades of rot and neglect. Many neighborhoods have seen a rebirth that at one time would have been unimaginable: the Williamsburg section of Brooklyn, including the Southside and particularly the Northside, are home to tens of thousands of new residents. The old warehouses have been turned into hip restaurants and chic boutiques. The glorious brownstones of Bed-Stuy and Harlem now command a princely sum, if you can even find one for sale. The frame buildings of Bushwick are being renovated, instead of rotting into piles of moldy debris. Even the poorest neighborhoods in East New York and the South Bronx have seen massive infusions of money, often from coalitions of religious groups, resulting in owner-occupied housing where once only "brick farms" flourished. There is still poverty, and crime, and fires. But it seems to me that the hopelessness that once gripped so many communities is easing. We will be all right. *We will be all right!*

Epilogue

In the preceding pages, I have told the story of my small part of the history of the Fire Department of New York, and of a few of the hundreds of great men and women that I was privileged to work with. It is not possible for me to write a story like what you have just read about every one of the 343 members of the FDNY who perished that terrible Tuesday, or the others who survived and continue to do great things for their fellow citizens. There is not enough room in this book for all of those stories, for one thing. But the main reason I can't tell you about all of these other brave men and women is that I didn't know them all. The FDNY is a huge job—over 11,000 fire officers and firefighters, and 2,500 emergency medical technicians and paramedics, and 1,500 civilians—who keep the big red machine running. It was simply not possible to know every one of them personally. But I do know what they are like. They are the same as Dennis, Mikey, Dave, and Joey. They are all awe-inspiring, each in their own way. They are like many of you and me. People who take their sworn oath to protect life and property seriously.

If they are called to lay down their lives for others, that is the price they will pay. Like thousands of Americans before them, they will have died doing their job, because it was *their job*. At the World Trade Center, no one else was going to go in there and pull people out of that flaming hell. It was going to take firefighters, and they were the ones to do it. Not one stopped at the front door and said, "I'm not going in there, it's too dangerous." The legions of civilians that they helped save, tens of thousands of people, tell over and over how the firefighters guided them down the stairs in their escape, using their flashlights in the darkened stairways, carrying those too seriously injured or infirm to proceed on their own, offering direction and words of encouragement to the exiting throngs, all the while plodding upward with their burden of protective clothing, masks, and tools. They stopped only to tend to a seriously injured person or momentarily for a brief respite. Upward to the 30th, 40th, and 50th floors, the youngest probies helped shoulder some of the burden of the older, more seasoned veterans, they continually surged upwards. Staff chiefs like Donald Burns and Gerry Barbara organized teams of companies, assigning groups of these teams to individual battalion chiefs with assignments to search groups of floors "You take 30 to 40! You take 40 to 50, Joe." "Orio, try to get as high as you can!" Upward they went. Heroes.

While I didn't know all these brave people personally, there is a saying that is repeated in firehouses throughout the city that epitomizes their lives. It says "Firefighters commit their greatest act of bravery when they take their oath of office. Everything else is part of the job." In the weeks after the attack, the spirit of the job was battered but not broken. Firefighters who had worked for hours digging through the smoldering pile or searching through the labyrinth-like below-grade area that extended six stories below ground, wrote messages of encouragement to the others that would follow them, using their fingers to scrawl through the layers of dust and grime clinging to vertical glass surfaces. Among the most heartwarming messages: "FDNY— still the greatest job on earth!" By the end of the second week, printed posters bearing the message sprouted everywhere on the site (fig. 34-1).

The job is still great. But now, my time has passed. I retired in 2007 after being injured at a church fire. Maybe it was "the Boss" telling me it was time to go and enjoy what He had planned for me in the future. Ah—the future.

On Friday afternoon, January 4th, 2019, I was hiking with my wife of 46 years, Jeanne, in the beautiful Adirondack Mountains of upstate New York, far from the hustle and bustle and danger that I had been associated with for over 49 years. It was a spectacular winter day and we were enjoying every second of our time in the woods. As we walked back toward our ATV, I stepped on a patch of ice and came down hard on my shoulders and

Figure 34-1. Despite all the death, hardships, and sorrow, it remains "The Greatest Job on Earth."

neck. I did not lose consciousness for more than a second or two, but when I could get my eyes to focus, I realized that while I could see my right hand, I could not feel it or move it. Jeanne was right there and came over attempting to find out what happened. Being a nurse, she began a patient assessment. I could not move or feel anything below my neck. This was bad.

We were in an area with no cell service, about 1½ miles off the road. There was no way Jeanne could safely move me by herself. She would have to take the ATV back down the icy logging road to our neighbors in Bay Shore Park to call 911. She did not want to leave me there lying on ice while she went for help, but there was no alternative. She covered me as best she could with several Mylar emergency blankets that we keep in a pack, and her own coat, then she tucked a can of bear repellent spray into my left hand, which she propped up on my chest, while she began the trip down the mountain to get help. I can say that lying there all alone, looking up at that beautiful blue sky through the trees was a very lonely and frightening period. We had seen plenty of coyote tracks as we hiked. Little good would the bear spray do me, since it was pointed at my face and I could not even move my hand if I had to. At the base of the mountain, Jeanne met our friends, Diana, Noel, and Jim, who called 911 for her and gathered a couple of blankets and gave Jeanne a coat to wear.

As I lay there motionless, I heard the Edinburg Volunteer Fire Department siren sounding. Oh, what a beautiful sound that was! I hoped that it meant that Jeanne had safely made it to Bay Shore (unless there was another emergency somewhere in town!). The EFD siren sounds so much like the old siren of the Inwood FD that I had jumped up at all hours to respond to for so many years. And now it was my turn to lie there helpless and silently wish the responders Godspeed. I will tell you that the apparatus siren does a lot more than warn other traffic to clear the way. When you are lying there as a patient, waiting for help, that siren is your connection to salvation. You hear the first distant warble as the units begin their response, and as they get closer and closer, the increasing volume helps you to hang on to whatever hope you can muster, and I did not have much hope at that point; being paralyzed from the neck down was not the way I wanted to live my remaining time on earth. The siren also tells you how seriously the responders are taking it: are they rolling like this is the real deal, or is this just a perfunctory occasional tap on the siren—just another "milk run?" If I ever get into a position to be able to use a siren again, I can tell you, I will lay on it and the air horns, long and hard, sending the message "Take it easy, guy, we're coming to get you!"

Jeanne waited for the emergency responders to arrive and drove the first two up to my location in the ATV, since none of their vehicles would make the climb over the ice. The first two on scene were Edinburg Fire Chief Wayne Seelow and my friend, Past Assistant Chief John Olmstead. Chief Seelow, on hearing the nature of the situation, had already requested mutual aid from the Northville Fire Department for their track-equipped 8-wheeled Argo ATV, which has patient transport capability. He also requested a life flight helicopter be launched to take me to the trauma center at Albany Medical Center. While waiting for EMS, Chief Seelow and Past Chief Olmstead placed additional space blankets and hand and toe warmers in vital areas to help ward off hypothermia. At this point, I had been lying on the ice for about 30 minutes, and even with my arctic weight Carhartt coat and fleece-lined pants, I was starting to get cold.

Jeanne made another run down the mountain with the ATV to bring up additional assistance, this time two volunteer EMTs from the Edinburg Emergency Squad, Kayla Milnyczuk and Walt Fitzgerald. Both were very professional, continuing patient assessment, prepping me for movement, coordinating with the troops at the staging area and requesting additional EMS resources in the form of mutual aid paramedics from the Fulton County Ambulance Service.

When the Northville FD arrived at my location, all hands placed me securely on a backboard and moved me to the Argo. This was done under very hazardous conditions over the ice on sloping terrain. Unfortunately, it was not done without incident, as past Chief Olmstead fell on the ice and broke his wrist. In the nearly 50 years that I had been responding to emergencies, I had never been responsible for having any of my people injured. (Okay, except for John Kiernan's broken nose at the plane crash.) Now this is my fault. When the backboard was secured to the Argo, all personnel boarded it and our ATV, and slowly made their way down to the staging area in Bay Shore Park where I was transferred to the Edinburg Emergency Squad ambulance. There I was treated by the Fulton County Paramedic, Dawn, who started an IV and administered medication for pain. I was taken to the Edinburg airport where the Edinburg FD had set up an LZ and transferred to Life Net for a 20-minute flight to Albany Medical Center. En route, the crew of Life Net 7-13, Andrew, Jess, and pilot Stevin, started an additional IV and administered more pain medication and gave me a very smooth flight over the beautiful Adirondacks (too bad I couldn't see any of the scenery). Past Chief Olmstead was transported by Fulton County Ambulance for treatment of his injury.

I cannot begin to thank all those who participated in this outstanding multi-agency effort: those doing the hands-on work, as well as all the support personnel that made it come off like it was a routine event, instead of a very complex wilderness rescue. It was truly a life and death situation. When I got to the ambulance my body was shivering uncontrollably; hypothermia was setting in. The spinal injury had the potential for serious consequences. On arrival at Albany Medical Center, the staff at that truly wonderful facility, from the Emergency Department doctors Hogan, Schuster, Chowdhury, and Fillion, to the Neurosciences Department of D5East (especially nurses Amy, whose husband is an Albany firefighter, and Isabella), conducted a series of tests and administered additional treatments that stopped the spasms I was experiencing. The care that all involved gave me helped to ensure that I will likely have no permanent effects (other than keeping me off the mountain when there is ice present).

As someone who spent nearly 50 years responding to all kinds of emergencies, big and small, I know how rare it is to ever have anyone say, "Thank you." Especially for those impacted by trauma, the need to put their lives back in some semblance of order, combined with the difficulty of figuring out who all the sea of swirling faces hovering over them belong to, often diverts their attention away from this task. I have made a conscious effort to try to remember the names and faces of all of you who were there for me and my family. Seemingly minor actions, like Mark Bomba's helping Jeanne to calm down and get the ATV put away, and give her directions to Albany Med, have a huge impact that you often don't think much about, but make life so much smoother for those in the middle of "the worst day in their life." You have all made a huge difference in our lives. Thank you!

Today, I continue to live the life of the FDNY, though I have to do it vicariously: through two of my sons, John and Conor, who fill me in on the latest jobs, letting me know the job is still great and in good hands. They and their coworkers are living through their own 9/11, now that Covid-19 is raging through our nation. Again the fire service, law enforcement, emergency medical services, nurses, truck drivers, food service workers, laboratory researchers, and so many others have stepped up when they have been needed, proving the meaning of what an essential worker is. As long as we have Americans who are willing to put the needs of their community and their nation before their own, we will be all right!

To all my friends who have devoted so much of your lives to helping others, please continue to do as much as you can. There may come a time when it is you or one of your loved ones who will need help. Then, hopefully you will be fortunate enough to have someone as caring as you are to help. I did!

May God bless and watch over all those who live by the words, "Greater love hath no man … "

MEMBERS OF THE FDNY SPECIAL OPERATIONS UNITS WHO MADE THE SUPREME SACRIFICE DURING MY FDNY CAREER 1979–2007

Ff. Lawrence P. Fitzpatrick	Rescue 3	6/27/80
Ff. Alfred E. Ronaldson	Rescue 3	3/5/91
Lt. Thomas Williams	Rescue 4	2/25/92
Ff. Peter F. McLaughlin	Rescue 4	10/9/95
Ff. Louis Valentino	Rescue 2	2/5/96
Ff. Brian D. Fahey	Rescue 4	6/17/01
Ff. Harry S. Ford	Rescue 4	6/17/01
Ff. Eric T. Allen	Squad 18	9/11/01
Capt. James Amato	Squad 1	9/11/01
Ff. Joseph J. Angelini Sr.	Rescue 1	9/11/01
Dedication, middle image; Fig. 29–1, 4th from left		
Ff. John P. Bergin	Rescue 5	9/11/01
Ff. Carl V. Bini	Rescue 5	9/11/01
Ff. Christopher J Blackwell	Rescue 3	9/11/01
Fig. 26–1, on right ladder		
Ff. Gary R. Box	Squad 1	9/11/01
Ff. Peter Brennan	Squad 288	9/11/01
Ff. Thomas M. Butler	Squad 1	9/11/01
Ff. Dennis M. Carey	Haz Mat 1	9/11/01
Ff. Peter J. Carroll	Squad 1	9/11/01
Ff. Tarel Coleman	Squad 252	9/11/01
Ff. Robert J. Cordice	Haz Mat 1	9/11/01
Lt. John A. Crisci	Haz Mat 1	9/11/01
Ff. Thomas P. Cullen III	Squad 41	9/11/01
Lt. Edward A. D'Atri	Squad 1	9/11/01
Ff. Martin N. DeMeo	Haz Mat 1	9/11/01
Lt. Kevin C. Dowdell	Rescue 4	9/11/01
DC Raymond M. Downey	SOC	9/11/01
Lt. Michael A. Esposito	Squad 1	9/11/01
Fig. 21–1, 3rd from left; Fig. 22–1, 2nd from right		
BC John J. Fanning II	Haz Mat Ops	9/11/01
Ff. Terrence P. Farrell	Rescue 4	9/11/01

Ff. Michael C. Fiore	Rescue 5	9/11/01
Ff. Andre G. Fletcher	Rescue 5	9/11/01
Ff. Thomas J. Foley	Rescue 3	9/11/01
Ff. David J. Fontana	Squad 1	9/11/01
Ff. Andrew A. Fredericks	Squad 18	9/11/01
Ff. Thomas Gambino, Jr.	Rescue 3	9/11/01
Ff. Thomas A. Gardner	Haz Mat 1	9/11/01
Ff. Matthew D. Garvey	Squad 1	9/11/01
Ff. Gary P. Geidel	Rescue 1	9/11/01
Ff. Ronnie E. Gies	Squad 288	9/11/01
Ff. David Halderman	Squad 18	9/11/01
Ff. Robert W. Hamilton	Squad 41	9/11/01
Lt. Harvey L. Harrell	Rescue 5	9/11/01
Ff. Timothy S. Haskell	Squad 18	9/11/01
Capt. Terence S. Hatton	Rescue 1	9/11/01
Lt. Michael K. Healey	Squad 41	9/11/01
Ff. William L. Henry	Rescue 1	9/11/01
Fig. 27–1, 3rd from right		
Capt. Brian C. Hickey	Rescue 4	9/11/01
Lt. Timothy Higgins	Squad 252	9/11/01
Fig. 14–1		
Ff. Jonathan R. Hohmann	Haz Mat 1	9/11/01
Ff. Joseph G. Hunter	Squad 288	9/11/01
Ff. Jonathan Ielpi	Squad 288	9/11/01
BC Charles L. Kasper	SOC	9/11/01
Lt. Ronald T. Kerwin	Squad 288	9/11/01
Ff. Thomas J. Kuveikis	Squad 252	9/11/01
Ff. William D. Lake	Rescue 2	9/11/01
Ff. Peter J. Langone	Squad 252	9/11/01
Ff. Daniel F. Libretti	Rescue 2	9/11/01
Ff. Michael J. Lyons	Squad 41	9/11/01
Ff. Patrick J. Lyons	Squad 252	9/11/01
Ff. William J. Mahoney	Rescue 4	9/11/01
Ff. Kenneth J. Marino	Rescue 1	9/11/01

Lt. Peter C. Martin	Rescue 2	9/11/01
Fig. 21–1, 2nd from right		
Ff. Joseph A. Mascali	Rescue 5	9/11/01
Lt. William E. McGinn	Squad 18	9/11/01
Ff. Raymond Meisenheimer	Rescue 3	9/11/01
Fig. 26–1, on ground at left		
Ff. Douglas C. Miller	Rescue 5	9/11/01
Capt. Louis J. Modafferi	Rescue 5	9/11/01
Lt. Dennis Mojica	Rescue 1	9/11/01
Dedication, left image		
Ff. Manuel Mojica	Squad 18	9/11/01
Ff. Michael Montesi	Rescue 1	9/11/01
BC John M. Moran	SOC	9/11/01
Ff. John P. Napolitano	Rescue 2	9/11/01
Ff. Peter A. Nelson	Rescue 4	9/11/01
Ff. Gerard Nevins	Rescue 1	9/11/01
Fig. 26–2, on scaffold under plank		
Ff. Patrick J. O'Keefe	Rescue 1	9/11/01
Ff. Kevin M. O'Rourke	Rescue 2	9/11/01
Ff. Jeffrey A. Palazzo	Rescue 5	9/11/01
BC John M. Paolillo	SOC	9/11/01
Ff. Durrell V. Pearsall	Rescue 4	9/11/01
Ff. Kevin M. Prior	Squad 252	9/11/01
Ff. Lincoln Quappe	Rescue 2	9/11/01
Ff. Edward J. Rall	Rescue 2	9/11/01
Ff. Adam D. Rand	Squad 288	9/11/01
Ff. Donald J. Regan	Rescue 3	9/11/01
Ff. Nicholas P. Rossomando	Rescue 5	9/11/01
Lt. Michael T. Russo	Squad 1	9/11/01
Ff. Dennis Scauso	Haz Mat 1	9/11/01
Ff. Gerard P. Schrang	Rescue 3	9/11/01
Ff. Gregory R. Sikorsky	Squad 41	9/11/01
Ff. Stephen G. Siller	Squad 1	9/11/01
Ff. Kevin J. Smith	Haz Mat 1	9/11/01

Ff. Joseph P. Spor	Rescue 3	9/11/01
Ff. Brian Sweeney	Rescue 1	9/11/01
Ff. Allan Tarasiewicz	Rescue 5	9/11/01
Ff. Richard B. Van Hine	Squad 41	9/11/01
Ff. Lawrence J. Virgilio	Squad 18	9/11/01
Capt. Patrick J. Waters	Haz Mat 1	9/11/01
Ff. David M. Weiss	Rescue 1	9/11/01
Ff. Timothy M. Welty	Squad 288	9/11/01

Glossary

FDNY RADIO CODE SIGNALS

10-18 Signifies that all units except the first-arriving engine company and ladder company can return to service.

10-30 A request for two engine companies, two ladder companies, and a battalion chief to respond to an incident. This signal is obsolete and no longer used. Usually indicated a moderately sized structural fire. (At the time, all engine companies were staffed with five firefighters plus an officer, allowing a single engine to operate at least one, and sometimes multiple, hoselines.)

10-45 Reports that a civilian has been seriously injured or killed in a fire. It is accompanied by one of three sub-codes, depending on the severity of the injury:

Code 1 Victim deceased

Code 2 Victim suffering serious injury (apparently life-threatening)

Code 3 Victim suffering serious injury (apparently not life-threatening)

10-75   The radio report from the first-arriving unit indicating a work-ing fire or other emergency: it results in a response of four engine companies, three ladder companies (one designated as the FAST team), one rescue company, one squad company, two battalion chiefs, and a deputy (division) chief.

10-76   Indicates a fire in a high-rise commercial building (over 75 feet in height). It is transmitted when, in the judgment of the inci-dent commander, conditions require a total response of the fol-lowing units:

4 engine companies	1 field communications unit
4 ladder companies	1 mask service unit
4 battalion chiefs	1 safety operating battalion
1 deputy chief	1 special operations battalion
1 rescue company	1 safety coordinator (fifth-due BC)
1 squad company	1 high-rise unit
1 CFR-D engine company	1 tactical support unit
1 "FAST" unit	1 RAC
1 command post company	1 public information officer

A second alarm for a fire in a high-rise building results in a second-alarm assignment of four engines, three ladders, one rescue company, two battalion chiefs, and one deputy chief, in addition to the 10-76 units above that have already responded. One of the two battalion chiefs shall be designated the commu-nications coordinator.

multiple-alarm fire   Indicates a fire or emergency that is beyond the capabilities of the first alarm units to control. Successive alarms result in additional units responding, bringing the total response to the levels shown in the following table.

	Engines	Ladders	Battalions	Deputy
10-75	4	3*	2	
Signal 75 (all hands)	4	3*	2	1
Second alarm	8	5	4**	1
Third alarm	12	6	5	1
Fourth alarm	16	7	5	1
Fifth alarm	20	8	5	1

*The third ladder shall be designated as the FAST unit.

**The third battalion chief shall be designated as the safety coordinator. The fourth battalion chief shall be designated as the communications coordinator.

10-84 Radio or data terminal signal that a unit has arrived at the assignment. Arrival times are recorded for reports and statistical analysis.

10-92 A malicious false alarm, one of the more frequently transmitted signals on the radio.

4.5 mask A self-contained breathing apparatus that contains air compressed to 4,500 psi.

"A-wing," "B-wing," etc. A designation by the FDNY to allow firefighters to have a common frame of reference to their location within a large, multi-wing building. When looking at the building from the street that the command post is located on, the far left wing is designated the "A-wing," the next to the right would be the "B-wing," and so on.

ADV Abandoned Derelict Vehicle, an official designation that an automobile or more often the hulk left after car thieves and strippers have stolen everything of value from tires to engines and transmission, is to be impounded and removed from city streets by the Department of Sanitation.

AFFF   Aqueous film-forming foam: a type of firefighting foam initially designed specifically for aircraft crash firefighting.

all hands A fire or emergency that results in all companies assigned on the first alarm being put to work. An All Hands usually precedes the transmission of a second alarm.

all hands chief   The second to arrive battalion chief at a working fire. Usually, this chief assumes responsibility for interior operations, while the first due chief remains outside in command until the arrival of the deputy chief.

BARS   Box Automatic Readout System, part of the department's Computer Aided Dispatch System (CADS) which automatically records and dispatches units to older style pull-type fire alarm boxes on street corners. These have been replaced by ERS Boxes which allow two-way voice communications with people reporting an emergency.

BLEVE   Acronym for Boiling Liquid, Expanding Vapor Explosion. A BLEVE is an explosion caused when a container that holds a liquid that is above its boiling point fails. It instantly allows the liquid to expand several hundred times in volume, releasing the contents, and sending the container rocketing. If the liquid is combustible, it ignites in a huge fireball.

booster tank A tank on the pumper that carries 500 gallons of water, which may be used to fight small fires such as rubbish or some car fires without connecting the pump to a hydrant or other water source. In the past it was most often used in conjunction with a one-inch diameter hard rubber hoseline stored on a reel, called a booster line. The hose and reel allowed the hose to be charged with water without pulling out all 200 feet of hose.

borough call   A multiple-alarm fire that is so large that additional units must be requested from another borough (county). Typically, it would be at least the equivalent of a six-alarm fire.

bucket or basket   A 3 ft × 5 ft platform attached to the end of the boom on a tower ladder that allows firefighters to stand and operate tools. It is equipped with a large nozzle that can deliver upwards of 1200 gallons of water per minute at heights of up to 95 feet above grade.

bulkhead   A structure that encloses a staircase, elevator, or dumbwaiter that extends above a building's roof. It has a door that opens onto the roof, and is often equipped with skylights on the top, allowing light down into the interior.

Chauffeur   The firefighter assigned to drive and operate the apparatus. They are usually among the senior members of the company, and unofficially act as "non-commissioned officers" within the unit. The Engine Company Chauffeur operates the pumps to supply water to hoselines, and is known as the ECC. Ladder Company Chauffeurs (LCCs) position and raise the aerial ladder and are responsible for search of the front apartments of a tenement building in addition to driving duties.

class 3 alarm   An automatic fire alarm or sprinkler water flow alarm. The name dates back to the era of the telegraph alarm system, when the alarm box number would be proceeded by a number 3, indicating an automatic alarm, and be followed by a number that designated a specific building or address, for example: telegraph signal 3-751-11 indicated a class 3 alarm has been received for a building designated terminal 11 at box 751. The companies had a record of the address for each terminal.

cockloft A void space between the top floor ceiling and the underside of the roof on wooden roofed buildings that provides insulation from sunlight beating down on the roof and ventilation for the wooden structural elements.

drop ladder A vertical steel ladder that connects the first level of a fire escape with the ground. It is held in a raised position above the reach of people at the street (to prevent thieves and others from climbing the fire escape to break in) by a hook that swings out of the way when the ladder is lifted slightly, then drops down like a guillotine to the extended position, allowing people on the fire escape balcony (platform) to reach the ground level. Firefighters can use a pike pole or other hook to lift the drop ladder and let it fall, in order to climb the fire escape. Drop ladders are often one of the first things stolen on vacant buildings.

eductor Eductors are appliances placed in the middle of a hoseline that allow the flowing water in the hoseline to "suck" another fluid into the hosestream by means of the venturi principle. Most often it is used to mix foam concentrate into the water stream allowing firefighters to float a layer of foam across the surface of a blazing flammable liquid like gasoline which cannot be extinguished with plain water.

ERS and ERS No Contact (NC) The Emergency Reporting System is an electronic communication system that replaced the old-style fire alarm boxes on street corners. ERS boxes allow the caller to speak to a dispatcher at fire alarm central offices (now 9-1-1 centers). They are activated by pushing a button and waiting for the dispatcher to acknowledge the call and question the caller. ERS provided greater flexibility to the fire department compared to the older manual pull boxes they replaced, since the dispatcher could question the caller about the type of incident needing assistance, the address of the incident and other important information, then select an appropriate response. The older pull boxes only sent a signal that there was an emergency at a general vicinity. Not knowing exactly what the problem was, the department sent two engines, two ladders and a battalion chief to every pull box activation in case there was a structure fire. In many cases, such as a rubbish or car fire, or auto accident, only one engine and one ladder might be required. The ERS box allowed only those needed units to be sent. If a person pushes the button and does not speak to the dispatcher it is known as an ERS NC. ERS NC results in only a single engine company response, just in case there is a deaf-mute person at the box who is unable to speak and there actually is some type of emergency to be checked out. ERS NC is nearly always a false alarm, often as schools are letting out, and kids walk past the box and push the button.

ETA (Estimated Time of Arrival) Usually requested of an agency or a unit in order to plan further actions.

exposure 1 The building or area directly in front of the fire building's main entrance; usually it is a street.

exposure 2 The building or area immediately to the left of the fire building; subsequent attached buildings in this direction would be indicated with a letter, e.g., exposure 2A is the second building to the left, 2B is three buildings to the left.

exposure 3 The building or area directly behind the fire building, often a rear yard.

exposure 4 The building or area immediately to the right of the fire building; as with exposure 2, subsequent attached buildings in this direction would be indicated with a letter, e.g., exposure 4A is the second building to the right, 4B is three buildings to the right.

Handie-Talkie   A lightweight, multi-channel, two-way radio worn by firefighters to provide on scene communications between personnel at an incident. They generally do not operate on the channel that is used for dispatching and apparatus communications and have a shorter range than apparatus radios.

Housewatch   The name given to the member who is responsible for monitoring the receipt and recording of alarms received, maintaining the security of the firehouse, and recording in the company journal the names and identity of all visitors to the firehouse. The duty is shared by all the firefighters on duty in three-hour rotations. It is conducted at a desk near the main entrance, dubbed the housewatch desk. Also used to refer to the small enclosure the desk is normally located in.

a job   A working fire, one that requires firefighters to stretch and operate a hoseline; as in, "looks like we have a good job."

the job   The FDNY, as in "the job sent down a directive"

"K"   A term used in radio transmissions to signify that the unit calling another is finished transmitting part of it's message, and is standing by for a reply, as in "Rescue 1 to Manhattan, K." The Manhattan dispatcher, hearing this will reply with "Manhattan to Rescue 1, go ahead with your message."

master stream   A large capacity nozzle, capable of delivering over 500 gallons of water per minute (GPM). Many flow over 1,000 GPM, with a reach that can exceed 100 feet. These are usually permanently mounted on a fire apparatus, which absorbs the huge forces, called nozzle reaction, that flowing such large quantities of water produces.

Mobile Data Terminal (MDT)   A computer screen and printer that sends and receives messages to and from the dispatcher

mope   A derelict, often drug addicted. AKA junkie, or skell.

mutts   A criminal or bad guy

off apartment   An apartment within the fire building that is located across the public hall from the fire apartment. This apartment is often used as a place of refuge if fire suddenly intensifies, causing firefighters to retreat. The designation may extend to all the other apartments in a vertical line above and below it, since it provides a common reference when you are unable to see the actual apartment number. The same is true in regard to the apartments above and below the fire apartment, which may be called the fire line apartments.

OV   Outside vent: a firefighter assigned to reach the area opposite the direction that the hoseline and interior firefighters are entering from, in order to vent that side, enter it, and search it for trapped occupants. A dangerous position that requires a skilled, experienced firefighter.

RAC unit   Recuperation and Care Unit: one of several ambulance-size vehicles staffed by light-duty firefighters that respond to multiple-alarm fires to provide cold bottles of water and cold wet towels in warm weather, and warm drinks and towels in cold weather, in order to minimize temperature-related injuries such as hyperthermia and frostbite.

Rabbit Tool   A small, powerful hydraulically operated ram that is used for forcible entry. It allows one firefighter to apply as much as 8,000 pounds of force to a door.

resuscitator   A mechanical device that was used to force pure oxygen into the lungs of injured victims. It has been replaced by the bag valve mask (BVM), which does a more efficient job at the same task.

skell   A derelict, often drug addicted. Also: junkie, or mope.

Special Operations Command (SOC)   SOC is the administrative unit that is responsible for all five rescue companies, eight squad companies, Hazmat Co. 1 and five hazmat technician companies (engine companies with additional training, tools and a second hazmat apparatus), the three large fireboats and several smaller fire/rescue boats, as well as numerous other specialized apparatus.

Stang   A large master stream nozzle, often fixed atop a pumper or on the basket of a tower ladder, which can flow upwards of 1,000 gallons per minute (GPM). Several very large versions on Satellite apparatus of the Maxi-water system can flow more than 5,000 GPM.

Stokes basket   A wire or hard plastic stretcher that holds and supports a human body so it can be transported over debris or raised or lowered.

taxpayer   A multi-occupancy commercial building, usually one-story, with all the stores under one common roof with little or no fire-stopping or other fire protection. The name derives from the 19th-century practice of erecting a quickly and cheaply built structure on a lot in order to pay the real-estate taxes until the land increases in value. New versions are called strip malls.

tiller   A tractor trailer aerial ladder with rear wheels that can be steered from the tiller seat at the very rear of the apparatus. Tillers are very maneuverable in tight streets but require a skilled team of driver and tiller operator.

torch   An arsonist: someone who intentionally sets a fire.

tour   A firefighter's work shift: either a nine-hour day tour from 9:00 a.m. to 6:00 p.m., or a 15-hour night tour from 6:00 p.m. to 9:00 a.m. the next day.

trussloft The space located between the structural elements of a truss. It can be from several inches high to tall enough for a person to stand in.

Venturi   A constricted opening through which a fluid is injected. In the fire service it is most often associated with foam eductors, where the Venturi effect is used to draw foam concentrate out of its container and mix it in with the water in a hose stream. It is also useful for ventilation where a hose stream operates out of a window or door and the flowing water sucks smoke along with the water spray to clear the area. The same process can worsen fire conditions if a hose stream is driven through a narrow open-ing into a fire-filled area, causing fresh air to be pulled into that space, intensifying the fire.

Vibralert The vibralert is the low air warning system on our SCBA that vibrates the mask facepiece and makes an audible sound to tell the wearer that they have about four minutes of air remaining.

Index

Symbols

9/11. *See* September 11, 2001

A

abandonment 145
ABC (acting battalion chief) 161
ABC status (airway, breathing, and
 circulation) 303
accelerants 170
access 72, 127
 stairs 6
accidents 70
 automobile 7, 94, 146–147
 scene 147
acting battalion chief (ABC) 161
actions
 corrective 102
 remedial 102
acts, meritorious 117
Addonna, Bert 367
ADV (derelict vehicle) 247
adz end 273
aerial ladders 128, 148–149, 152, 174
aerosol 245
AFFF (aqueous film forming
 foam) 275–276
AFID (Apparatus Field Inspection Duty).
 See Apparatus Field Inspection Duty
 (AFID)
aggressiveness 118
aides 202
aid, mutual 268, 446
AIDS 94, 100, 156
air 67
 alarms 111
 bags 73, 82, 90, 220, 225
 bottles 3, 23
 cart 220, 225
 cylinders 110, 113, 128
 horn 131, 172
 hose 47
 in hoseline 243

 monitoring 310
 pressure 224
 shaft 67
 supply 110–111
aircraft carrier 308, 338
airplanes 129, 267. *See also* jets
 cabin 139
 construction 139
 crashes 120, 133, 268, 270
 crash rescue 267
 door 139
 fire rescue 133
 fuselage 133, 139
 impact 268
 interior 277
 landing gear 273
 mechanics 278
 perimeter 134
 runway 133
 tail 139
 wings 134, 138
 wreckage 134, 139
airports 132. *See also* John F. Kennedy
 International Airport
 runway 133
alarms 5, 146
 air 111
 all hands 75
alarms (*continued*)
 assignment 4
 automatic fire 109
 bells 77
 boxes 27, 41, 160
 Class 3 109, 154
 false 7, 47, 66, 95, 126
 fill out 171
 first 75
 fourth 81, 163
 general 18
 message 77
 multiple 27
 second 56, 61, 83, 148

telegraph system 27
third 129, 148
units 154
verbal 42, 155
voice 96, 166
water motor 166
Albany Medical Center 446–448
alerts, fog horn 19
Alitalia crash 268
all hands 66, 75, 81, 147
chief 66, 190
aluminum 139
ambient noise 87
ambulances 48, 85, 202
service 45
size 440
American Airlines flight 587 289
American LaFrance 172, 180, 242
America's Heroes xi
Amtrak 80
Angelini, Joseph 2, 295, 303–306,
336–342, 411
animals 326
encounters with 328
anthrax 436
apartments
conditions 100
doors 149
fires 58, 150
H-type 180
searches 167
apparatus 1, 27, 33, 65, 130
booster tank 79
breathing 9
cab 165
doors 41, 94, 153
fire 41
floor 297
inventory 363
layout 32
lights 191
loss 435
preparation 82
siren 446
Apparatus Field Inspection Duty
(AFID) 45
apparitions 80
appliances 212
aqueous film forming foam
(AFFF) 275–276
architects 6
areas
cellar 85
lean-to 83

prisoner 220
of refuge 149, 249–250
sheltered 84
staging 132, 447
top safety 86
of work 87
Argo ATV 446
arrests 170
arson 18, 35, 58, 145
arsonists 59
epidemic 35
assessments, patient 445, 447
assignments 56, 66, 70, 95
personnel 76
of positions 180
sought-after 53
of tools 180
of units 179
variety 102, 120
wish list 118
Athanas, Bob 300, 302
Atlantic Avenue 141
Atlantic Beach 18
attack crew 29
attack, interior 190
attic fires 127
attitudes 207
Atwell, John 104
auditions 65
automobile accidents. *See* cars, accidents
aviation gasoline (AvGas) 276
awards 116
Class I 51
Class II 51
Class III 51
A-wing 57
axes 190, 205, 230
flathead 250

B
Baal, Joe 22–23, 26, 32, 270–271
backboard 447
backdraft 195
back stretch 8
backup 24, 54
man 9, 14
Bag-Valve mask 202
Baker, Tom 336
Baldwin, Paul 321, 323
Ballenger, Dave 30
balusters 175
Balzano, Frank 25
Banker, Mike 410

Barbagallo, John 146, 220, 222, 225, 241–246
Barbara, Gerry 422, 444
Baretto, Maria 413
barred windows 197. *See also* windows
barriers 190–191
    fire resistive 100
bars, drop-in 155
baskets 111, 127, 139, 163, 242
    ladder 86
    stokes 85, 305
Battalion
    3 66
    7 300
    13 65, 75, 81, 119
    14 78
    19 70–71
    26 83
    28 161, 163
    31 172–173
    32 165
    33 147
    35 98, 109
    38 194
    40 166, 228
    42 126
    44 35
    57 147, 152, 244
    58 48, 202
    Special Operations 298
battle dress uniform (BDU) 369
BC (buoyancy compensator) 279
BDU (battle dress uniform) 369
Bedford-Stuyvesant 119
Bellevue Hospital 224
bells 27, 77
Bengston, Russ 55
Bensonhurst 126
Bergen Street 120, 125, 155, 172
BI (building inspection). *See* building inspection (BI)
Big Blue 54, 76, 81
the Big House 75, 81
black water rescue diving 131. *See also* diving
Blackwell, Chris 138–139, 300–302
blades 57
    carborundum 87
    change 87
    metal cutting 87
    steel cutting 82
blankets 445
bleeding 85
blips 287

blockbuster 47
blocks, cement 97
Bloomberg, Michael 437
Board of Merit 51
boats 134, 138
    fiberglass 284
    inflatable 130
bodegas 128
bodies, removal of 90
boiler rooms 36
boiling liquid expanding vapor explosion (BLEVE) 197, 244–246, 271
Bomba, Mark 448
bombers 5
bombings
    dirty 347, 348
    Oklahoma City 351
    World Trade Center 351
Bondy, Pete 121–124, 152–154, 220–221, 222–223, 244–246, 265
booby-traps 95, 98
boom, weight of 243
booster lines 32, 232
booster tank 13–14, 79
boots 100, 242
    steel sole 100
Borfitz, Butch 23, 367
Borst, Joe 123
bottles 130
boxes
    crash 272
    malfunction of 66
    pull 209
    transmission 76
brake shoes 71
break-ins 155
    car 156
breathing apparatus 9. *See also* SCBA (self-contained breathing apparatus)
Brett, Walter 355–356, 361, 403, 405, 418
bricks
    farms 441
    wall 86
Broadway Rescue 293. *See also* Rescue Company 1
Bronx 53, 58, 66, 69, 76–78
    Communications Office 70
    dispatcher 78
Bronx (*continued*)
    jobs 82
    radio 76
Brookdale Hospital 51, 203
Brooklyn 66, 75–76, 119, 140, 152
    Communications Office (CO) 148

District Attorney's Office 170
fire radio 160
House of Detention 219
Navy Yard 307
Union Gas Company 102
Brooklyn-Queens Expressway 93, 131
Brown, Patty 292
Brown, Steve 311–312
brownstones 242, 246
fire in 120–121
Brownsville (Brooklyn) 145
bruisers 303
Bryant, Joe 76–77
Bucca, Ronny 4
bucket 11, 112, 121, 174
seat 19
buddy-breathing 309
budgeting 17, 439
building inspection (BI) 45, 102, 105
buildings
collapses 83, 120
construction 54
cost 179
entrance 180–181
factory 93
fire 102, 142, 166
frame 15, 213
gas supply 84
H-type 55, 149
layout 245
line 9
sealed-up 99
steel-plated 190
vacant 7–9, 14, 94, 101, 365
bulkheads 11
bullet holes 84
Bullock, Jim 285, 318, 321
bumpers, car 70
bunker gear 3
buoyancy compensator (BC) 279
Bureau of Fire Investigation 38
burglary deterrents 83
burners 35
burns 54, 195, 269. *See also* injuries
inhalation 264
kit 202–203
second degree 186
unit 203
Burns, Donald 86–91, 275, 362, 422, 444
Burns, Philip 85
Bush, George W. 440
Bushwick 208, 211
Bushwick Avenue 161
Byrne, Susan 76

C
cables 73, 139
Callan, Joe 423
CAMs (chemical agent monitors) 336
Canal Street (NYC) 81
Canarsie 201
Pier 48, 54
candidate
criteria 53
skills 54
can man 106
cans 182
Cape Lambert (motor vessel) 308
capsules, thermal recovery 130
captains
promotion 291
rank 360
responsibilities 359
test 359
carbon monoxide (CO) 9, 82, 110
indicator 84
carborundum blades 87
cardiovascular illnesses 3
cargo jets 274
Caribbean Day Parade 189
Carlson, Howie 289–290
Carney, Jimmy 365
cars. *See also* fires, cars
accidents 7, 70, 94, 146–147
break-ins 156
bumpers 70
thieves 94
train 79
wreckage 72
Casani, Steve 45–52, 53, 69, 76, 293
Cassano, Sal 362–363, 439
cats 325
catwalks 338
cautions 92
cave-ins 120
ceilings 87
bays 100–101
crackling 142
lath-and-plaster 88
plaster 95
pulling 100, 142, 177
stripping 101
tin 89
cellar
area 85
fire 82, 98, 127
grate 121
opening 85
Cerato, John 130

2 84, 126, 141, 142, 173
3 126, 173
4 83, 108, 126, 141, 173
4A 211–213
    to cold water 130
    fire 163–164, 164
    suit 136
extension 12
    cord 220
    of fire 248
    ladder 234
exterminators 326
extinguishment 164
    extinguishers 106, 108, 316
    extinguishing system 341
extrication
    gear 335
    of vehicles 247

F

facades 294
facepiece 196
factories 102
    buildings 93
Fahey, Brian 430
false alarms. *See* alarms, false
Fanning, Jack 349
Fanning, Richie 190–191
Farnsworth 194, 196, 198
Father McGrath 409
fatigue 92
FDIC (Fire Department Instructors
    Conference) 364
Feehan, Bill 120, 394
Fellini, Frank 386, 390–391, 401, 423
Ferran, Mark 396
Ferrante, Pete 69–72
field communications unit 2
fieldstones 85
Fillipelli, Steve 12, 214
fines 102
fins 136
fireballs 47, 70, 251, 269, 289
fireboats 131, 280
firebombing 58
fire companies 28, 53, 58, 75–76, 179
Fire Department Instructors Conference
    (FDIC) 364
fire departments
    cuts 45
    doctor 51
    Edinburg Volunteer Fire
        Department 446

Inwood Fire Department 18, 21–22,
    267, 446
Jersey City Fire Department 298
Lawrence-Cedarhurst Fire
    Department 19, 270
Northville Fire Department 446
Stillwater Fire Department 28–31, 32
Suffolk County Volunteer Fire
    Department 370
fire escapes 9, 55, 62, 67, 107–112
    location 248
    steps 112
    window 55
firefighters 1
    age 54
    career 17, 26
    efforts 89
    light-duty 155
    mission ix
    promotion 115
    salary 94
    tasks 102
    volunteers 17
fireground 168, 254
firehouse 7, 19, 77, 153, 179
    block 38
    break-ins 155
    inspection 81
    robbery 154
    single 179
    watchman 77
fire inspector 104
Fire Laws 104
firelight 198
fire marshals 65, 155, 166, 170
fire officers 68, 105
Fire Officer's Handbook of Tactics ix
fire protection 28
    classes 28
    engineer 54
    systems 159, 170
    technology 28
Fire Protection Department 30
fire-rated material 149
fire room 97
fires 6, 71. *See also* The Winter Offensive
    10-alarm 82
    activities 28, 37
    apartment 58, 150
    apparatus 41
    attic 127
    blowing 121
    box 36
    Brownstone 120

brush 26
building 102, 142, 166
car 47, 48, 66, 71
cellar 82, 98
classes 28
cockloft 176, 177, 182
companies 75, 76, 179
conditions 141, 177
duty 119, 159, 359
escape 55
experience 361
exposure 163, 164
extension 79, 217, 248
extent 78
extinguishment 84
fireproofing 6
floor 112
fires (*continued*)
garage 228
grease duct 82
hazards 102
hidden 101
indication 148
intensity 251
interior cabin 273
isolation 149
keep-warm 159
knockdown 14, 30, 128, 143, 192
line 248
main body 79
mattress 47
minor 7
movement 99, 150
no impact 165
oil burner 35
overhead 142, 198
phenomenon 159
prevention 102
problem 58
radio 160
real 7
record 193
reports 37, 170
resistive barrier 100
rubbish 26, 47, 66
safety 58
safety improvements 58
ship 120
spread 30
storm 47
traffic 76
train and 78, 82
traps 27
traveling 100

units 83
fire service
career 26
history 104
firestorm 38
fireworks 46
first aid hose 164
first-alarm units 82
first contact 89
first due 24, 65, 78
first due engine 78, 167, 210
First Line Supervisors Training Program
(FLSTP) 116
Fitzgerald, Walt 447
Fitzpatrick, Larry 67
fixtures
electrical 95
light 100
flames, venting 79
flashlights 72, 98, 136
Sunlance 264
flashover 110, 195, 198, 200
flashpoint 276
Fleet Week 338
floodlight trucks 19
floor
apparatus 297
fire 110
holes 98
joists 214
floorboards 245
flooring 86
layers 88
Floor Warden Stations 317
flotation equipment 136
flotation suits 130
cold water 132
FLSTP (First Line Supervisors Training
Program) 116
Flushing Bay 133
foam 267, 272
blanket 71
concentrate 36
coordinators 361
educator pickup tube 36
eductor 36
handline 36
fog horn alerts 19
fog nozzles 29, 273
Foley, Butch 281
forcible entry 60, 93, 161, 249
tools 59
Ford, Harry 430
forklift 278

Forsyth, Warren 300, 302, 315, 319
foundation walls 85. *See also* walls
fourth alarm 163
Fox, John 250–251, 375
Fox, Patty 122–124
Fox, Tom 336
Foy, Brian 306, 356
fractures, compound 85. *See also* injuries
frame building 15, 213
frame house 12
frames 307
    vacant 15
    wood 27
Frank, Robby 346–347
Fredericksen, Neil 278–279, 279
freight trains 80. *See also* trains
Frisbee, Jerry 67
Fuchs, Warren 229, 231
fuel 140
    behavior 46
    load 24
    oil 46
    tank 70
Fulton County Ambulance Service 447
fumes, exhaust 86
funding 2
fuselage 133, 139, 272

G
gadgets 336
Galdi, Alfred 83, 85
Galione, Bobby 191, 194–195, 198, 248–253, 259–266
Ganci, Peter 4, 363, 394–395, 422, 429–430
Gander, George 165–166
gangs 58
garages
    fire in 228
    one-car 228
gas 71, 84, 95, 210
    company 84
    detection paper 336
    leak 155
    meter 84
    natural 84
    odor 84
    supply 84
gasoline 23, 25, 46–47, 70, 174
    canteen 57
    powered chain saws 82, 87
    powered tools 86
    truck 26

gates
    roll-up 84
    steel security 83
Gausintu, Milde 198–200
gear 140
    landing 273
    rescue 130
Geidel, Gary 295, 411
general alarm 18. *See also* alarms
generators 82, 275, 363
George Washington Bridge 65
Geraghty, Ed 4, 244
Giamo, Sam 108
Gies, Ronald 429, 431
Gill, Dwayne 122
Giordano, Nick 76, 84–87, 90–92, 300, 302
Giovina, Serge 250–251
Giuliani, Rudolph 309
glass 36, 61
    bulletproof 257
gloves 197, 269
Gorup, Frank 31–32
graffiti 65
Grand Concourse 145
grants 440
grass highway 370
grates 297
Grawin, Ray 63, 375
Gray, Larry 147–148, 150, 154, 167–170
grease duct fire 82
Greene, Robert 84–85, 88–91
griphoists 170, 297–299, 301, 327
Grosso, Joe 314–317, 320, 327–328, 361
guide man 89
Gumby Suit 133
guns, Lyle 330, 332–334

H
Haddock, Mike 400
Halligan 175
    hooks 54, 56, 99, 295
    mini 156
    tools 22, 54–56, 150–156, 175, 250
handie-talkies 2
hand lights 22, 191
handlines 14, 59, 142, 167, 210
    foam 36
hand tools 151
Hannon, Dick 54–57
harbor launch 48
hardware, panic 107
Haring, Bobby 279

Haring, Tom 375, 377–380, 383
Harlem (NYC) 53, 58, 65, 76, 365
   housing stock 58
   tenements 58
harnesses 252, 261
Harris, Josephine 397
Hashagen, Paul 137–138, 278–280, 286,
   336, 350–351
Hatton, Terry 2, 237–239, 279, 308–309,
   376
hauling system 296
Hay, Al 391
Hayde, Donald 119, 382, 435
Hayden, Peter 414, 423
Hazardous Materials Company 83
hazards 66, 102
   target 343
hazmat 84, 138, 436
   Company No. 1 102
   instructors 438
   teams 436
   technician 438
   units 54
H design 149
Healy, Tom 300
heart attacks 3, 5. *See also* injuries
heat 67, 79, 210
   condition 197
   level 162
   mushrooming 249
   pattern 249
   radiant 79, 164
   tendencies 249
heaters
   electric 159
   kerosene 37, 159
   space 37
heating
   creative heating 37
   plants 159
   sources 159
heating, ventilating, and air conditioning
   (HVAC) system 296
heavy axes 59
heavy rescues 75
helicopters 293, 338, 446
helmet earflaps 177
Hendries, John 76
Henry, Billy 316–317, 323, 342, 349, 411
hepatitis 94, 100. *See also* injuries
heroes xi
   loss of xi
heroin 93, 94, 95, 156. *See also* drugs
   junkies 93

   overdose 94
Herold, Hugo 76, 79
Hewitson, Billy 190–199, 212, 214, 228,
   230–232
hidden fire 101
Higgins, Tim 110, 167, 220–221, 259–
   261, 329–334
high-rises 2
   collapse 5
highways 147
   divider 71
   Interstate 878 18
   Major Deegan Expressway 70
   Ocean Parkway 147–148
Hitter, Thomas 42
hoists 295, 297
holes 25, 97, 175, 212, 244
   bullet 84
   in floor 98
   inspection of 56, 57, 212
   mouth 89
   vent 57
Holland Tunnel 298
homicide rate 38
hoods
   Nomex 177, 194, 197
   ventilation 341
hooks 65, 176, 183, 230, 236
   Halligan 54, 295
   hook and ladder truck 19
horses 326, 327, 328
Hose Company 24
hose, first aid 164
hoselines 56, 84, 167–168, 174, 180
   advance 127
   air 243
   charging 243
   coupling 97
   foam 23
   small 32
   stretch 243
   uncharged 150
   withdrawal 163
hoses 97, 141, 148
   air 47
   bed 8, 25, 30, 271
   carrying 79
   folds 8
   hand-stretch 165
   hydraulic 59
   lengths 30
   streams 24, 57, 99, 101, 163
   stretching 8, 36
   suction 20

hosestream 167
hospitals 271
   Bellevue Hospital 224
   Brookdale Hospital 51, 203
   Kings County Hospital 203
   Peninsula General Hospital 271
   St. Mary's Hospital 154
Hostage Negotiating Unit 344, 347
house line 164
housewatch 46, 76
   booth 76
housing projects 37
Howard, Bruce 127–139, 141–142,
   220–224, 248–251, 253
H-type 56, 58
   apartment 180
   buildings 55, 149
Hudson River 298
hurricanes 441
Hurst
   cutters 72
   tools 71, 139, 147, 247
HVAC (heating, ventilating, and air
   conditioning 296
hydrants 8, 13, 180
   bonnet 8
   lack of 78
   supply line 33
hydraulic
   pads 243
   pump 156
   rescue tools 82, 139
   spreaders 147
hypothermia 447
   cold water exposure 130

I

IAFF (International Association of Fire
   Fighters) 438
I-beams 58, 86–87, 88
Ielpi, Lee 127–142, 171–177, 204–206,
   233–239
IG (Inspector General) 104
ignition 30, 84
   source 84, 258
image, public 118
immigrants 103, 145
impact 6, 133, 268
   site 2
incentives 437
incidents
   command 90
   life-threatening 53
   unusual 76, 120
income, disposable 93
indicators
   of carbon monoxide 84
   combustible gas 84
   of fire 148
   of oxygen 84
Infanzon, Lloyd 303–305, 305–306, 323,
   336–337
inflatable
   boat 130
   raft 138
   suit 137
Ingram, Bob 437
inhalation burns 264. *See also* injuries
initial
   contact 91
   report 130
   response 171
injuries 85, 88, 140. *See also* burns
   cardiovascular illnesses 3
   compound fractures 85
   heart attacks 3, 5
   hepatitis 94, 100
   inhalation burns 264
   report 128
inlet 29
inquiries, status 79
inside team 100, 180
inspections 81–82, 106
   activity log 103
   cycle 105
   of firehouse 81
   holes 56–57
   routine 109
   visit 111
Inspector General (IG) 104–105
inspectors 108
   fire 104
instructor 183
   Hazardous Materials 54
   Hurst Tool 54
insulation 95
intelligence agencies 343
intensity of fire 251
interior attack 190
interior team 210
International Association of Fire Fighters
   (IAFF) 438
intersections 147
Interstate Highway 878 18
Intervale Avenue 76
interview 116
investigator 105

Inwood (Long Island) 267
  Inwood Fire Department 18, 21–22,
    267, 446
irons 65, 104, 180, 205
  man 59, 62
isolation of fire 149

J

jackhammers 230, 299
jacks 243
  trench 86, 90
jaws 59, 60, 63, 139
  hydraulic 59
Jersey City (NJ) 298
  Jersey City Fire Department 298
jets 130–131. *See also* airplanes
  cargo 274
  fuel 6, 140, 276
jobs, man-in-machine 81
job spirit 444
Johannesen, Thor 323
John F. Kennedy International Airport 129,
    267
  runaway 268
John Paul II 335
joists 58, 236–238
  floor 6, 214
  overload 237
  of roof 87
  second-floor 83
  support 83
Jonas, Jay 396, 398
Jove, Joe 115
Judge, Michael 366, 429
jumpers 5, 121, 172–173
jump seat 8
junkies 37, 95, 98, 101, 156
  heroin 298

K

Kane, Alan 284–285
Kasper, Charlie 2, 298–299
keep-warm fires 159
Kelly, Kerry 140
Kelly, Tim 303
kerosene heaters 37, 159
Kiernan, John 174–177, 229–231,
    248–253, 273–278, 330–334
Kings County Hospital 203
Kingston 189
Kingston Avenue 189, 193, 201
Kleehaus, Jack 120
Kobes, Wicker 110

Kopetz, Fred 7
Kroth, Kevin 311–312, 319
Krowl, Teddy 122–123

L

ladder companies 29, 41, 87, 163, 208
  operations 364
Ladder No.
  13 142
  14 59, 60–64, 61, 63, 64, 76
  16 61
  17 79
  19 83
  21 82
  23 59
  28 59, 62, 65, 67
  30 59, 61, 62
  31 96
  33 56, 71
  40 61, 62
  45 65, 75, 81, 82
  46 71, 74
  55 79, 83
  59 56
  102 65, 162, 242
  103 35, 46, 96, 118, 191
  103-2 41
  104 94, 96–97, 99–101, 105–106,
    110–111
  105 174, 176
  108 99–101, 106, 162–163
  109 232
  110 174
  111 119–120, 141–142, 180–181,
    242–244
  112 210
  113 194
  114 167
  117 133
  119 111, 112
  120 12, 35, 190, 191, 192
  122 165
  123 124, 148, 180, 194
  124 161
  132 124, 142, 152, 244
  149 228
  170 48, 49, 201, 202
  176 121, 124, 191
  290 118
  330 128
ladders 83, 134. *See also* aerial ladders; *See
    also* tower ladders
  assignments 70

basket 86
drop ladder 9
extension 234
pinnacle 56
portable 194
rope 141
straight 194
LaFemina, John 362–363
LaGuardia Airport 130, 131
Laird, Bobby 366
Lake, Billy 174–178, 180–181, 186, 259–262, 265–266
Lakiotes, Artie 228, 232
Landa, Emilio 85, 90, 91
landing gear 273. *See also* airplanes
landing zone 339
landmarks 343, 349
land value 326
LaRocca, Bobby 121–123, 211–212, 214–218, 259–260, 264
lath-and-plaster 88
ceiling 88
Lawrence-Cedarhurst Fire Department 19, 270
Leach, Jay 356
leadership 437
leaks 137
lean-to 83. *See also* void, types
leave, medical 140, 141
lessons 6
levers 297
lieutenant, tasks of 117
life-saving
efforts 325
rope 54, 55, 67, 133
lifting 73
subway cars 73
train cars 73
lights 99, 196
emergency 126
fixtures 100
hand 191
out 258
Lim, David 396–397
lines
booster 232
charging 150
fire 248
preconnected 32
liquefied natural gas (LNG) 102
load-bearing wall 6
loading operation 204
load replacement 90
load up the box 70

Local Law 11 of 1990 294, 305
locomotives. *See* trains
Loftus, Mike 84, 88, 89
logjam 174
Long Beach (NY) 18
Long Island (NY) 145
lookouts 288
loss of heroes xi
lots, vacant 9
loudspeaker 77
Lucas, Precious 265
lumber, tongue-and-groove 88
Luna, Norberto 85–91
Lund, Pete 59–65, 120
Lyle gun 330, 332–334

M
Mack 83
CF Pumper 7, 131
R model 66, 77
maintenance
cost 179
duties 118
Major Deegan Expressway 70–71
Mako Alley 279
maneuvers, reduced profile 274
Manhattan 76, 93, 103
Central Office 77
man in machine 81, 219
man-made openings 100
manpower 230
marble 145
Marchbanks, Joe 400–401
Marchese, Sal 382
Mari, Dominic 24–25
Marino, Kenny 411
Marriott Hotel 4
Martin, Peter x–xi, 229–231
mascots 207
masks 3, 8, 24, 444
AGA Mark II 131
Bag-Valve 202
cylinder 97, 168
facepiece 11, 98, 110, 142, 169
microphones 131
speakers 131
master stream 14, 32, 127, 176
material, fire-related 149
Maxi Water System 163
mayday 110–111, 251
McBride, Neil 83
McCross, Vinnie 270–271
McDonald, Mickey 246

McFadden, Paul 83–87, 91–92, 120, 125
McGlinn, Jim 396
McGrath, Father 409
McIntosh, Celestine 123
McNeala, Seamus 374
McTigue, Marty 69, 120
McTigue, Tommy 76
MD (multiple dwelling) 180, 247
medals 51
    Medal Board 52
    Medal Day ceremony 51, 52, 56
    for valor 91
media, interaction with 293
medical evaluation 88
medical leave 51, 140–141, 224
medical officers 140
medical technician 203
Meisenheimer, Ray 300–302, 431
members, rotation of 92
mental condition 92
meritorious acts 117
Merrick Fire Department 352
meter man 302–303
Metro-North 78. *See also* trains
    power control office 80
    third rail 80
Metz, Neal 346
midnight movers 38
Midtown Manhattan 86
militants 42
Milner, Mike 84–85, 139
Milnyczuk, Kayla 447
mobility 237
Mojica, Dennis 2, 161–170, 327, 376,
    412–413
Molle, Hank 278, 315, 316–319, 321
monitoring stations 348
Montagna, Frank 202–203
Montesi, Mike 411
mooring bollard 134
Moran, John 2
Moriches Inlet 288
Morris Avenue 77
Morris, Bob 78, 364, 436–437
Morstadt, Joe 76
motorcade 336
motor replacement (boiler room) 36
movement of fire 150
Mulroy, John 66, 69, 76, 78
multiple dwelling (MD) 180, 247
Murphy, Billy 67
Murphy, Denis 121–124, 127–134,
    137–144, 427–428
Murphy's Law 57, 135

Murtha, Jerry 64, 138–139, 300–301
mushrooming, heat 249
mutual aid 268, 446
Myslinski, Ed 305

N
Narbutt, Tom 244
NASA 9
Nassau County (Long Island)
    Fire Academy 165, 270
    Police 25, 370
National Fire Academy (NFA) 116
National Transportation Safety Board
    (NTSB) 140–141
natural gas 84
    odor of 84
Naudet, Jules 402
NBC defense (nuclear, biological, and
    chemical attacks) 2
needles 94, 98–100
neglect 145
Nevins, Gerry 438
Nevins, Jerry 306, 411
Nevins, Steve 438–439
New York City 18
    Fire Department (FDNY) xi, 7, 9, 26,
      75
    Police Harbor Unit 282
    Transit Museum 225
New York Fire Department (FDNY) xi, 7,
    9, 26, 75
New York Harbor 279, 281
New York State Labor Law 103
New Yorker Multiversal Nozzle 210
NFA (National Fire Academy) 116
Nigro, Dan 363, 422–423
no impact fires 165
noise
    ambient 87
    level 87
Nomex hoods 177, 194, 197
Noonan, Danny 386
Norgrove, Idela 124
Norman, Butch 268, 339
Norman, Conor 367, 373
Norman, David 283, 367, 377, 422
Norman, Jeanne 351, 367, 373, 413,
    444–447
Norman, John W. 18, 21, 91, 351, 367
Norman, Joseph 283
Norman, Patrick 351, 354, 356–357,
    367–370
Norman, Warren 270, 283

North Tower (World Trade Center) 3. *See also* Tower 1 (World Trade Center)
   collapse 1
Northville Fire Department 446
Notice of Violation (NOV) 108–109, 112
nozzleman 25. *See* nozzles, operators
nozzles 8, 36
   controlling 97
   fog 29, 273
   New Yorker Multiversal 210
   operators 8, 33
   playpipe 32
   tip 29
   of tower ladders 165
NTSB (National Transportation Safety Board) 140–141
nuclear, biological, and chemical attacks (NBC defense) 2

O

oath of office x
obstructions 182
occupancy 208
occupants 211
Ocean Parkway 147–148
O'Connell, John 65
odors 20
off apartment 250–251
Office of Emergency Management (OEM) 382
officers
   covering 109
   overtime 109
   police 86, 138
   rescue 120
   selection process 116
   tools 221
O'Flaherty, Brian 4, 291, 298–299, 395
oil 46
   burners 36, 46
   chain saw bar 73
O'Keefe, Patty 300, 411
Oklahoma State University (OSU) 28
   campus fire station 28
Oliva, Vito 96–97, 100
Olmstead, John 446, 447
openings
   man-made 100
   vertical 101
open red 223
operating
   mechanism 135
   nut 8

operations 193
   loading 204
   relay 79
   report 193
orders 102, 448
   compliance 102
   promotion 119
   transfer 119
   violation of 102, 104
ordinances, city 294
outboard 79
outriggers 243
overdoses 41. *See also* drugs; heroin
overhauling 25, 177, 217
overhead 198
overload 10
overtime 436
   average 119
oxygen 82
   cylinders 3
   indicator 84
   supply 3, 262

P

packaging 305
paddles 137
pads, hydraulic 243
Palmer, Orio 4, 363
Pampalone, Frank 119, 121
pancake section 84
panic hardware 107
Paolillo, John 2, 280–281
paperwork 359
paramedics 1, 447
parapet 67
   wall 56
partition 89–90
Partner K-12 circular saw 54
passenger trains 78. *See also* trains
   cars 79
patient assessment 445, 447
pavement breakers 82
PCBs (polychlorinated biphenyls) 321
Pelligrino, Tommy 13, 48
Pena, Mike 125–134, 140–141, 211–217, 375–379, 412
Peninsula General Hospital 271
performance
   categories 118
   evaluations 118
permanent assignment 101
personal escape rope 67
personality clash 118

personal rope 67, 261
personnel
    assignment 76
    decrease 182
Peterson, Don 83, 86–89, 115
Pfeiffer, Joe 402
Picciotto, Rich 397
Piegari, Benny 13, 43
piers 136–139
pikes 182
pilasters 299
piles. *See* debris, piles
pilings 136–137
Pinsent, Mike 44
pipes 25, 100
    brass play 29
    chase 196
    drain 101
    play 29, 32
    sprinkler 159
plants
    heating 159
    storage 267
plaster 101
plastics, burning of 80
platforms, raised 79
plating 166
    roof 87, 89
    switch 100
plumbing 95
poles, pike 182
police
    cars 48
    harbor launch 48
    launch 139
    officers 86, 138
    Police Accident Investigation Squad 74
    Police Commissioner 86
    truck 137
policy 9, 227
    response 75
polychlorinated biphenyls (PCBs) 321
polyvinyl chloride (PVC) 95
portable ladder 194. *See also* ladders
Port Authority 140, 268
    crash truck 133
    police officers 1
posts 72
    command 138
    newel 181–182
powder, Purple K 345
power
    equipment 46

saws 180
shutdown 78
source 80
tools 87, 153
precautions 84
Precinct 77 155
preliminary reports 78, 300
premonitions 80
pressure
    of debris 85
    direct 85
primary search 177, 181, 202, 207, 216
prisoner area 220
probes 63
Probie School 12
projectiles 330
promotions 115–119, 291, 359, 360
    delays 118
    order 119
    process 116
    system 115
property 101
protection 163, 195
    clothing for 444
    crash rescue 335
    flash 336
protocol 223
Prunty, Richard 397
public
    housing 24
    image 118
    relations 325
    service officers, murders of 42
pull box 209
"Pull 'em up!" 7
pumpers 8, 130, 132, 141
    discharge 30
    discharge gate 97
    first attack 32
    inlet 14
    Mack 19
    third alarm 132
pumper trucks 95
pumps 60, 63
    hydraulic 59–60, 63, 156
punishment 177
Purple K powder 345
PVC (polyvinyl chloride) 95
pyrotechnics 46

Q
Quatrochi, Phil 279
Quinn, Patrick 12, 214

R

Rabbit tool 59–65, 112, 156, 250, 253
radiant heat 30, 79, 164. *See also* heat
radiation 350
  detection equipment 351
radio 76
  contact 3
  fire 160
  frequency 272
  reports 2, 130–131
  scanner 2
  traffic 150–151, 201
rafts 134–140
  inflatable 137–138
  inflation of 135
  leaks 137
  Switliks 134
railroads 78. *See also* trains
  ties 79
rails 79
  power 80
  yard 78
raised platforms 79
rams 147
Rand, Adam 431
Randall's Island (NY) 362
rating 118
rear
  exposure 12
  extension 12
reciprocating saw 89
recognition 117
recovery 111
  of bodies 287
recreation complexes 257
reduced profile maneuvers 274
reference points 110
refuge 251
registered nurse (RN) 220
Regler, John 101–102
regulators 220
Reichel, Tommy 87
Reidy, George 7
reinforcements 136, 214
relay operations 79
  hose 79
relief 11, 191, 264
  early 126
  valve 224
relief groups 101, 119
  captain 292
relocating companies 153–154
remedial action 102
removable parts 94

removal of debris 230. *See also* debris
repairs 36
repeater systems 3
reports 2
  10-75 77
  eyewitness 51
  fire 170
  initial 75, 130
  injury 128
  operations 193
  preliminary 78, 300
  radio 2, 130–131
  victim 83
reputation 207
rescue. *See also* rescue companies
  aircraft crash 267
  captain 53
  collapse 439
  Company No. 1 53, 81–87, 130,
    136–138, 293
  Company No. 2 53, 75, 119–120, 130
  Company No. 3 53, 75–76, 81–82
  Company No. 4 53
  Company No. 5 130, 139, 147
  dive 132, 135
  diving 131
  duration 86
  effort 113, 139
  equipment 82, 147
  fire 52
  gear 130
  heavy 75
  hydraulic tools 82
  immediate 83
  liaison 130
  Liaison Unit 291
  officer 120
  operations 120
  result 86
  rig winch 72
  school 293
  site 87
  truck cab 132
  trucks 82, 130, 147
  water 52
rescue companies 53, 76, 130, 146, 153,
    208
  candidates 53
  jobs 120
  quarters 153, 172
rescuers 83, 86, 202
  monitoring of 92
resource man 1
resources 325

respiration, spontaneous 111
responders, emergency 139, 446
response
    area of 41, 48
    initial 171
    message 78
    policy 75
    times 179
Response to Terrorism program 116
rest and recuperation (R&R) 154
resuscitators 122, 178, 199, 202–204
retirements 115, 241, 436
revenues 145
Richardson, Tommy 180–183, 186–187
riding list 125
Riechel, David 91
riggers 303
rigs 65, 72, 141, 300
    winch 72
risers 127, 174, 176
risk 51
River Avenue 78
R-model Mack 66, 77
RN (registered nurse) 220
Roby, Tom 368, 377
Rocco, Louis 85
Rockaway Peninsula (Queens) 271
Rockefeller Center 314
rods 330
Rogers, Harry 141, 142, 147, 152, 211
Rogers, Jimmy 293
Rogers, Vincent 91
Rohan, Glen 396
roll call 95, 126, 128, 229
rollers 222
rollover 182
Rolon, Mark 349, 367
roof
    boards 57, 87, 191
    bulkhead 11, 62, 149
    chief 149
    collapse 229
    cutting 191
    cutting methods 190
    firefighters 142
    holes 212
    joists 57, 87
    layers 57
    level 165
roof (*continued*)
    men 56
    plates 87, 89
    purlins 238

    removal 89
    roofing nails 99
    rope 55, 57, 68
    skylight 11
    team 54
    timber truss 227, 234, 235
    ventilation 128, 190, 250
    wooden 87
roofman 165
ropes 67, 140, 252. *See also* search ropes
    ladder 141
    life-saving 54, 55, 67, 133
    personal 67, 261
    roof 55, 57, 68
    tugs 132
Rose, David 31
rotary saws 82, 87, 89. *See also* saws
rotation of members 92
routes, escape 141, 211
routine for standby 336
R&R (rest and recuperation) 154, 194
rubber boat 138
rubbish fires 66
rubble 86, 230
Ruland, Donald 213, 216, 217
runs 15, 26, 47, 96
runways 267, 268
    JFK Runway 22 272
Rush, Steve 439–440
Ruvolo, Phil 279, 289
Ryan, Bill 54, 69

S
Safe-T-Firsts 346
safety
    division 68
    improvements 58
Safety Operating Battalion 4
salary 94
Salka, John 115
SAR (surface-supplied air respirators) 310
satellite units 163
saws 56, 161, 170, 197, 296–297. *See also* chain saws
    bag 57
    blades 57, 190
    power 180
    reciprocating 89
    rotary 82, 87, 89
Sawzall 220–221
scaffolds 294, 297, 304
    jobs 295
    unattended 294

SCBA (self-contained breathing
    apparatus) 48, 86, 264, 309
  bottle 274
  strap 274
scene
  of accident 147
  control 90
Scheffold, Freddy 400–401
Scheider, Roy 284
Schenectady Avenue 180
Scherer, Gene 71–73
Schildhorn, Bob 361
Schwartz, Al 376
Scoppetta, Nicholas 437
scrap metal 37
  dealers 37
scraps 95
SCUBA 48, 130. *See also* diving
  divers 131, 180
  equipment 153
  flashlight 136
  gear 134, 137–138, 140
  lights 139
  police unit 438
  set-ups 130
  training 438
  underwear 140
Search and Rescue Manager 435
searches 22, 78
  for access 83
  of apartments 167
  primary 177, 181, 202, 207, 216
  secondary 138, 216–217, 229
  training 22
  underwater 134
  for victims 83
  for voids 83
search groups 444
search ropes 111, 113, 161–162, 165. *See also* ropes
second alarm 148
second due 362
  engines 79
Secret Service 335, 337, 339–340
security 155
  gates 83
  measures 87, 93
Seelig, Bill 441
Seelow, Wayne 446–447
self-contained breathing apparatus (SCBA).
    See SCBA (self-contained breathing
    apparatus)
seniority 116–117
  credit 117

September 11, 2001 1, 289, 367, 436
  aftermath ix
  attack ix
  loss xi
Serpico, Frank 96
service road 147
services, lack of 145
Seventh Avenue 59, 61, 65
shafts 88, 162–163
  conditions 89
Shea, Kevin 375
sheets 89
  coverings 97
  metal 99
  steel 83, 87
Sheffield Avenue 76, 241
Shelley, Craig 120, 209–210, 220, 223, 280–283
shells 46
sheltered area 84
shifts 101
shipboard 438
shootings 42–45
shoring 90
shutters 110–111
Siamese connection 166
siding 192
Siegel, Bill 441
Sifton, Charles 262
signals
  10-75 10, 75, 77, 82, 141
  all hands 75, 81
  "signal ten" 19
silence 168–169
single houses 179
single units 179
sirens 126, 131, 446. *See also* alarms
sites
  crash 131, 137
  of rescue 87
size-up of conditions 83, 179
Skillings, Emmanuel 51
skin 199
skycams 293
skyhooks 305
skylights 11
sledgehammers 9, 97
sling-pak 22
smoke 8, 63, 95, 110, 162
  conditions 64, 78, 141, 211
  detector 200
  lift 178
  odor 82
  pumping 121

wood 8
SOC. *See* Special Operations Command (SOC)
SOC Battalion 2
sockets 83
Southside (Brooklyn) 93–96, 101, 107, 109, 113
   reputation 96
   units 96
South Tower (World Trade Center). *See* Tower 2 (World Trade Center)
space heaters 37, 159. *See also* heaters
spaces
   confined 86, 307
   limits 85
   void 229
sparks 191
Special Operations 292
   Battalion 298
   Headquarters 2
   Unit 291, 308
Special Operations Command (SOC) 1, 278, 285, 291, 435
   personnel 436
   recovery 441
   specialization pay 438
   support ladders 440
   training 438
speed (of travel) 85, 182
Spillane, Jack 436
Spinelli, Ronnie 25
spirit 444
spontaneous respiration 111
Spor, Joe 69–73
sprinklers 6, 164. *See also* sprinkler systems
   heads 164
   pipes 159
   valve 165, 166
sprinkler systems 6, 108, 110, 166, 200
   engineer 108
   inspector 108
squad companies 1
Squad Company No.
   1 173
   3 27
   54 28
squatters 15, 63, 159
stables 327
Stabner, Cliff 278–279, 288, 336–337
Stack, Larry 4, 395–396
Stackpole, Timmy 149, 151–152, 154, 181
staffing cuts 181
staging area 132, 447
stairs

fold down 79
noncombustible 58
risers 150, 176
stairs (*continued*)
   staircases 141
   stairways 14
   stringer 175–176
staking a claim 305
standbys 335
   presidential 340
   routine 336
standpipes 165
   control valves 164
   systems 164
Stang 229–230
   nozzles 12
Staten Island (NY) 76, 147
stations, monitoring 348
status inquiry 79
steam 57, 63, 98, 236, 241
   momentum 99
   production 99
steamer connections 8
Steam Team 241
steel 79, 145
   security gates 83
   sheets 83, 87
   shutters 110
   sole shoes 100
Stefandel, Harry 97–101
Steyert, Richie 10, 195, 199
St. George, Bill 440
Stillwater Fire Department 28–31, 32
St. John's Place 190, 201
St. Mary's Hospital 154
stock market 439
stokes
   basket 85, 305
   stretcher 90
storage plants 267
St. Patrick's Cathedral (NYC) 351, 354
straight ladder 194. *See also* ladders
streams 101
   master 127
   reach 98
   tower 142
   tower ladder 149, 213
streets, network of 146
stress 182
stretcher, stokes 90
stretching 14
stringers 174–176
strips, commercial 145
structural engineers 6

structure collapse 190
style of trains 79
subways
    attacks 440
    wrecks 120
Suffolk County (Long Island)
    Police 280
    Volunteer Fire Department 370
Suhr, Danny 5
suits
    cold water flotation 132
    diver 279
    dry 131
    exposure 136
    flotation 130
    inflatable 137
Sullivan, Irene 440
Sullivan, Tom 323
summons 102
    book 104, 108
    criminal 103
    for failure 108
supervision 236
supply lines 22
supply locker 46
support, psychological 86
surface-supplied air respirators (SAR) 310
surface victims 83
surveys 100
survivors 83, 131, 139
Sutton, Robin Mundy 440
Sweeney, Brian 411
switch plates 100
Switliks 134. *See also* rafts
syndrome, crying wolf 313
systems
    extinguishing 341
    fire protection 170
    fire sprinkler 30
    hauling 296
    hoist 297
    sprinkler 166, 200
    standpipe 164
    telegraph 27
    tracking 125

T
T-32s 301
Tactical Support Unit 2
tailboard 96
take up 231, 254
tank farms 102
tape measures 87

target hazards 343
Tate, Chip 350
teams
    hazmat 436
    inside 100, 180
    interior 210
technicians, medical 203
telegraph systems 27
teleprinters 96, 126, 166, 179, 242
    glitch 78
television coverage 1
temperatures 150, 159, 171, 197
tenacity 177
tenements 58–59, 61, 67, 93, 102
    design 248
    new law 58
    old law 58
tenure 118
terrorism
    anthrax 436
    counter-terrorism 335
    dirty bombs 347
    training 438
tethers 134, 137
Theobald, Jack 248
thermal balance 24
thermal imagers 112–113
thermal imaging camera (TIC) 111
thermal recovery capsules 130
thieves 154
    car 94
third alarm 132, 140, 142, 148
Third Avenue 91
third due engines 79
throats (of apartments) 55
TIC (thermal imaging camera) 111
tightrope 175
timbers 84–87, 131–133, 136, 168, 300
timber trusses
    construction 227
    roof 227, 234, 235
tin ceiling 87, 89
tin snips 89
tires, pulling 73
tongue-and-groove lumber 88
tools 130–131, 205, 230, 305, 444. *See also* Halligan, tools
    air-powered 220
    assignments 180
    bag 56
    box 69
    capabilities 69
    change 89
    cutters 139, 147

forcible entry 59
gasoline-powered 86
hand 122, 151
Hurst 71, 139, 147, 247
hydraulic rescue 82, 139
hydraulic spreaders 147
jaws 139
kit 220
officer 222
power 87, 153
Rabbit 59–65, 112, 156, 250, 253
rams 147
truck 54
top area 86
safety 86
torches 82, 170
cutting 155
torching 59
total structural failure 5
tours 15, 82
day 90
exchanges 101
night 90
Tower 1 (World Trade Center)
south wall 4
west wall 4
Tower 2 (World Trade Center) 3
collapse 4
tower collapse 3
tower ladders 14, 59, 148–151, 170,
230–231. *See also* ladders
basket 86, 163, 202
nozzles 165
stream 142, 149
streams 213
tracking system 125
tracks 80
tractor trailer 70–73
traffic 71, 126, 145, 179
circle 166
fire 76
lights 147
tragedy
national xi
personal xi
trailer, chassis 72
training
essentials 28
search 22
trains 41, 80. *See also* passenger
trains; Metro-North
Amtrak 80
cars 79
commuter 77

ConRail 80
diesel locomotives 80
engineer 80
fire 78, 82
freight 80
trains (*continued*)
Metro-North 80
operation 78
outboard 79
style 79
tracks 80
transfer order 119
transmit the box 228
trauma 3, 54, 448
treads 174
trenches 57
jacks 82, 86, 90
Triangle Shirtwaist Company 103
fire 103–105
Tri-County Fire Equipment 268
Troy Avenue 205
Truck 45 66
truck companies 9, 11, 54, 76
truckies 63, 212, 216
trucks
axles 72
collapse 300
company 76
crash 268, 273
fire 71
floodlight 19
gasoline 26
hook and ladder 19
police 137
rebreather 439
rescue 82, 130, 132, 147
tools 54
trusses 236. *See also* timber trusses
trussloft 234
tunnels 78, 260–261
conditions 79
eight-pack 79
six-pack 79
Turi, Al 361–362, 362
turnout bells 43
turnout coats 197, 242
Tuttlemundo, Frank 14–15, 214
Twin Towers. *See* World Trade Center

U

UFOA (Uniformed Fire Officers
Association) 370
UFO (until further orders) 119–120, 209

Uniformed Fire Officers Association
(UFOA) 370
unions 438
United Nations 344
    fire brigade 344
units
    apparatus 1
    assignment 179
    assistance 79
    burn 203
    Collapse Utility 298
    development 68
    field communications 2
    fire 83
    first-alarm 82
    hazmat 54
    Hostage Negotiation 344, 347
    Rescue Liaison 291
    research 68
    return 179
    satellite 163
    single 179
    Special Operations 308
until further orders (UFO) 119–120, 209
urban
    mining 95, 160
    renewal project 27
U.S. Air Flights
    737 130
    5050 271
USS *America* 339
USS *Constellation* 308
USS *Grasp* 282
USS *Intrepid* 338
U.S. Urban Search and Rescue Teams xi

V
vacancies 115, 291, 292
vacant
    buildings 7–14, 59, 84, 94–101, 365
    frame 15
    houses 25
    lots 9, 46
    theater 26
vacations 101
Vacchio, David 367
Valenzano, Pete 275
value of land 326
valves
    relief 224
    replacement 36
vandalism 95
vantage points 88

Van Vorst, Dave 220, 222–223, 244–246,
443
vehicles. *See also* trucks
    extrication of 247
vending machines 219
vent, enter, and search (VES) 173
ventilations 96, 127, 176, 190, 202
    hood 341
    lack of 190
    of roof 250
ventilators 174
venting 249
    flames 79
Venturi effect 142
verbal alarms 42, 155
VES (vent, enter, and search) 173
vibralert 237, 254
V.I.B.s (Very Important Buildings) 343
victims
    bond 90
    buried 85
    call out 84
    discovery 90
    packaging 305
    reassurance 91
    reply 84
    searches for 83
    surface 83
    trapped 83
Vigiano, John 292
violations 104, 108
    of orders 102, 109
violence 42
Visconti, Nick 394
visibility 60, 98–99, 110, 139, 199
voice alarms 96, 166
voids 84, 86
    access to 84
    conditions in 86
    entrance 87
    mouth 87
    rescuers 85
    spaces 100, 229
    stabilization of 86
    types of 83
volunteers 17
Von Essen, Thomas 404, 423, 437

W
Waldbaum's Supermarket 227
walls
    brick 86
    cinder blocks 11

collapses 82, 165, 229
  foundation 85
  knee 127
  parapet 56
  plaster 95
  sockets 83
wash-down 101
Washington, Al 127–128, 131–138, 142,
  152–153
Washington Bridge 78
Washington Heights 77–78, 81, 119
watchman 27, 42, 47, 96
water 215
  amount 101
  can 106
  current 134–136
  flotation suits 132
  flow 36, 109
  hot 251
  lack of 165
  leaks 66, 155
  level 139, 216–217
  polluted 51, 131
  rescue gear 130
  sterile 202
  streams 29
  supply 78–79, 95
  wait for 79
Water Monster 254
water motors 166
  alarm 166
weapons of mass destruction (WMD) 438
weight 199
  belt 137
Weiss, David 341, 411
welders 82–83
welding equipment 153
Wendell, Gary 100–101
Wendling, Jim 361
Weston, Larry 162, 165, 167–170
Whittaker, Viola 123
Williamsburg (Brooklyn) 93, 101
winches 72–73

windows 110
  barred 197
  glass 173
  guard gate 67
  HUD sealed 171
  reinforcement 155
  sash 173, 195
  screen 154
  venting 64
Wingert, Clem 13, 43–44, 45, 47
wings 55
The Winter Offensive 37, 159
wires 25
  electrical 279
  overhead 80
wish list 118
WMD (weapons of mass destruction) 438
wooden roofs 87
work areas 87
workbenches 154
working fires 19
World Trade Center ix, 289, 368–434, 443
  5 WTC 383
  1993 bombing 3, 351
  architects of 6
  collapse of 435
  site 435
  towers 435
wreckage 71, 131, 285
  car 72
wrenches 57

Y
Y2K
  event 441
  project 362
  task force 362

Z
Zazulka, John 198–199
Ziegler, Ray 315–316
zone of collapse 213

About the Author

John Norman began his fire service career as a volunteer firefighter in his hometown of Inwood, New York in June of 1970. In September of that year, he enrolled in the Fire Protection Technology program at Oklahoma State University, living in Fire Station #2 of the Stillwater Fire Department as a student firefighter. John left OSU in December of 1972 for work as a fire protection systems designer for a New York City sprinkler contractor. One of his first jobs was laying out portions of the high-pressure standpipe risers being installed in the new twin towers of the World Trade Center, then under construction in lower Manhattan. He worked for nearly seven more years at fire protection system design, during which he designed and supervised installation of many types of systems throughout the metropolitan New York area.

On his return from college, John resumed an active role with the Inwood Fire Department, being elected to lieutenant, and then captain, deputy chief, and eventually chief of the department in 1991. The Inwood FD provided a variety of rescue services, ranging from water rescue to high angle and confined space, as well as extrication services. In June of 1979, John was hired as an instructor for the Nassau County Fire Service Academy on Long Island and taught a variety of firefighting and rescue subjects.

In October of 1979, he was appointed to the New York City Fire Department, and after training was assigned to Engine Company 290 in the East New York section of Brooklyn, the busiest unit in the city at the time. There he met one of the many mentors that would shape his career, Lt. Stephen Casani, recently promoted from Rescue Co. 1. After three years in Engine 290, John transferred across the floor to Ladder Co. 103, where he worked for a year. In the interim, Lt. Casani was transferred to Rescue Co. 3, which covered the Bronx and Harlem. When that unit had a vacancy for a firefighter, Casani recommended Norman for that position. John worked in Rescue 3 for about a year when the department opened a dedicated hazardous materials company. Since he was a state-certified hazmat instructor, John was selected for assignment to the new unit. He worked there for 15 months, helping to train the members before returning to Rescue Co. 3.

In August of 1987, he was promoted to lieutenant and assigned to cover in the 11th Division in northern Brooklyn. After a year of covering, he was detailed to Ladder 111 in Bedford Stuyvesant, where he ran in with Rescue 2 on a daily basis. This is where he first met Ray Downey, the Captain of Rescue 2. After a year at Ladder 111, Norman got a call from Captain Downey. One of Downey's lieutenants was due to get promoted in a few months—would John be interested in working in Rescue 2? The answer took some time to think over—about half a second! Thus began a working relationship that continues to this book. John worked with Downey and others to establish the FDNY's first formal rescue training school in 1992. In 1993, John was promoted to captain and was briefly assigned to the 15th Division in southern Brooklyn before being detailed back to Rescue Operations. There he served in the Rescue Liaison Unit, and as a covering captain rotated through the five rescue companies and Squad Company 1. In 1994, he was recommended by then Chief of Rescue Operations Downey to be the company commander of Rescue Co. 1 in Manhattan, and that was where he would spend the next five years. That year, FEMA conducted its first rescue specialist "train the trainer" program for the 26 USAR task forces. Norman and Downey served together as instructors in that and subsequent programs. In 1999, John was promoted to battalion chief, and covered in Manhattan and the Bronx before being assigned to the 16th Battalion in Harlem.

On September 14th, 2001, after the depth of the FDNY's losses in the World Trade Center attack was fully recognized, Chief Norman was given a dual role: the Search and Rescue Manager for the Trade Center site, a job that Ray Downey most certainly would have held had he survived, and command of the Special Operations Command, Downey's prior assignment. He also served as Rescue Squad Officer, Technical Team Manager, and Task Force Leader of New York City's FEMA USAR Task Force, NY-TF-1. In June 2003, John was promoted to Deputy Chief and remained in SOC. A year later, he was promoted to Deputy Assistant Chief, assuming the duties of Citywide Command Chief in addition to his responsibilities as Chief of Special Operations. He retired from the FDNY in January 2007.

John lives in New York with Jeanne, his wife of more than 45 years.